LOW-CARBON TECHNOLOGY TRANSFER

Low-carbon technology transfer to developing countries has been both a lynchpin of, and a key stumbling block to, a global deal on climate change. This book brings together for the first time in one place the work of some of the world's leading contemporary researchers in this field. It provides a practical, empirically grounded guide for policy-makers and practitioners, while at the same making new theoretical advances in combining insights from the literature on technology transfer and the literature on low-carbon innovation. The book begins by summarizing the nature of low-carbon technology transfer and its contemporary relevance in the context of climate change, before introducing a new theoretical framework through which effective policy mechanisms can be analysed. The North–South, developed–developing country differences and synergies are then introduced together with the relevant international policy context. Uniquely, the book also introduces questions around the extent to which current approaches to technology transfer under the international policy regime might be considered to be 'pro-poor'. Throughout, the book draws on cutting-edge empirical work to illustrate the insights it affords. The book concludes by setting out constructive ways forward towards delivering on existing international commitments in this area, including practical tools for decision-makers.

Dr David Ockwell is a senior lecturer in the Geography Department at the University of Sussex, UK. He is also a senior fellow in the Sussex Energy Group and the Tyndall Centre for Climate Change Research. David's research and teaching focuses on transitions to a low-carbon economy with particular interest in low-carbon development, public engagement with climate change, and reflexive climate and energy policy appraisal. He provides regular policy advice to a range of developed and developing country governments and several intergovernmental bodies.

Dr Alexandra Mallett is an assistant professor at the School of Public Policy and Administration (SPPA), Carleton University, Ottawa, Canada, and visiting fellow at SPRU (Science and Technology Policy Research), Sussex Energy Group, University of Sussex, UK. Alexandra also worked for the Canadian government in the area of international energy and environmental policies. Her research focuses on the development, production, cooperation and adoption processes involved in low-carbon energy technologies. She is also interested in debates involving low-carbon energy transitions (e.g. the role of policies and appropriate pathways for developing countries).

LOW-CARBON TECHNOLOGY TRANSFER

From Rhetoric to Reality

Edited by David Ockwell and Alexandra Mallett

Routledge
Taylor & Francis Group

LONDON AND NEW YORK

First published 2012
by Routledge
2 Park Square, Milton Park, Abingdon, Oxon OX14 4RN

Simultaneously published in the USA and Canada
by Routledge
711 Third Avenue, New York, NY 10017

First issued in paperback 2017

Routledge is an imprint of the Taylor & Francis Group, an informa business

The opinions expressed by the contributing authors to this volume are the personal views of the authors of individual chapters and do not necessarily reflect the views of the editors or their institutions.

British Library Cataloguing in Publication Data

A catalogue record for this book is available from the British Library

Library of Congress Cataloging in Publication Data
Jones, Amelia.
  Low-carbon technology transfer : from rhetoric to reality / edited by David Ockwell and Alexandra Mallett. -- 1st ed.
    p. cm.
  Includes bibliographical references and index.
  1. Climate change mitigation. 2. Green technology--Developing countries. 3. Technology transfer--Developing countries. I. Ockwell, David G. II. Mallett, Alexandra.
  QC903.L88 2012
  363.738'74--dc23
                                                                2011043779

ISBN 13: 978-1-138-11001-4 (pbk)
ISBN 13: 978-1-84971-269-9 (hbk)

Typeset in Bembo and Stone Sans
by MapSet Ltd, Gateshead, UK

CONTENTS

List of Figures and Tables

Figures

Tables

List of Contributors

Editors

David Ockwell is a senior lecturer in the Geography Department at the University of Sussex, UK. He is also a senior fellow in the Sussex Energy Group and the Tyndall Centre for Climate Change Research. David's research and teaching focuses on transitions to a low-carbon economy with particular interest in low-carbon development, public engagement with climate change, and reflexive climate and energy policy appraisal. He provides regular policy advice to a range of developed and developing country governments and several intergovernmental bodies.

Alexandra Mallett is an assistant professor at the School of Public Policy and Administration (SPPA), Carleton University, Ottawa, Canada, and visiting fellow at SPRU (Science and Technology Policy Research), Sussex Energy Group, University of Sussex, UK. Alexandra also worked for the Canadian government in the area of international energy and environmental policies. Her research focuses on the development, production, cooperation and adoption processes involved in low-carbon energy technologies. She is also interested in debates involving low-carbon energy transitions (e.g. the role of policies and appropriate pathways for developing countries). Alexandra completed her PhD dissertation in Development Studies (DESTIN) at the London School of Economics and Political Science (LSE), examining the uptake of renewable energy technologies in Mexico and Brazil.

Authors

Ahmed Abdel Latif is senior programme manager for innovation, technology and intellectual property with the International Centre for Trade and Sustainable Development (ICTSD) in Geneva Switzerland. He has taken an active part in global discussions on intellectual property, public policy and development. His chapter was written in his personal capacity. As an Egyptian career diplomat, he previously worked at the Permanent Mission of Egypt to the United Nations and the World Trade Organization (WTO) in Geneva (2000 to 2004) where he was a delegate to the TRIPS Council and to the World Intellectual Property Organization (WIPO). He is a graduate of the London School of Economics and Political Science (LSE), the American University in Cairo (AUC) and the Institut d'Études Politiques de Paris (Sciences-Po Paris).

Martin Bell is an historian and economist at SPRU (Science and Technology Policy Research), University of Sussex, UK. His research interests centre on policy and management issues concerned with the development of innovation capabilities in industrial firms in African, Asian and Latin American countries. He currently focuses on the roles of these capabilities in the structural diversification of these economies, in the long-term evolution of their innovation systems, and in their transitions towards lower carbon futures.

Thomas L. Brewer is a senior fellow at the International Centre for Trade and Sustainable Development (ICTSD) Geneva and former Schoeller Foundation senior research fellow at Friedrich-Alexander University in Nuremberg, Germany. He has also had a recurring appointment as a visiting senior research fellow at Oxford University in the Smith School for Enterprise and the Environment, UK. He is an associate fellow of the Centre for European Policy Studies (CEPS) in Brussels, and an emeritus faculty member of Georgetown University in Washington, DC.

Rob Byrne has been working in the energy field for 15 years as an engineer, a project manager and a policy analyst. Five years of this experience have been in an African context – in Botswana, Tanzania and Kenya. His research focuses on the role of low-carbon energy in development.

Ben Castle has worked on UK climate change policy since 2004. His areas of focus have included emissions trading, consumer behaviour change, domestic energy efficiency and renewables. Ben has recently started work for the Evaluation Office at the Global Environment Facility, based in Washington, DC. He holds a BSc in Environmental Management and Policy from the London School of Economics and an MSc in Climate Change and Development from the University of Sussex, UK.

Joy Clancy is an associate professor in development studies at the University of Twente, The Netherlands, specializing in technology transfer. Dr Clancy's research has focused, for more than 25 years, on small-scale energy systems for developing countries, including the technology transfer process and the role that energy plays as an input for small businesses, and the potential it offers entrepreneurs through the provision of a new infrastructure service.

Andreas Falke is a professor of international studies at the School of Business and Economics of the University of Erlangen Nürnberg, Germany. A political scientist by training, he received his PhD from the University of Göttingen. From 1992 to 2002 he worked as principal economic specialist at the American Embassy in Bonn/Berlin. He specializes in trade policy, the politics of globalization, the relationship between trade and climate change, transatlantic relations and American politics, generally. He is also director of the German–American Institute in Nürnberg.

Tim Forsyth is reader in environment and development at the London School of Economics and Political Science. His work focuses on the implementation of global environmental policy within developing countries via local stakeholder participation and deliberation. He is the author of *Critical Political Ecology: The Politics of Environmental Science* (Routledge, 2003) and *International Investment and Climate Change: Energy Technologies for Developing Countries* (Earthscan, 1999). He was a co-author of the chapter on climate change in the *Millennium Ecosystem Assessment* (Island Press, 2005), and is on the editorial boards of the journals *Global Environmental Politics*; *Conservation and Society*; *Progress in Development Studies*; and *Social Movement Studies*.

Erik Haites is president of Margaree Consultants Inc. in Toronto, Canada. He advises clients on economic aspects of climate change, including the design of market mechanisms and international financial mechanisms. He was a consultant to the United Nations Framework Convention on Climate Change (UNFCCC) Secretariat during the negotiation of the rules for the Kyoto mechanisms. He is an author of several peer-reviewed articles on technology transfer in the Clean Development Mechanism (CDM), linking emissions trading systems and international financial support to address climate change. He holds a BSc in Mathematics from the University of Alberta, an MBA from McGill University and an MS and PhD in Economics from Purdue University.

Rüdiger Haum is a research analyst at the German Advisory Council on Global Change, Berlin. He is also a research associate to the Environmental Policy Research Centre of the Free University of Berlin, Germany. He holds postgraduate degrees in media studies as well as in science and technology policy, and has previously worked as a journalist and policy consultant.

Merylyn Hedger has led research projects across both the mitigation and adaptation agendas and delivered a range of products, including as convening lead author for the Intergovernmental Panel on Climate Change (IPCC) *Special Report on Technology Transfer*. She has extensive experience in public-sector decision-making at all scales of governance and direct responsibilities as a practitioner for policy-making on climate change, and the provision of advice on regulation. She has provided guidance, and developed tools and vision for frameworks on new ways of working on the cross-cutting issue of climate change at international, national and sub-national levels. Merylyn has degrees from the London School of Economics and Liverpool University, UK, and holds a PhD in Energy Policy from Imperial College, London. In 2002 she was awarded the OBE for services to climate change assessment.

Peter S. Hofman is an assistant professor in corporate governance and corporate social responsibility at the Nottingham University Business School China. His research interest lies in understanding how progress towards sustainability is shaped through the interaction of business, government and civil society, with a specific focus on Asian countries, China and Thailand, in particular. His research focuses on topics such as cleaner production and environmental management, governance for energy innovation, and more collaborative policy strategies to induce sustainable innovations.

Grant A. Kirkman is responsible for CDM and Joint Implementation (JI) project analysis at the Secretariat of the UNFCCC, in support of the climate change negotiations and the regulatory bodies of these market mechanisms. Prior to joining the UNFCCC, Grant worked in the private sector and academia on industrial air quality, environmental compliance issues and, finally, research on boundary-layer air chemistry in Southern Africa, South America and in Europe. Grant received his BSc in Agricultural Economics from the University of Natal, his MSc in Environmental Sciences from Witwatersrand University and his PhD in Biogeochemistry from the University of Mainz.

Annemarije L. Kooijman-van Dijk is assistant professor at the University of Twente, The Netherlands, specializing in energy and development, diffusion of energy innovations, and the socio-economic context of energy technologies. As of 2012, she will also take on the position of adviser on the sustainability of biomass for the Faculty of Science and Technology, University of Twente. Her PhD study, in 2008, was on the role of energy in poverty reduction. Her prior professional experience, since 1998, was at the Energy Research Centre of The Netherlands ECN, the United Nations Development Programme (UNDP) and the University of Twente in the fields of energy in developing countries and renewable energy policy.

Jon C. Lovett is professor of sustainable development in a North–South perspective at the University of Twente, The Netherlands, and leader of the sustainability strategic research orientation at the Institute for Innovation and Governance Studies. His main interest is the institutional economics of natural resource management with a focus on both adaptation and mitigation to climate change. He led a study on market-based approaches to technology transfer for the Dutch Ministry of Foreign Affairs and was recently part of the team that prepared the Africa Energy Strategy for the African Development Bank.

Muthukumara S. Mani is a senior environmental economist in the Sustainable Development Department of the World Bank's South Asia region, based in Washington, DC. He primarily works on climate change mitigation and adaptation issues. Prior to this, Dr Mani led the World Bank's work on assessing the environmental implications of development policy reforms, and was earlier an economist in the Fiscal Affairs Department of the International Monetary Fund, where he analysed the environmental implications of macro-economic policies and programmes. He has a PhD and an MA in Economics from the University of Maryland, College Park.

Karlijn Morsink is a PhD student at the University of Twente, The Netherlands, researching the diffusion of innovations in developing countries in the context of poverty reduction. She focuses on the influence of the international, national and local context on the adoption of innovations. She has worked with the non-governmental organizations (NGOs) HOPE and HESCO in India on the development of projects to promote the uptake of technologies and has collaborated with several universities in Kenya and India.

Kevin Murphy is professor of economics at Oakland University in Rochester, Michigan, where he has taught since 1985. He was also assistant professor of economics at the University of Tennessee from 1981 to 1985, and he held a Fulbright Fellowship to University College Cork, Ireland, in 1987 to 1988. He received his PhD in Economics from Michigan State University in 1981. His areas of research and teaching expertise are applied econometrics, statistics and labour economics.

Prosanto Pal is a senior fellow in the Industrial Energy Efficiency Division at The Energy and Resources Institute (TERI) in New Delhi. The division undertakes research projects on energy-efficiency improvement and cleaner technologies promotion among industries. Prosanto has been involved in cleaner production studies; in research, development, demonstration and diffusion (RDD&D) of energy-efficient technologies among small- and medium-sized establishments (SMEs); and in studies on issues related to technology transfer from developed to developing countries. He leads several multidisciplinary research projects in these fields and has published numerous articles in international and national journals. He has authored a book on cleaner technologies in the foundry industry. Prosanto has been educated in India and the UK, and received his Bachelor of Technology in

Chemical Engineering from the Indian Institute of Technology, Delhi; his MSc in Manufacturing Systems Engineering from the University of Warwick, Coventry; and his MSc in Process Engineering from the University of Strathclyde, Glasgow.

Stephen Seres is an economist working on climate change policy issues in Montreal, Canada. As a consultant to the Canadian government, Stephen participated in the negotiations on the decisions to the Kyoto Protocol at the annual Conferences of the Parties (COP). He is an expert on the Clean Development Mechanism and has provided his expertise to the Canadian government, the private sector, and to the UNFCCC. Stephen Seres has published several peer-reviewed papers on the subject and remains on the faculty of the Greenhouse Gas Management Institute in Washington, DC, where he teaches a course on the Kyoto mechanisms.

Girish Sethi is a senior fellow and director of the Industrial Energy Efficiency Division of TERI, with more than 26 years of experience in the field of energy conservation and environment improvement in the industrial sector. He has been working at TERI for the last 18 years. He is a chemical engineer and holds an MSc in Energy Studies from the Indian Institute of Technology, New Delhi. He has also completed a multidisciplinary Master's course on Technology in the Tropics from University of Applied Sciences, Cologne, Germany. His direct responsibilities include providing strategic direction and coordinating the activities related to industrial energy efficiency. These involve energy audit assignments, capacity-building programmes and projects concerning the rational use of energy in various industrial subsectors. He has led multidisciplinary research teams in action research projects involving development/adaptation of energy efficient and environmentally benign technologies. He is currently managing a large seven-year programme funded by a bilateral organization that focuses on the holistic development of a few energy-intensive small-scale industry sectors in India. Other areas of interest include inventorization of corporate-level greenhouse gas (GHG) emissions and aspects related to the transfer and promotion of low-carbon energy technologies in the context of climate change.

Adrian Smith is a researcher at SPRU (Science and Technology Policy Research), University of Sussex, UK, and focuses on the politics and governance of innovation for sustainable development. His research draws upon analytical frameworks from innovation studies, the sociology of technology and political science.

Krinshna Ravi Srinivas received his PhD at the National Law School, Bangalore, and is currently conducting research on, *inter alia*, intellectual property rights, climate change, biodiversity and biotechnology, and science, technology and society issues at the Research and Information System for Developing Countries (RIS), New Delhi, India. His chapter was written in his personal capacity. He has been a visiting scholar at University of Pennsylvania (on a Fulbright Fellowship), Indiana University Bloomington, and a post-doctoral fellow at South Centre, Geneva.

Some of his publications are available at http://papers.ssrn.com/sol3/cf_dev/AbsByAuth.cfm?per_id=290086.

Mahesh Sugathan is programme coordinator for economics and trade policy analysis with the International Centre for Trade and Sustainable Development (ICTSD) in Geneva, Switzerland. He was also engaged as a consultant to the World Bank. At ICTSD, he has worked on a broad range of issues, including trade, development, environment and agriculture. His areas of expertise are trade in environmental goods and services, and trade, climate change and sustainable energy. He has published as well as spoken extensively on these issues at numerous international conferences. At the World Bank he was engaged in a pilot project on Assessing Investment Climates for Doing Climate Business, with a focus on South Asia. Mahesh has an MA in International Law and Economics from the World Trade Institute in Bern, Switzerland. He completed undergraduate and graduate studies in Economics and International Relations at the University of Kerala and the Jawaharlal Nehru University in New Delhi, India. He also holds an MPhil from Jawaharlal Nehru University.

Andy Sumner is research fellow in the Vulnerability and Poverty Reduction Team at the Institute of Development Studies, Sussex, UK, and a visiting fellow at the Centre for Global Development, Washington, DC. He is a cross-disciplinary economist. His research relates to poverty, inequality and human well-being, with particular reference to poverty concepts and indicators, global trends and Millennium Development Goal (MDG)/post-MDG debates.

Frauke Urban is lecturer in environment and development at the School of Oriental and African Studies (SOAS) at the University of London. She used to be research fellow in climate change and development at the Institute of Development Studies (IDS) at the University of Sussex before joining SOAS in September 2011. She is an environmental scientist and specializes on the links between energy, climate change and development, particularly in relation to China.

Anne-Marie Verbeken, an independent consultant, has extensive international experience in the field of the environment, energy and technology policy. She worked nearly ten years for the Global Environment Facility (GEF), supporting its Scientific and Technical Advisory Panel, and developing and managing climate change projects. She has also worked as a consultant for SPRU and the UNFCCC Secretariat on energy and technology transfer issues. She holds an MSc degree in Science and Technology Policy from the University of Sussex.

David Vincent's career in the energy, energy efficiency and renewable energy technology fields dates from the first oil price shock of the 1970s. He has had over 35 years of experience in various government departments managing research, development and demonstration (RD&D) programmes and developing policy. He helped to set up the Carbon Trust in 2000 to 20001 and worked there as technology director and projects director until 2011. He is now an independent consultant working with business, universities and governments on strategic aspects of energy efficiency and low-carbon technology.

Jim Watson is director of the Sussex Energy Group and a research fellow with the Tyndall Centre for Climate Change Research. He has over 15 years of research experience on energy, climate change and innovation policy issues – particularly in the UK and China. He regularly advises UK government departments, parliamentary committees and international organizations. He is the 2011 chair of the British Institute for Energy Economics.

Acknowledgments

The editors would like to extend their thanks to the chapter authors for their valuable contributions to this book. We would also like to thank our colleagues in the Sussex Energy Group at the University of Sussex and other academic colleagues, including several contributors to this book, from whom we've benefited from working with on climate and energy policy issues over the years. We have also learned from interactions with a wide range (too numerous to list here) of policy-makers, practitioners and academics across a range of developing countries whose commitment and drive in addressing climate change, sustainable development and a more equitable share of development benefits have been a source of both inspira-tion and motivation, not to mention a source of invaluable insight on the reality of these issues in the everyday lives of poor and marginalized people across the world. We would also like to acknowledge the UK Department for Energy and Climate Change (DECC), the UK Foreign and Commonwealth Office (FCO), the Organisation for Economic Co-operation and Development (OECD) Environ-ment Directorate, the United Nations Framework Convention on Climate Change (UNFCCC) Secretariat, the University of London, and, in particular, the London School of Economics and Political Science (LSE), the Commonwealth Secretariat and the UK Economic and Social Research Council (ESRC) for financial support to several of our research efforts in this field, all of which have contributed empiri-cal and theoretical insights, some of which are reflected in this book.

Last, but by no means least, we would like to thank our partners and families for the most important contribution of all.

List of Acronyms and Abbreviations

ADB	Asian Development Bank
4AR	IPCC *Fourth Assessment Report*
ARGeo	African Rift Valley Geothermal Development Facility
AWGLCA	Ad Hoc Working Group on Long-Term Cooperative Action
BASIC	Brazil, South Africa, India and China
BOOT	build–own–operate–transfer
BOP	best operating practice
BRIC	Brazil, Russia, India and China
BYD	Build Your Own Dreams
CC	continuous casting
CCS	carbon capture and storage/sequestration
CDM	Clean Development Mechanism
CER	Certified Emissions Reduction
CET	clean energy technology
CFL	compact fluorescent light/lamp
CFR	coke feed ratio
CGIAR	Consortium of International Agricultural Research Centres
CIF	Climate Investment Fund
CIRI	Climate Investment Readiness Index
CLASP	Collaborative Labelling and Appliance Standards Programme
CO_2	carbon dioxide
CO_2e	carbon dioxide equivalent

COP	Conferences of the Parties
CSD	Commission for Sustainable Development
CSI	Cement Sustainability Initiative
CSP	concentrated solar power
CSP	cross-sector partnership
CTC	Climate Technology Centre
CTCN	Climate Technology Centre and Network
DBC	divided blast cupola
DECC	UK Department for Energy and Climate Change
DFID	UK Department for International Development
DNA	Designated National Authority
DOE	designated operational/operating entity
DRI	direct reduction of iron
ECLA	European Classification System
ECN	Energy Research Centre of The Netherlands
EE	energy efficient/efficiency
EGTT	Expert Group on Technology Transfer
EPC	European Patent Convention
EPO	European Patent Office
ESCO	energy service company
ESL	electron-stimulated luminescence
ESRC	UK Economic and Social Research Council
EST	environmentally sound technology
ETS	Emissions Trading Scheme
EU	European Union
EV	electric vehicle
FCB	fuel cell bus
FCO	UK Foreign and Commonwealth Office
FDI	foreign direct investment
FSF	Fast Start Fund
FTC	Federal Trade Commission
GATT	General Agreement on Tariffs and Trade
Gcal/day	gram calories per day
GCF	Green Climate Fund
GDP	gross domestic product
GEF	Global Environment Facility
GEI	Green Economy Initiative
GGND	Global Green New Deal
GHG	greenhouse gas
GNI	gross national income
GNP	gross national product
GSK	GlaxoSmithKline
GW	gigawatt
GWh	gigawatt hours

HEM	high-efficiency motor
HEV	hybrid and electric vehicle
HFC	hydrofluorocarbon
HS	Harmonized System
IA	implementing agency
IADB	Inter-American Development Bank
IBRD	International Bank for Reconstruction and Development
ICC	International Chamber of Commerce
ICE	internal combustion engine
ICEEE	International Conference on Electrical and Electronics Engineering
ICT	information and communication technology
ICTSD	International Centre for Trade and Sustainable Development
IDA	International Development Association
IDS	Institute of Development Studies
IEC	International Electrotechnical Commission
IEG	Independent Evaluation Group
IFC	International Finance Corporation
IGCC	integrated gasification combined cycle
IGO	intergovernmental organization
IP	intellectual property
IPC	International Patent Classification
IPCC	Intergovernmental Panel on Climate Change
IPP	Independent Power Producer
IPR	intellectual property right
IRENA	International Renewable Energy Agency
ISCC	Integrated Solar Combined Cycle Power Plant
ISO	International Organization for Standardization
IWRM	integrated water resources management
ITT	international technology transfer
JBIC	Japan Bank for International Cooperation
JGI	joint geophysical imaging
JI	Joint Implementation
JNNSM	Jawaharlal Nehru National Solar Mission
kgce	kilograms coal equivalent
kW	kilowatt
kWh	kilowatt hour
LCCRED	Low Carbon, Climate Resilient Development Programme
LCD	low-carbon development
LCDP	Low Carbon Development Plan
LCDS	Low Carbon Development Strategies
LCR	local content requirement
LDC	least developed country
LDCF	Least Developed Countries Fund

LECRED	Low Emission Climate Resilient Development Strategy
LED	light-emitting diode
LEDS	Low Emission Development Strategies
LEG	Least Developed Countries Expert Group
LESI	Licensing Executives Society
LIC	low-income country
Li-ion	lithium-ion
LSE	London School of Economics and Political Science
LULUCF	Land Use, Land-Use Change and Forestry mechanism
MDG	Millennium Development Goal
M&E	monitoring and evaluation
MEA	Multilateral Environmental Agreement
MIC	middle-income country
MNRE	Indian Ministry of Non-Renewable Energy
MRV	monitoring, reporting and verification
MW	megawatt
N_2O	nitrous oxide
NAMA	Nationally Appropriate Mitigation Action
NAPA	National Adaptation Plan of Action
NDRC	National Development and Reform Commission
NGO	non-governmental organization
NREA	New and Renewable Energy Authority
NSP	new suspension pre-heater kiln
NTB	non-tariff barrier
ODA	official development assistance
ODA	overseas development assistance
OECD	Organisation for Economic Co-operation and Development
OEM	original equipment manufacturer
OP	Operational Programme
OSDL	Open Source Development Labs
PATSTAT	EPO/OECD Worldwide Patent Statistics Database
PCT	Patent Cooperation Treaty
PDD	project design document
PFC	perfluorocarbon
PPA	Power Purchase Agreement
ppm	parts per million
PRI	policy, regulation and incentive
PV	photovoltaic(s)
PVMTI	Photovoltaic Market Transformation Initiative
R&D	research and development
RD&D	research, development and demonstration/deployment
RDD&D	research, development, demonstration and diffusion
REEEP	Renewable Energy and Energy Efficiency Partnership
REDD	Reduced/Reducing Emissions from Deforestation and Forest Degradation

REDP	Renewable Energy Development Project
REN21	Renewable Energy Policy Network for the 21st Century
RPO	renewable purchase obligation
SBI	Subsidiary Body for Implementation
SBSTA	Subsidiary Body for Scientific and Technological Advice
SC	strip casting
SC	supercritical
SCM	Subsidies and Countervailing Measures Agreement (of the WTO)
SD	sustainable development
SDC	Swiss Agency for Development and Cooperation
SEG	Sussex Energy Group
SGP	Small Grants Programme
SHS	Solar Home Systems
SIDS	small island developing state(s)
SiO_2	silicon dioxide
SME	small- and medium-sized enterprise/establishment
SPV	solar photovoltaic technology
SSRM	steel re-rolling mill
STAP	Scientific and Technical Advisory Panel
STEPS	Social, Technical and Environmental Pathways to Sustainability
TBT	Technical Barriers to Trade Agreement
TEC	Technology Executive Committee
TERI	The Energy and Resources Institute
TM	Technology Mechanism
TNA	Technology Needs Assessment
toe	tonnes of oil equivalent
tpd	tonnes per day
TPRI	Thermal Power Research Institute
TRIMS	Trade Related Investment Measures Agreement
TRIPS	Trade Related Aspects of Intellectual Property Rights Agreement
TSC	thin slab casting
TT	technology transfer
TTA	Technology Transfer Agreement
TTZ	Taj Trapezium Zone
UK	United Kingdom
UNAGF	United Nations High-Level Advisory Group
UNCED	United Nations Conference on Environment and Development
UNDP	United Nations Development Programme
UNEP	United Nations Environment Programme
UNFCCC	United Nations Framework Convention on Climate Change
UNICEF	United Nations Children's Fund
UNIDO	United Nations Industrial Development Organization

US	United States
USC	ultra-supercritical
VAT	value-added tax
WBCSD	World Business Council for Sustainable Development
WBG	World Bank Group
WETC	Regional Wind Technology Centre
WIPO	World Intellectual Property Organization
WSSD	World Summit on Sustainable Development
WTO	World Trade Organization

PART I

New Analytic Approaches

1

INTRODUCTION

Low-carbon Technology Transfer – From Rhetoric to Reality

David Ockwell and Alexandra Mallett

Climate change is considered by many to be one of the most pressing issues facing society today (IPCC, 2007). Potential solutions to the problem are complex and multi-faceted, involving a range of possible changes to lifestyles, ways of doing business, growing food, using land, and producing and consuming energy. Indeed, some commentators argue that it challenges the desirability of commitments to sustained economic growth that lie at the very heart of neo-liberal thinking (e.g. Jackson, 2011).

Due to the significant proportion of global emissions that are related to burning fossil fuels for energy—estimated to be 56.6 per cent (IPCC, 2007)—and the difficulty in decoupling energy from economic growth (Ockwell 2008), attempts to mitigate future climate change have prioritized the development and deployment of low carbon technologies. Low carbon technologies are defined in this book as technologies that aim to minimize greenhouse gas (GHG) emissions, especially carbon dioxide emissions, relative to those technologies currently in use in a particular context (including, for example, renewable energy generation technologies, energy efficient end use technologies and more efficient fossil based energy generation technologies). The deployment of low carbon technologies is as much a priority for developed as for developing countries. But developing nations have attracted particular attention due to potential opportunities to industrialize on the basis of low carbon technologies as opposed to conventional ones. The unprecedented rate of industrialization and current and future increases in energy consumption of countries such as China and India highlights the significance of this opportunity to mitigate future GHG emissions. China and India's share of global CO_2 emissions, for example, grew from 13 per cent in 1990 to 26 per cent in 2007 (EIA, 2010). But the issue is equally significant for smaller developing countries. The fact that many of these new technologies were traditionally owned by companies based in developed countries meant that the issue became characterized as one of 'technology transfer', reflecting the idea that technologies could be transferred from where they were owned to new country contexts. This characterization of developed–developing, North–South transfer is now outdated with many developing countries, particularly the BRIC countries (Brazil, Russia, India and China), having become leading manufacturers and developers of low carbon technologies (Brewer, 2008). Nevertheless, the term 'technology transfer' is still widely used within policy and development discussions, and captures the idea that there are many cases where internationally owned technologies can be beneficially brought to bear within different developing country contexts, with subsequent environmental, economic and human development benefits.

A tension might be assumed between the adoption of often more expensive low carbon technologies and the achievement of more immediate development concerns, including aggregate economic growth and other priorities, such as wider access to modern energy services to meet basic needs (e.g. heating, cooking and lighting), particularly for poor and marginalized people. But this tension may be artificial. Low carbon technologies have the potential to make a strong contribution to human as well as economic development aims. In many cases, rural energy access could be better met – from an economic, environmental and social perspective – by, for example, solar home systems that use combinations of solar photovoltaics, batteries to store energy and highly efficient light-emitting diode (LED) light bulbs. Solar hot water systems have been shown to be an excellent way of providing hot water in urban and rural developing country contexts (Rodrigues and Matajs, 2005; Mallett, 2009), and the provision of energy efficient cleaner burning cook stoves has potential to reduce reliance on scarce wood fuel resources and to reduce the health impacts of fumes from indoor wood burning for cooking and heat. The integration of low carbon technologies within a country's energy matrix can also reduce its reliance and exposure to energy imports. Together with the increase in the diversity of the energy matrix, this can also serve to increase energy security in the face of both future price fluctuations for fossil resources and, in the long term, reductions in fossil-fuel availability (Biswas et al., 2001; Renewables, 2004).

Moreover, access to technologies has been a key priority for developing countries as a way in which to spur innovation and, thus, economic development (traditionally correlated with levels of economic development) (UNCTAD, 2007; Maskus and Okediji, 2010). Developing countries' access to new low carbon technologies and the knowledge that underpins them can therefore potentially benefit economic competitiveness. The transfer of low carbon technologies to developing countries thus stands to contribute to both mitigating future climate change and to sustainable economic and human development in these countries. It is also worth noting here that, while low carbon technologies for the purpose of climate change mitigation is the main focus of this book, the transfer of technologies of relevance to adaptation to climate change is of equal importance. Many commentators thus now prefer to refer to the transfer of 'climate technologies' as opposed to 'low carbon technologies', the former now tending to characterize international policy discussions on the issue.

In large part due to the historical responsibility of developed nations for current atmospheric GHG concentrations and the high per capita emissions of these countries, low carbon technology transfer has formed a high profile issue within international climate change negotiations. Developed nations, at the time of writing, have obligations under both the United Nations Framework Convention on Climate Change (UNFCCC) and the Kyoto Protocol to facilitate the transfer of low carbon technologies to developing nations. The promise of access to new technologies is widely viewed as the carrot that brought developing nations on board with these multilateral environmental agreements. However, technology transfer is also widely viewed as one of the areas where these agreements have failed to deliver (Khor, 2008), creating difficult tensions and often bringing negotiations on a post-Kyoto climate change agreement to a standstill. Any post-Kyoto deal will have to directly address this issue and do so in such a way that it stands to deliver more effectively than previous policy efforts.

It is against this backdrop of the failure to deliver low carbon technology transfer to developing countries in practice and the significant environmental, economic and human development benefits that it can yield which this book seeks to contribute. Most importantly, the book engages with the lack of empirical analysis that has been available to support policy efforts in this area. Political rhetoric on the issue often takes on a life of its own, reflecting more the idea of 'received wisdom' (Leach and Mearns, 1996) than careful analysis based on empirical evidence. This has, for example, led to a failure to acknowledge insights from the field of innovation studies that emphasize the importance of the knowledge, or software, component of technology as opposed to simply the hardware – an issue attended to further below and by several of the contributors to this book (see, for example, Chapters 2, 4, 6 and 7 to 10).

A key concern, therefore, is that policy rhetoric on low carbon technology often fails to reflect the reality of how technology transfer can be achieved, not least in ways that will yield maximum development benefits for developing countries. This edited collection brings together leading contemporary thinkers on low carbon technology transfer to provide new empirical evidence, theoretical advancements and policy recommendations in one place for the first time, providing an overview of some of the key issues and emerging thinking in this field in an accessible policy relevant way. The book is targeted at policy makers, practitioners, researchers and students, and includes contributions from the research and practitioner communities in both developed and developing countries.

Several fundamental objectives have underpinned the development of this book. The first objective is to provide evidence based on 'real world' examples; the second is to revisit the theoretical underpinnings through which policy decisions are made;

the third is to provide examples of ways to move policy forward in constructive new directions. We (the editors) have tried as much as possible to take a light touch when editing chapters and allow authors space for their own voices, views, opinions and interpretations of empirical evidence. Low carbon technology transfer is very much an emerging field with much scope for varying and contrasting views on theory and practice. We hope that readers will consult individual chapters for insights into the informed views of the various contributors and begin to develop their own empirically informed views as a result.

As a backdrop to this edited collection, in this introductory chapter we seek to introduce some of the issues that we see as critical in this emerging field and how they link to the different chapters of the book. These can be characterized around four key issues – namely:

- Low carbon technology transfer is unique from conventional technology transfer.
- Drawing on the innovation studies literature, innovation capacities can be understood to be what supports sustained uptake of new technologies, and these are developed via access to qualitatively different strands of knowledge, as opposed to simply the hardware component of technology. Innovation capacities also encompass recognition of the importance of adaptive and incremental innovation in the context of developing countries.
- Drawing on the socio-technical transitions literature, the socio-technical nature of technological transitions and the way that technologies and society co-evolve and shape one another can be identified as fundamental to understanding how low carbon development might be catalysed.
- Building on insights from the innovation studies and socio-technical transitions literatures, the context-specific, spatially situated technology needs of different people in different spaces (physical, cultural, economic and so on) can be understood to play a critical role in defining both what technologies are appropriate and how likely they are to be successfully adopted.

The coverage in this chapter is, however, by no means intended to be exhaustive and readers are urged to consult the individual chapters of this book for further insights on priority contemporary concerns in this field. The text below draws on our experience in researching technology transfer, as well as our interactions with the pioneering unconventional approaches to low carbon technology transfer of many of the academic and policy thinkers whom we have interacted with over the years – many of whom are represented in this book – and combines these insights with promising strands of contemporary research on climate change, energy and development. A range of chapters in this book adopt some or all of these analytical components. In the text below we provide a brief overview of each, referring readers, where relevant, to chapters that deal with them in more detail. We conclude the chapter by providing an outline of the structure of the book and an overview of the content of the individual chapters therein.

Why is low carbon technology transfer unique?

Technology transfer from one country to another has long been a subject of interest for scholars from a number of disciplines – for example, economic geographers analysing spatial configurations of various industries and their value chains, innovation studies scholars interested in technological development within different industries and firms, and development studies scholars interested in the introduction of technological innovations in developing countries to benefit poor and marginalized people. These efforts have yielded many insights of relevance to low carbon technology transfer. However, there are several unique characteristics of low carbon technologies that mean these insights need to be revisited and often demand new theoretical stances and empirical analysis.

The first point to consider is the fact that low carbon technologies are of interest due to an urgent global problem to which they form a part of the potential solution – namely, the problem of climate change. This immediately introduces a temporal concern in the form of urgency. So rather than being interested in the transfer of technologies that naturally occur via market processes over unspecified timescales, in this case we are interested in the transfer and uptake of technologies within a timescale that is fast enough to mitigate or adapt to future climate change impacts.

Secondly, policy aimed at facilitating transfer of *low carbon* technologies is essentially attempting an intervention to speed up a process that might or might not be delivered by the market, but is doing so for the purposes of delivering a global public good (i.e. climate change mitigation or adaptation) (Mowery et al., 2010). This implies a need to intervene to create conditions that incentivize transfer in the absence of an obvious market.

Thirdly, low carbon technology transfer is further problematized by the fact that many of these technologies are at pre-commercial stages in their development. Policy is therefore not simply concerned with their 'horizontal' transfer from one country context to another, but also their 'vertical' transfer from early research and development (R&D) stages through demonstration, early pre-commercial deployment, to commercially viable stages of development. This introduces a whole range of additional considerations, such as how to overcome high levels of investor risk for technologies at early stages of development, or how to adapt technologies to new contexts – whether economic, environmental or socio-technical (see below for discussion of the relevance of context specific considerations).

Hence, there are several reasons why low carbon technology transfer demands a reappraisal of the relevance of existing insights on technology transfer. There is a need to revisit and reinterpret existing theory and empirical evidence and to carry out new analysis targeted directly at understanding the considerations that characterize low carbon technology transfer. One particularly useful theoretical stance stems from the innovation studies literature on innovation capacities to which we now turn.

Innovation capacities: Distinguishing between technology hardware and types of knowledge

Low carbon technologies have a key role to play in boosting economic growth and productivity more generally in developing countries due to the important link between technology ownership, technological capacities (more correctly articulated as 'innovation capacities' – see below) and economic growth. A North–South gap historically characterizes technology ownership (Missbach, 1999) with developed countries having a clear technological advantage. As stressed by UNCTAD (2007), the key to sustained economic growth and poverty reduction in developing countries is increased productive capacities, a core part of which relies on technological progress. So 'unless the LDCs [least developed countries] adopt policies to stimulate technological catch-up with the rest of the world, they will continue to fall behind other countries technologically and face deepening marginalization in the global economy' (UNCTAD, 2007, pI). This is not to say that developing countries would or should necessarily become technology leaders. Rather that, as emphasized by UNCTAD (2007), access to new technologies and the development of related capacities will play a key role in helping all developing countries (including LDCs) to develop their productive capacities and sustain economic growth and poverty reduction.

A central insight from the literature in the field of innovation studies is that the way in which these productive capacities are developed is via the development of 'innovation capacities' (c.f. Bell, 1990, 2009; Bell and Pavitt, 1993). This refers to the capacities to adopt, adapt, develop, deploy and operate technologies effectively within specific contexts. As noted above, this issue is dealt with in detail in several chapters in this book (including Chapters 2, 4 and 6 to 10), but it is worth highlighting some key points here in the introduction. Critically, work in the field of innovation studies has shown that the development of innovation capacities, particularly via international transfers of technology, relies on recognizing the fundamental knowledge component of technology. The construction of 'technology' within most international policy discussions assumes that technology consists solely of 'hardware' (physical equipment). Of equal, if not more, importance is the 'software' element of technology, including the knowledge and processes which underpin technology development, refinement, operation and new innovations on the basis of existing technologies (Bell, 1990; Bell and Pavitt, 1993), including adaptations to suit new physical and cultural contexts (Bell, 2009; Ockwell et al., 2010b). The knowledge component of technology includes both codified knowledge (e.g. engineering and manufacturing processes) and tacit knowledge (human-embodied knowledge acquired by doing, e.g. applied engineering and systems integration skills). Furthermore, this dynamic recognizes nuances to do with knowledge – that facets include know-how skills (e.g. the ability to operate and maintain 'hardware') and know-why skills (or the ability to understand the principles behind how such 'hardware' works) (Lall, 1995). The accumulation of tacit knowledge is particularly important in the context of developing countries seeking to catch up with technological frontiers and places particular emphasis on the value of international partnerships within technology transfer projects, as well as around softer capacity building activities, such as international personnel exchanges and training activities.

Some clarifications are, however, important. First, the argument here is not necessarily that all developing countries, particularly LDCs, for example, should aim efforts at developing, producing and/or using 'radical' innovations. Radical innovations are those with a higher degree of new knowledge – often occurring as a result of directed efforts into R&D, providing a major point of departure from previous ways in which energy is being generated, developed and/or used (Ockwell et al., 2008). Such an idea rests on an incomplete understanding of the idea of innovation. As per the Organisation for Economic Co-operation and Development (OECD) *Oslo Manual* (emphasized by Bell, 2007; UNCTAD, 2007), innovation should be understood not just as instances when a technology or process is new to the world, but also instances when they are new to a firm or new to a market – the latter two categories of innovation being most relevant to LDCs, but often equally relevant to other developing countries. It is often slow, incremental processes of technology acquisition, adaptation and improvement that best characterize the pathways that developing countries follow as they catch up with, and sometimes surpass, technological frontiers – as was, for example, observed with the Korean steel industry. But these incremental, adaptive processes should be considered no less innovative than the more radical innovation processes that describe the development of products and processes that are new to the world. In this vein, opportunities exist where the ability for countries, regions, communities and organizations to become technology innovators becomes broader, as opposed to orthodox frameworks which centre on a privileged few (large scale firms with headquarters in industrialized nations or BRIC countries – Brazil, Russia, India and China).

Such instances of incremental and adaptive innovation are widespread in developing countries. During the implementation and later operation of new technologies in these countries, significant opportunities arise to increase capacities by learning from the systems and procedures put in place to install and run these technologies and to subsequently improve and modify them within the specific circumstances where they are implemented (Bell, 2007). This implies two things: first, that the implementation of new low-carbon technologies can present significant learning and capacity-building opportunities; and, second, that existing 'innovation capacities' are important in facilitating this kind of process and ensuring its contribution to overall productive capacities in developing countries, including LDCs. Innovation capacities therefore do not refer exclusively to capacities to undertake more radical innovation.

The socio-technical nature of technology

The software/hardware distinction made within some of the innovation studies literature is clearly important to assisting energy policy in effectively facilitating the development of indigenous innovation capacities. However, the approach to innovation capacities within the innovation studies literature has been criticized as requiring revision when broader processes of technological change, such as to lower carbon development trajectories, are being considered (see Byrne et al., 2010, and Chapter 7 in this volume for a more detailed exploration of the relevance of a socio-technical transitions approach to low-carbon development). First, any approach stressing low-carbon innovation capacities needs to respond directly to the specific challenges raised by climate change – namely, the need for time-constrained, rapid transitions to low-carbon development pathways and, hence, the need to roll out technologies often at early stages of development in countries and communities with little related experience and capacity to work with or innovate (technically or socially – e.g. via new energy-related work practices such as pooled social arrangements for sharing labour) around these technologies. Second, it needs to be broadened to deal more fully with the demand-side aspects of innovation systems, including the political and institutional contexts of these systems, and the need to ensure that technology development proceeds on a self-determined, needs-led basis. Demand-side signals – users' wants, needs and demands communicated through social structures, markets and other means – help to shape innovation processes by indicating profitable and socially acceptable directions for technological trajectories. Political and institutional contexts influence innovation processes by shaping supply-side resource allocation decisions and by constraining or enabling demand. The innovation systems literature recognizes much of this but tends to focus on supply-side actors and their interactions. Where the demand side is understood, it is strongest in regard to user-firms rather than final consumers. Moreover, analysis tends to emphasize the introduction to the market of new technologies and concentrates less on how changes to innovation systems themselves have occurred in the past and might be incentivized (e.g. in low-carbon directions) in the future. One final critique is that while the innovation systems literature is extensive in terms of understanding institutional contexts, it is weak in terms of understanding political influences on innovation, and hence leaves little to work on in terms of how policy might influence directions of development.

A socio-technical transitions-based understanding of technological change can add value here (see Rip and Kemp, 1998; Geels, 2002; Smith et al., 2010). A socio-technical approach recognizes that technologies are embedded interdependently in existing social practices and reflect knowledge of these practices, as well as knowledge of technical principles. A firm or farm might be the locus of innovation and technological learning (UNCTAD, 2007), but they are embedded within a broader framework of institutions which define and are defined by these processes. The spatial development of towns and cities, the nature of transport infrastructure and the institutional norms governing travel, for example, have all developed around the use of the internal combustion engine, and continue to influence and constrain the direction and nature of current innovations in transport modes and efficiency. An important hypothesis that flows from this perspective is that technologies will be widely adopted not simply because they successfully harness technical principles, but also if their form and function are 'aligned' with dominant social practices, or offer opportunities to realize new practices that are attractive in particular social and geographical settings. This has important implications for the kinds of low-carbon energy approaches that are likely to be adopted and work effectively within the specific contexts of small developing countries (see the discussion of context-specificities below). It also emphasizes a more dynamic systems-level approach to policy analysis.

Context-specificities: A needs-based approach

A final insight that can be teased out of several strands of climate and energy policy research and reverberate within this volume is the current lack of a needs-based approach to policy, coupled with, and related to, a need to attend to the context specificities that define the appropriateness of low-carbon energy options in any given situation. Hulme (2008, 2009) alludes to the importance of context specificities through his emphasis on how the idea of 'climate change' has been dominated by certain constructions of the issue which ignore the multiple spatially and culturally contingent understandings and meanings of climate, and hence (by implication) potentially undermine constructive ways forward for society to both interpret and decide how to respond to a changing climate.

When examined at the level of considering whose needs international climate policy (particularly regarding low-carbon energy technology transfer) is geared towards meeting, the need to attend to context specificities becomes especially acute. Important questions therefore must be asked regarding the context specificity of technology needs, both in their applicability within different socio-technical circumstances and their applicability within different physical, cultural and economic contexts. For example, the technological needs of communities with different wealth levels should be understood – poorer communities, for example, might perhaps have a greater need for technologies related to subsistence needs; wealthier communities might have priorities around transport or processing goods to add value. Questions must be asked as to what extent the flows of low-carbon technologies facilitated under existing international policy mechanisms are pro-poor (for a useful point of departure, see Urban and Sumner, 2009, and Chapter 12 in this volume). In poor rural areas, for example, it might be more viable to explore adaptive innovation around low-maintenance configurations of solar and light-emitting diode (LED) technology, as opposed to clean options for centralized energy generation which might better suit urban industrial interests. And in adapting to climate change, technologies such as drought-resistant strains of crops or knowledge regarding new farming methods in increasingly flood-prone areas might be of more relevance to poor people than advanced engineering solutions for strengthening coastal flood defences.

Context specificities in relation to technology needs are also likely to extend to differences between the needs of rapidly emerging economies and those of small developing countries – and a whole spectrum of contexts between the two. For example, while rural electrification and poverty alleviation is still an issue for both emerging economies and small developing countries, a distinct need exists to understand and chart the distribution of capacities for absorbing and working with different low-carbon technologies (referred to in the innovation literature as 'absorptive capacities'), across and within different country contexts. For example, to what extent do different developing countries, regions, firms or communities have the capacities to work with technologies at different stages of commercial development (e.g. dealing with higher investor risk at earlier stages of technology development), or to work with the hardware and software components involved? One example of this would be a technology such as carbon capture and storage (CCS), which involves more complex systems management capabilities than small-scale solar photovoltaics (Ockwell et al., 2010b), although it should be noted that CCS provides an illustrative example here as opposed to being an option under consideration for all developing countries. Such nuanced understandings of relative technological capabilities internationally and intra-nationally have a key contribution to make to better orienting international policy efforts.

The issue of context specificities in relation to technology needs also emphasizes a necessity to attend to the widely varying suitability and viability of different low-carbon technological options within different spatial contexts. For example, carbon capture and storage technologies will need to be adapted to suit both local fuel sources and geological storage options (Tomlinson et al., 2008); energy efficiency or clean decentralized energy options need to work within the context of existing cultural (behavioural) practices and existing infrastructure; and so on.

The structure of this book

Having introduced some key insights of relevance to low-carbon technology transfer, we now turn our attention to outlining the structure of this book and the contributions that the individual chapters make across and beyond the themes outlined above. The first part of the book focuses on introducing some thoughts on new analytic approaches to low-carbon technology transfer. As well as the analysis in this introductory chapter, this includes an important contribution from Martin Bell in Chapter 2, arguably one of the most influential thinkers in the field of technology transfer more broadly – not just in relation to low-carbon technology. Bell's chapter focuses, in particular, on areas that remain to be addressed by work in this area and includes constructive criticism of approaches adopted to date, which includes highlighting some weaknesses in how other authors (including several contributors to this book, the editors included) attending specifically to low-carbon technology transfer have operationalized insights from innovation studies. This focus on new analytic approaches is not exclusive to this opening section of the book and is taken up and developed further by several other chapters. For example, Byrne et al., in Chapter 7 pick up in more detail the ideas explored above in relation to socio-technical transitions via their critical analysis of the Clean Development Mechanism (CDM) and alternative framings of development pathways. Several authors also provide insightful analyses and empirical demonstrations of the benefits of an analytic approach based on insights from innovation studies (see, for example, Chapters 4, 6 to 8 and 10).

The book then moves on to give some examples of grounded country and technology-specific analyses of the barriers faced by developing countries in accessing low-carbon technologies. These analyses also highlight areas where policy and other initiatives have worked well, such as Pal and Sethi's analysis (Chapter 3) of the uptake of fuel-efficient technologies by small- and medium-sized enterprises (SMEs) in India, and Watson and Byrne's (Chapter 4) detailed analysis of various low-carbon energy technologies in China. These chapters provide exemplary examples of how such context-specific empirical analysis can provide the basis for effective policy recommendations that move away from a reliance on the received wisdom of existing policy rhetoric.

A key challenge regarding low-carbon technology transfer relates to the polarized, entrenched negotiating positions between the 'North' and the 'South', which permeate the atmosphere of international climate negotiations. Arguably, nowhere are these divisions as pronounced as they are within the debates surrounding intellectual property rights (IPRs). IPRs have become a pivotal area that has been subject to much controversy within international political debates on low-carbon technology transfer. For example, by analysing empirical evidence on IPRs in relation to low-carbon technology transfer, Ockwell et al., (2010a) offer an example of how political rhetoric can result in the same evidence being interpreted to support very different policy positions. They argue that opposing policy positions can be explained in part by the different motivations of developed and developing country actors for coming on board with an international climate change agreement. Developed countries were motivated by finding solutions to a global environmental problem – climate change – and their interest in low-carbon technology transfer is therefore primarily concerned with the rapid diffusion of low-carbon technologies in developing countries. Developing countries, on the other hand, were motivated by the potential role that access to new technologies could play in underpinning their economic development – access to technology traditionally having been linked to levels of industrialization. In examining the available evidence, Ockwell et al., (2010a) found that the picture is far more complex than the two opposing framings of development versus diffusion would suggest.

The third section of this book also engages directly with the IPR debate and offers new empirical analysis on this subject by Srinivas (Chapter 6) and Abdel Latif (Chapter 5). Krishna Ravi Srinivas provides readers with some of the historical and current debates which have characterized the discussions surrounding low-carbon technology transfer and IPRs within the high-level climate change political discussions. He also highlights interesting open-innovation concepts from other sectors, including information technology, health and environment, which can help to move forward from the current impasse between North and South divisions with respect to IPRs and low-carbon technologies. Ahmed Abdel Latif sheds further light on this subject through his chapter on research conducted with the European Patent Office (EPO), the United Nations Environment Programme (UNEP) and the International Centre for Trade and Sustainable Development (ICTSD)'s project on assessing the landscape of clean energy technology patents and licences. Here, Abdel Latif avows (similar to other studies) that patenting activity tends to be concentrated in industrialized countries, but that when clean energy technology patenting data is assessed *vis à vis* total patenting activity in a country, some developing countries are emerging as leaders (e.g. India on solar photovoltaics (PV) and Brazil and Mexico on hydro/marine). Furthermore, 'policy triggers' do play an important role in spurring innovation in the area of low-carbon technology, as manifested by the rise in global patents in this sector after the conclusion of the 1997 Kyoto Protocol. Finally, organizations and firms interested in licensing to developing countries were concerned about intellectual property (IP) protection, but 'attached greater weight' to other attributes, including having a favourable market, investment climate, human capital and scientific infrastructure. Many firms also indicated their receptivity to providing flexible terms to developing countries with more constrained capacities.

The fourth section of the book focuses on empirical critiques of existing policy mechanisms, particularly the United Nations-based mechanisms of the CDM and Global Environment Facility (GEF). The UN-backed climate discussions are viewed by many as an important channel through which to influence actions (e.g. the investment actions of multinational corporations, the programming efforts of international and local non-profit organizations, and national and subnational public policies) at a more macro level and are the key vehicles through which policy might deliver on UNFCCC commitments on low-carbon technology transfer. These mechanisms have tended to reflect the dominance of environmental economic framings of environmental problems as needing to pay the incremental costs of low-carbon technologies to reflect their social benefits. Within much of the literature, debates seem to gravitate either towards optimistic accounts of what the CDM and GEF have done to increase technology transfer and/or economic and social development (e.g. Duic et al., 2003), or accounts (much more prominent) which are critical of these entities (Forsyth, 2007; Pearson, 2007). Our book adds to these discussions. Haites et al., (Chapter 9) provide enhanced statistical analysis of technology transfer within projects in the CDM pipeline through examining the notion of time lags, knowledge stock and overall technology transfer. Their analysis yields interesting results, including the fact that technology transfer tends to happen more frequently among larger projects, but that overall it is decreasing over time and is less common in host countries with a larger 'technological base'. Furthermore, Haites et al., unpack the CDM project process further, noting that countries tend to purchase credits from those projects which involve firms from their countries, but that in the technologies they examined, there was no one 'dominant supplier' preventing choice.

Both Verbeken (Chapter 8) and Haum (Chapter 10) respond to a lack of empirical analysis of the GEF (including by the agencies tasked with implementing it) and present some detailed empirical analysis of the experiences of, and possible lessons to learn from, the GEF. Adopting a theoretical stance based on an insightful discussion of innovation capacities (drawn from the innovation studies literature), Haum undertakes empirical analysis of the GEF-funded Photovoltaic

Market Transformation Initiative (PVMTI) in India. He presents a critical assessment of the extent to which the market transformation approach favoured by the GEF contributed to industrial development and capacity development around photovoltaics in India. His findings show that the contribution of the PVMTI was limited and highlight a need for the GEF to integrate a more targeted focus on knowledge transfer as part of its approach. Also adopting an innovation studies based stance, Verbeken is critical of the lack of reflective analysis by GEF-implementing agencies of what works and what does not in terms of its technology transfer activities. She contributes to addressing this gap by tracing the development of financial support for technology under the UNFCCC and the issues that remain to be addressed. By examining four case studies of technologies that have received support under the GEF, she illustrates the challenges and opportunities for technology transfer funded by multilateral institutions. Her analysis highlights limited understanding and experience of technology transfer within multilateral agencies, while noting some positive moves under emerging international policy negotiations to address this.

Byrne et al., (Chapter 7) take analysis of the CDM in an entirely new direction. Drawing on critical new approaches from the emerging development literature, particularly the STEPS Centre's[1] Pathways Approach (Leach et al., 2007) and integrating this with insights from socio-technical transitions research, they highlight the limitations of policy based on limited political framings of the problem of low-carbon technology transfer which have failed to deliver in practice. They demonstrate how policy has tended to frame the issue as a need for hardware financing to cover the incremental costs of low-carbon technologies. This has resulted in policy mechanisms such as the CDM, which provides opportunities for developed countries to meet their emissions reduction targets under the Kyoto Protocol at reduced cost by investing in emission-reducing projects in developing countries. Byrne et al.,'s empirical analysis shows how this has led to a limited number of technologies being invested in, in a limited number of countries. For example, China has received two-thirds of the finance available to date, India 14 per cent and Brazil 2 per cent, leaving the remaining participating countries with just 17 per cent between them. And over 75 per cent of registered CDM projects are implemented using just five types of technology – hydro, methane avoidance, wind, biomass energy and landfill gas (based on data available from UNEP Risø, 2010). Byrne et al., argue that reframing the problem as a need to develop indigenous innovation capacities in developing countries, as well as understanding the socio-technical nature of technological development, could lead to more effective policy solutions that better deliver against the development needs of developing countries. CDM reform and its potential contribution to development and poverty reduction constitute a focus of several other chapters in this book, including Chapters 12, 18 and 19.

The fifth section of this book tackles an issue that is all too often absent in international discussions of low-carbon technology transfer and which, to date, is under-represented in research in this field – namely, the link with poverty reduction (readers should note that the broader development benefits of low-carbon technology transfer are also dealt with elsewhere in the book, most significantly in Chapters 18 and 19 by Forsyth and Castle). The inadequacy of low-carbon technology transfer in meeting the needs of developing countries is not a new story, but it is an important one to reiterate: on the one hand, current practices continue – with CDM projects being skewed towards large-scale initiatives in emerging economies (especially China and India), while, on the other, a new opportunity exists in which to undertake a major overhaul of mechanisms, shifting climate tools to align more closely with poverty tools. Urban and Sumner (Chapter 12) present new theoretical thinking on what might constitute 'low-carbon development'. This is one of the first published attempts to theorize on a development model which simultaneously addresses human development, economic development and climate change mitigation concerns. Their contribution provides us with valuable heuristic approaches to conceptualize and operationalize low-carbon technology transfer within a broader development context.

Hedger (Chapter 11) charts the extent to which technology transfer initiatives under the UNFCCC address poverty issues, examining in detail emerging climate finance issues under the UNFCCC and situating these discussions within an emerging green economy agenda that links the global financial crisis to the broader development agenda. The chapter benefits from Hedger's extensive experience of the UNFCCC since its inception and offers some timely insights into the emerging dynamics at the international policy level. Kooijman-van Dijk (Chapter 13) also focuses on the links between climate technologies and the poor with an interesting focus on SMEs. Through detailed empirical analysis, she highlights the significance of SMEs in a developing country context, both in terms of their contribution to the economic welfare of poor people and their potential contributions to reducing carbon emissions, and shows how the needs of SMEs are not currently addressed by climate change policy.

Low-carbon technology transfer and the problem of climate change do not exist in a vacuum. They are situated within policy space which is also characterized by a wide range of other urgent concerns. For many governments, a primary concern is the interaction between transferring technologies to other countries and current and future trade policy regimes. Part VI of this book contains contributions which situate low-carbon technology transfer within the context of these broader global policy concerns. Sugathan and Mani (Chapter 14), through empirical evidence gleaned from a World Bank project in South Asia, purport that governments can play a key role in not only policy engagement (as echoed by numerous studies), but also through catalysing effective implementation efforts on the ground. They further remind audiences about the importance of engaging the private sector and, in so doing, the need to ensure that in pursuing the goal of addressing climate change effectively, efforts dovetail with the broader trade and investment regime.

By the end of the book, the critical empirical analysis and new theoretical directions offered by the chapter authors will leave readers in no doubt that effective policy in this area requires new thinking and new directions. All of these experiences call out for a fundamental shift regarding how technology transfer is practised and discussed – from the hallways of the UN negotiating meetings in Bonn,

Technology Transfer Agreements (TTAs) agreed upon by firms in the North and South, to communities in LDCs seeking an appropriate less carbon-intensive pathway.

The final section therefore provides several chapters that engage with this need for a new direction and offer a range of empirically grounded insights on new or revised institutional approaches to policy delivery in this field. Vincent (Chapter 16) draws on extensive personal experience in the field of low-carbon technology policy practice to reflect on key contemporary challenges and the potential value of the emerging idea of a network of low-carbon innovation centres in developing countries. Castle (Chapter 19) explores the extent to which a reformed CDM might better deliver on broader development benefits. His commentary includes a systematic review of the potential shape and benefits of alternative options for CDM reform in this regard. Forsyth (Chapter 18) and Lovett et al., (Chapter 17) both explore new institutional approaches to facilitating technology transfer based on deeper engagement with developing country stakeholders. Forsyth's insights regarding low-carbon technology experiences from South-East Asia suggest that cross-sector partnerships (i.e. partnerships between local communities and investors/developers/governments) can be a way in which to bring about effective technology transfer and benefits at the local level. They can do this by considering local development needs (sometimes referred to as the 'development dividend') concurrently with low carbon mitigation – through fostering community discussion and active participation in the project. Echoing, to some extent, Forsyth's emphasis on the benefits of cross-sector partnerships, Lovett et al., assess existing technology financing mechanisms and conclude that neither finance nor technology access are the real limiting factors to broader uptake of low-carbon technologies; rather, it is a lack of 'enabling environments' which, they suggest, might be fostered via a new approach characterized by 'multi-stakeholder partnerships'.

Conclusion

People generally agree that the magnitude of low-carbon technology transfer required to tackle the problem of climate change has failed to materialize since the UNFCCC's creation in 1992. Furthermore, low-carbon technology transfer raises problems and challenges that are unique and require a new and nuanced empirically grounded approach to tackling the problem. This requires a fundamental shift in understandings of technology transfer within the broader context of innovation; the intertwined, mutually defining nature of society and technologies; and the context-specific, spatially situated technology and development needs of different people in different developing countries. These complexities demand policy built on grounded empirical analysis and developed around institutional approaches that can build innovation capacities in developing countries while being responsive to their reflexively defined needs.

This book serves as a platform through which to present the work of mindful scholars, researchers and practitioners, constituting alternative views, which while starting to gain traction and a stronger voice, nevertheless tend to be drowned out in the cacophony of rhetorical, circular, regurgitated discussions which surround this issue within international policy forums. The rapid advancement of policy discussions, both international and bilateral, and the increasing number of funded initiatives in this field make this a critical time to appraise and act upon the insights available. We hope this book will provide a valuable resource that can contribute to such appraisal amongst policy-makers, practitioners and researchers alike.

Note

1   STEPS – Social, Technical and Environmental Pathways to Sustainability; see http://www.steps-centre.org/.

References

Bell, M. (1990) *Continuing Industrialisation, Climate Change and International Technology Transfer*, SPRU, University of Sussex, Brighton, UK

Bell, M. (2007) *Technological Learning and the Development of Production and Innovative Capacities in the Industry and Infrastructure Sectors of the Least Developed Countries: What Roles for ODA?*, UNCTAD, The Least Developed Countries Report 2007, Background Paper, SPRU, University of Sussex, Brighton, UK

Bell, M. (2009) *Innovation Capabilities and Directions of Development*, STEPS Working Paper 33, STEPS Centre, Brighton, UK

Bell, M. and Pavitt, K. (1993) 'Technological accumulation and industrial growth: Contrasts between developed and developing countries', *Industrial and Corporate Change*, vol 2, pp157–210

Biswas, W. K., Bryce, P. and Diesendorf, M. (2001) 'Model for empowering rural poor through renewable energy in Bangladesh', *Environmental Science and Policy*, vol 4, no 6, pp333–344

Brewer, T. (2008) 'Climate change technology transfer: A new paradigm and policy agenda', *Climate Policy*, vol 8, pp516–526

Byrne, R., Smith, A., Watson, J. and Ockwell, D. G. (2010) 'Energy pathways in low-carbon development: From technology transfer to socio-technical transformation', STEPS Centre Working Paper, STEPS, University of Sussex, Brighton, UK, http://www.steps-centre.org/PDFs/Energy_PathwaysWP.pdf

Duic, N., Alves, L. M., Chen, F. and da Graca Carvalho, M. (2003) 'Potential of Kyoto Protocol Clean Development Mechanism in transfer of clean energy technologies to Small Island Developing States: Case study of Cape Verde', *Renewable and Sustainable Energy Reviews*, vol 7, pp83–98

EIA (Energy Information Administration) (2010) *International Energy Outlook 2010*, US Energy Information Administration, Washington, DC

Forsyth, T. (2007) 'Promoting the "development dividend" of climate technology transfer: Can cross-sector partnerships help?', *World Development*, vol 35, no 10, pp1684–1698

Geels, F. (2002) 'Technological transitions as evolutionary reconfiguration processes: A multi-level perspective and a case-study', *Research Policy*, vol 31, pp1257–1274

Hulme, M. (2008) 'Geographical work at the boundaries of climate change', *Transactions of the Institute of British Geographers*, vol 33, pp5–11

Hulme, M. (2009) *Why We Disagree about Climate Change*, Cambridge University Press, Cambridge

IPCC (Intergovernmental Panel on Climate Change) (2007) *Fourth Assessment Report, Climate Change 2007: Synthesis Report Summary for Policymakers*, IPCC, Geneva, Switzerland

Jackson, T. (2011) *Prosperity without Growth: Economics for a Finite Planet*, Earthscan, Abingdon, UK

Khor, M. (2008) *Access to Technology, IPRs and Climate Change*, Day One Session, European Patent Forum, http://www.epo.org/about-us/events/epf2008/forum/details1/kohr.html

Lall, S. (1995) *Science and Technology in the New Global Environment: Implications for Developing Countries*, United Nations, Geneva, Switzerland

Leach, M. and Mearns, R. (1996) *The Lie of The Land: Challenging Received Wisdom on the African Environment*, James Curry, Oxford, UK

Leach, M., Scoones, I. and Stirling, A. (2007) *Pathways to Sustainability: An Overview of the STEPS Centre Approach*, STEPS Approach Paper, STEPS Centre, Brighton, UK

Mallett, A. (2009) *Technology Adoption, Cooperation and Trade and Competitiveness Policies: Re-Examining the Uptake of Renewable Energy Technologies (RETs) in Urban Latin America Using Systemic Approaches*, PhD thesis, London School of Economics and Political Science (LSE), London

Maskus, K. and Okediji, R. (2010) *Intellectual Property Rights and International Technology Transfer to Address Climate Change: Risks, Opportunities and Policy Options*, Issue Paper No 32, International Centre for Trade and Sustainable Development (ICTSD), Geneva, www.ictsd.org/downloads/2010/12/maskusokedijiitests.pdf

Missbach, A. (1999) *Das Klima zwischen Nord und Süd: Eine regulationstheoretische Untersuchung des Nord-Süd-Konflikts in der Klimapolitik der Vereinten Nationen*, Westfälisches Dampfboot, Münster, Germany

Mowery, D. C., Nelson, R. R. and Martin, B. R. (2010) 'Technology policy and global warming: Why new policy models are needed (or why putting new wine in old bottles won't work)', *Research Policy*, vol 39, pp1011–1023

Ockwell, D. G. (2008) 'Energy and economic growth: grounding our understanding in physical reality', *Energy Policy*, vol 36, pp4600–4604

Ockwell, D. G., Watson, J., MacKerron, G., Pal, P. and Yamin, F. (2008) 'Key policy considerations for facilitating low carbon technology transfer to developing countries', *Energy Policy*, vol 36, pp4104–4115

Ockwell, D. G., Haum, R., Mallett, A. and Watson, J. (2010a) 'Intellectual property rights and low carbon technology transfer: Conflicting discourses of diffusion and development', *Global Environmental Change*, vol 20, pp729–738

Ockwell, D. G., Watson, J., Mallett, A., Haum, R., MacKerron, G. and Verbeken, A. (2010b) *Enhancing Developing Country Access to Eco-Innovation: The Case of Technology Transfer and Climate Change in a Post-2012 Policy Framework*, OECD Environment Working Papers, No 12, OECD Publishing, doi: 10.1787/5kmfplm8xxf5-en

Pearson, B. (2007) 'Market failure: Why the Clean Development Mechanism won't promote clean development', *Journal of Cleaner Production*, vol 15, pp247–252

Renewables (2004) *Conference Report for Renewables 2004*, Renewables 2004, International Conference for Renewable Energies

Rip, A. and Kemp, R. (1998) 'Technological change', in S. Rayner and E. Malone (eds) *Human Choices and Climate Change, vol 2: Resources and Technology*, Battelle, Columbus, OH

Rodrigues, D. and Matajs, R. (2005) *Brazil Finds Its Place in the Sun: Solar Water Heating and Sustainable Energy*, Vitae Civilis and the Blue Moon Fund, Sao Lourenço da Serra, Brazil

Smith, A., Voß, J. and Grin, J. (2010) 'Innovation studies and sustainability transitions: The allure of the multi-level perspective and its challenges', *Research Policy*, vol 39, pp435–448

Tomlinson, S., Zorlu, P. and Langley, C. (2008) *Innovation and Technology Transfer: Framework for a Global Deal*, E3G and Chatham House, London

UNCTAD (United Nations Conference on Trade and Development) (2007) *The Least Developed Country Report 2007*, United Nations, Geneva

UNEP (United Nations Environment Programme) Risø (2010) *CDM/JI Pipeline Analysis and Database*, http://cdmpipeline.org/

Urban, F. and Sumner, A. (2009) *IDS in Focus: Policy Briefing. After 2015: Pro-Poor Low Carbon Development*, Issue 9, Institute for Development Studies, Brighton, UK

2

INTERNATIONAL TECHNOLOGY TRANSFER, INNOVATION CAPABILITIES AND SUSTAINABLE DIRECTIONS OF DEVELOPMENT

Martin Bell

Introduction

This chapter locates the issue of international transfer of lower-carbon technologies at the interface between two issues: one concerned with responses to climate change, involving both mitigation measures and adaptation activities; the other about achieving per-capita income growth and wider forms of socio-economic development in middle- and low-income economies.[1] At a general level, this perspective involves nothing new. International transfer of technology has been widely seen as an important link between the two issues. Very roughly, the argument is that the greater the transfer of lower-carbon technologies, especially if their financial costs are reduced, the better will middle- and low-income economies be able to pursue growth in more environmentally sustainable directions. But that relationship between transfer and growth is highly variable, as is the extent to which transfer influences wider dimensions of development.

Aspects of that variability have been explored in previous studies. Some of these have identified institutional and other barriers or obstacles to transfer, such as prevailing intellectual property rights regimes (e.g. Schneider et al., 2008). However, such barriers – if, indeed, that is what they are – have been seen primarily as constraints on access to technology rather than as constraints on the outcomes that follow from having acquired it. In other words, the main problems about the relationship between transfer and development are identified as problems about variability in the *quantitative scale* of international technology flows in the first place, and much less about variability in the growth and development effects that arise thereafter. This chapter concentrates primarily on that second kind of variability.

A number of studies have begun to address that issue by highlighting the importance of the *qualitative content* of international knowledge flows, not just their quantitative magnitude (Ockwell et al., 2008, 2010; Doranova et al., 2011). They have suggested that differences in this aspect of international transfer influence its outcomes. In particular, these views focus on the extent to which transfer results in only:

- additions to the *production capacity* of the acquiring economy (e.g. a new coal gasification plant or an up-to-the-frontier energy-efficient steel casting-rolling mill, plus the skills to operate them efficiently); or also
- additions to the *innovative capacity* of the economy (new competences to innovate in these areas of technology).

However, these studies have an important limitation. The analysis usually stops at the point where additional innovation capability is identified as a developmentally positive outcome from international transfer. But beyond the rather obvious idea that this will contribute to innovation in some unspecified way, what is the developmental purpose of that capability? In particular, what kinds of innovation is it expected to contribute to? With Ockwell et al., (2009) as an exception, such questions have rarely been addressed, let alone answered. But a starting point for this chapter is that seriously misleading answers are implicit and sometimes explicit in much of the discussion of international transfer and climate change. This arises because those discussions are deeply embedded in two kinds of widespread views about innovation.

One view is about the types of innovation that are necessary for achieving low-carbon trajectories of growth – 'radical' or 'breakthrough' innovations. These usually dominate lists of policy priorities for global technology development to address climate change, and they are central to discussions of national-level priorities in the advanced economies (e.g. Slocum and Rubin, 2008, with respect to R&D priorities in the US). Such policy-centred emphases have been buttressed by more academic studies – for example, by literature over the last decade or so about the nature of socio-technical transitions towards sustainable futures. This commonly identifies 'radical' innovations as the basis of correspondingly radical shifts to more sustainable technology systems developed initially in niche contexts, while 'incremental' innovations are identified as those that contribute only to perpetuating the existing regimes of non-sustainable technological development (e.g. Geels and Schot, 2007).

A second view centres more specifically on the way in which the innovation process is often discussed in connection with the development of low-carbon technologies: as an 'innovation chain' running from research and development (R&D), via demonstration to deployment and diffusion. Such chains are described as delivering the necessary novel technologies as one-off entities that are deployed and diffused to an increasingly wide range of users (e.g. UNFCCC, 2009, pp15–20). Within this schema the technologies are considered more or less 'fixed' by the time of the deployment stage, and diffusion is seen largely as a matter of innovation-free selection and replication. The technologically creative components of innovation are therefore confined to the initial research, development and demonstration (RD&D) steps used to develop significantly novel technologies. Consequently, discussion about developing countries engaging in low-carbon innovation tends to slip past a few brief remarks about 'improvement' and 'adaptation' and slide into a focus on the engagement of only a few of them in such initial RD&D stages of technology development.[2]

The *World Development Report 2010: Development and Climate Change* (World Bank, 2010) illustrates the combination of these views in connection with policy discussion. Chapter 7 ('Accelerating innovation and technology diffusion') emphasizes the importance of 'breakthrough' technologies for addressing both mitigation and adaptation. These are described as passing through the standard research, development, demonstration and diffusion (RDD&D) innovation chain (p295); the early stages of this are the province of high-income countries. Middle-income countries (should/do?) focus on investment in existing technologies, and low-income countries (should/do?) concentrate on identifying, assessing and adapting and improving them (p288). In discussing technology transfer, the emphasis on capacity development should be placed on absorbing existing technology (p289). There are brief comments about improving and adapting what has been acquired; but there is no elaboration of these ideas and they become lost in the emphasis given to breakthrough technologies – in particular, energy efficiency, carbon capture and storage, next generation renewables and nuclear power (p289). The tabulated summary of suggested policy priorities across the three groups of countries (p303) reflects these perspectives.

This chapter argues that these kinds of perspectives, by highlighting selected features of low-carbon innovation while underemphasizing or neglecting others that may be just as important, lead very easily towards imbalanced discussions about what to do in practice about international transfer of such technologies to developing countries. In particular they diminish the attention given to the importance of efforts to shift the qualitative content of transfer in ways that add much more substantially to the innovation capabilities of *all* types of developing economy.

The chapter seeks to redress that imbalance a little. It does so with primary reference to innovation and technology transfer in industry – broadly defined to include both energy 'supply' industries and energy-'using' industries, ranging from mining through various kinds of manufacturing. Although the experience in these kinds of industry also seems applicable in many of the infrastructure service industries (rail, road, sea and air transport) and construction, slightly different perspectives seem necessary in industries such as agriculture and services. The second section elaborates upon the distinctions made above between different types of qualitative content of transfer and the differing capabilities they may augment. The third section then further explores the types of innovation to which augmented innovation capabilities might contribute, highlighting the importance of incremental as well as radical types. The fourth section uses a case study of long-term low-carbon technology development in the steel industry to illustrate how these different kinds of innovation interact over time in generating paths of disruptive change in the industry. The fifth section returns to international transfer and raises questions about its qualitative content – in particular, what forms of knowledge and skill contribute to increased innovative capabilities? How are these acquired? In what kinds of organization is their acquisition and accumulation likely to be most effectively located?

The qualitative content of international transfer and the capabilities acquired

The qualitative content of transfer is identified in some discussions in terms of the extent to which the technology is 'complex/simple', 'advanced/mature', 'high-tech/low-tech', and so forth. For example, in discussing technology transfer via different kinds of projects under the Clean Development Mechanism (CDM), one study notes that, as between projects, the 'technological content varies in its complexity and its relative performance compared with state-of-the-art technology in the field' (Schneider et al., 2008, p2932).

However, those are not the senses in which the notion of the 'qualitative content' of transfer is used here. Instead, it refers to a spectrum of different forms of technology that may (or may not) be transferred *within* any transfer project, regardless of whether the overall package of technology is 'high-tech' or 'low-tech', 'advanced' or 'mature'. Such differences in the intra-project composition of internationally transferred technology have been examined in several studies of CDM projects. For example, Dechezleprêtre et al., (2008) and Seres et al., (2009) use a simple distinction between two kinds of content: 'equipment', referred to sometimes as 'hardware'; and 'knowledge', referred to sometimes as 'software'. For both of the different samples of CDM projects in the two studies, international transfer was involved in less than half of the cases (43 and 30 per cent, respectively). Within these the composition of the content of transfer varied quite widely: the transfer of 'equipment' alone was present in 21 and 37 per cent; the transfer of only 'knowledge' in 35 and 15 per cent; and the transfer of both in 44 and 53 per cent.

But such observations do not tell us very much because the distinction between 'equipment' and 'knowledge' is very unclear and does not identify many forms of technology that may be important. For example, technology for the 'hardware' of new production facilities may not be transferred only in the form of equipment. Instead, technology-importing organizations may acquire the preceding designs and specifications for the equipment via the purchase of engineering and related services for the design and construction of facilities, or through the licensing of some of the process and/or product designs and specifications to be incorporated within local engineering for new production facilities. It is not at all clear whether such design and informational 'precursors' of the tangible technology that finally comes to be used in production should be included in the category of 'equipment/hardware' or 'knowledge/software'.

At the same time, as well as overlapping in such ways with 'equipment/ hardware', the highly heterogeneous 'knowledge/software' category includes a wide range of other forms of knowledge and skills: for carrying out routine operational tasks; for initial trouble-shooting and problem-solving in the new facility; for maintenance of the facility and, hence, for sustaining its performance over time; or for various types of subsequent improvement and development of the technology that was initially acquired.

One could, of course, go further with such comments and elaborate extensive lists of the different components of the content of transfer – quickly proceeding beyond the point of practical utility. Instead, the discussion here keeps things fairly simple and follows closely the framework used in Ockwell et al., (2010) and subsequently drawn on in Mitchell et al., (2011) (see Figure 2.1). This distinguishes between only three forms of technology within the content of what may be transferred in particular projects:

- Flow A: capital goods (i.e. equipment and other constructed facilities), capital services (i.e. design, engineering and related services that are used in creating capital goods) and 'ready-made' designs and specifications for products or processes that may be purchased or licensed.
- Flow B: various types of skill, knowledge and know-how for operating new production facilities and for undertaking associated maintenance activities.
- Flow C: a bundle of many kinds of knowledge and skill for adapting, improving and further developing the technology initially acquired.

Obviously, there are fuzzy boundaries between these categories; but one merit of this simple framework is that, almost by definition, it allows one to move on to distinguish between two broad kinds of developmental outcomes of international transfer:

1. Outcomes that, drawing on elements of Flow A and/or Flow B, contribute additions to the production capacity of the technology-importing organization and its home economy (e.g. a new coal gasification plant, a new solar panel and refrigeration system in a rural health centre, or a more energy-efficient compact strip mill in an existing steel plant);
2. outcomes that draw on Flow C and add to the technology-importer's innovative capability (e.g. capabilities to contribute to subsequent innovation in coal gasification technologies, solar-based refrigeration systems for health centres, or compact strip mill technology).

To oversimplify somewhat, the underlying argument here is that very large quantities of technology, including current vintages of lower-carbon technology, are transferred to developing economies in the forms of Flows A and B. These flows occur via imports of goods and services – sometimes linked to foreign direct investment or donor-supported schemes, but to a great extent simply via arm's-length trade.

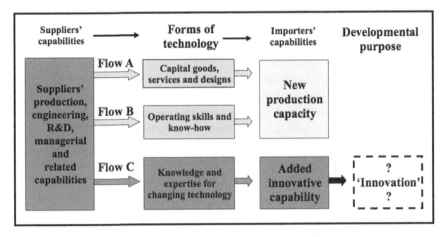

FIGURE 2.1 The content of technology transfer and the capabilities required
Source: Adapted from Bell, 1990, and Okwell et al., 2010

In contrast, associated transfer of technology in the form of Flow C is limited. But, the contention is, the larger the components of Flow C in the international transfer of lower-carbon technologies (and, hence, the greater the additions made to innovative capabilities and not just to production capacities), the larger are likely to be *both* the consequent shifts towards lower-carbon growth paths *and* the wider developmentally positive outcomes.

But what kinds of innovation act as the link between the qualitative content of international transfer and these outcomes – or, in other words, what is the content of the 'Innovation' box on the right of Figure 2.1?

Innovation capabilities: Capabilities for what?

As noted earlier, discussions about moving towards low-carbon growth paths commonly emphasize the importance of 'radical' rather than 'incremental' innovation as the basis for such transitions. The suggestion here is that such an emphasis is highly questionable. This is partly because there are considerable problems about what these categories actually mean. But, more importantly, it is also because it is misleading to dismiss the significance of types of innovation lying towards the 'incremental' end of the spectrum.

Problems about the meaning of 'radical' and 'incremental' innovations

The common distinction between radical and incremental innovations is one among many that attempt to bring some structure to discussions about the huge array of different types of innovation. The distinction in this case is usually drawn in terms of two kinds of criteria – essentially about the knowledge inputs to innovation and the effects arising from it:

- *The novelty of the knowledge base underlying the innovation.* Roughly, the idea is that 'radical' innovations involve a considerable discontinuity between the types of knowledge that they incorporate and the types that underpinned previous technical systems (products, processes and so forth). Incremental innovations rest on knowledge that reflects much greater continuity with the knowledge embodied in previous systems. This idea about the novelty of the knowledge *underlying* the innovation frequently blurs into an emphasis on the novelty of the technological configuration of the innovation itself.
- *The scale and significance of the economic (and other) consequences of the innovation.* The idea here is that the impacts and effects of radical innovation are much greater and probably more 'disruptive' than in the case of incremental innovations.

In trying to elaborate upon these kinds of distinctions, some studies have focused more on the novelty of knowledge inputs (e.g. Mokyr, 1990), some have concentrated on the impacts and consequences (e.g. Abernathy et al., 1983) and some have attempted to encompass both (e.g. Freeman and Perez, 1988; and Freeman, 1992, 1994) It is perhaps not surprising that none of these has been able to develop a sound basis for the distinction. Freeman, for example, perhaps did most to try and elaborate upon the distinction. He emphasized the importance of the novelty of the underlying knowledge, but offered little by way of explanation of what that meant.[3] Instead, he relied on trying to base the distinction on the consequences of innovations. Radical types were those that give rise to distinct new industries, recognizable as new entries in statistical industry classifications – such as new rows/columns in an economy's input/output table, while incremental innovations change economic activities and performance *within* existing industries. But this approach does not help a great deal – for example, his two most frequently used illustrations of radical innovations, nylon and nuclear power, never (as far as I am aware) attained the status of separate rows/columns in national input/output tables.

Others have also contributed to elaborating upon the effects-based approach to classification. In particular, Abernathy et al., (1983) emphasized the importance of dealing separately with the technical novelty and the effects axes of distinction. They reserved the 'radical' term as a basis for discussing the former, and focused mainly on the latter, where they differentiated between 'conservative' and 'disruptive' types of effect with respect to both production systems and input/product markets. They assessed the combined conservative and disruptive effect ('transilience', in their terms) in considerable detail with respect to an exhaustive compilation of 631 innovations in the US automobile industry between 1893 and 1981 – weighting the transilience of individual innovations on a seven-point scale. The vast majority of these (613, or 97 per cent) fell into the lowest five categories, and only nine innovations were allocated to the highest (7) – listed in Table 2.1. However, questions arise about the compatibility of this perspective and others: for example, while Ford's Highland Park assembly line is widely recognized as leading to a 'radical' transformation of the industry, it is hard to see the others in this list as being in a similar 'radical' ball-park as Freeman's illustrative example of nuclear

TABLE 2.1 High 'transilience' innovations in the US automobile industry (1893–1981)

Company	Innovation	Date
Oldsmobile	First mass-produced vehicle	1901
Ford	First branch assembly plant	1910
Cadillac (GM)	First large-scale production V-8 engine	1914
Ford	Elevated moving chassis assembly line (Highland Park)	1914
Hudson	Inexpensive closed car	1922
Ford	Continuous V-8 engine casting line	1934
Oldsmobile	'Hydra-matic' automatic transmission	1940
Budd	All-steel single-unit body	1941
Ford	Cleveland engine plant	1951

Source: Abernathy et al., 1983

power. Indeed, since none of the cases in Table 2.1 gave rise to new industries, they would all be 'incremental' in Freeman's terms.

The difficulty of aligning similar things in similar categories has been compounded by the proliferation of other classification schemes that use different terms but still try to address differences with reference to similar criteria about technological novelty and the outcomes or impacts of innovations. Two of these are particularly interesting in the way they identified novelty in the knowledge base of innovations:

- Enos (1958, 1962) examined innovation in the petroleum industry between 1913 and 1955, and distinguished between what he called the Alpha and Beta phases of innovation. Alpha phase innovations involved the initial invention and commercial introduction of technologically novel processes that were based on new underlying technological principles. Beta phase innovations involved subsequent improvements to the Alpha phase innovations. These were not based on significantly novel knowledge inputs, but involved incrementally improved designs of new plants together with incremental improvements to existing plants.

- Hollander (1965) examined innovation in rayon production by the DuPont company between 1929 and 1960. He distinguished between 'major' and 'minor' technical changes (or innovations) that rested on different knowledge bases. An innovation was considered 'major' if its development 'was considered "difficult" to accomplish by men skilled in the pertinent arts before the development program'; and they were 'minor' if their development involved 'a relatively simple process' (p195).[4] Hence, the latter were achieved on the basis of the firm's existing stock of knowledge, and the former required new knowledge beyond what was already available.

Again, however, it is not at all clear how these categories line up with each other or with those of Freeman. For example, although Enos describes his Alpha phase innovation as involving 'radically different' technologies, all of them occurred as improvements *within* a single industry; hence, Enos's idea of 'radical' differed from Freeman's.

The literature on innovation management has taken further the ideas of Abernathy et al., (1983) on disruptive types of innovation – in particular, Christensen (1997). But rather than helping to clarify the meaning of radical and incremental technologies, this strand of work has reinforced the point raised earlier by Abernathy and colleagues: differences between innovations in terms of their consequences may align poorly with differences in terms of their technological novelty, and these two axes of definition should be kept separate. For example, with reference to a series of highly disruptive innovations in the computer disk drive industry through the 1970s and 1980s, Christensen noted that:

> Generally disruptive innovations were technologically straightforward, consisting of off-the-shelf components put together in a product architecture that was often simpler than prior approaches{…} They offered a different package of attributes valued only in emerging markets remote from, and unimportant to, the mainstream.
>
> (Christensen, 1997, p15)

In other words, if one can generalize from these observations to other possibilities, disruptive innovation steps towards lower-carbon growth paths might *not* necessarily require innovation that is 'radical' in the sense of involving novel discontinuity in the underlying knowledge base incorporated within innovation.

But, less speculatively, the main point is that significant empirical studies in the innovation management field highlight three issues:

- There is a considerable problem about what exactly (or even approximately) is meant by 'radical' and 'incremental'.[5]
- There are considerable merits in using such terms to refer to distinct 'units' of innovation rather than clusters or constellations of many innovations, while also keeping analytically distinct the ideas of differences in the novelty of innovations and difference in their consequences.
- There is a more significant problem about underestimating the importance of the kinds of innovations that lie towards what is commonly thought of as the incremental end of the spectrum. I take further the last of those points.

The importance of 'incremental' types of innovation

Most of the discussion about radical and incremental innovations is about the characteristics of identifiably distinct innovation events – for example, along the lines of those included in the list of 631 automobile innovations compiled by Abernathy and colleagues. But the focus on individual innovations is somewhat beside the point in connection with questions about transitions to lower-carbon growth because the technological discontinuities underpinning such wide-ranging economic and social transformations do not rest on single innovations, however radical they may be. They arise from constellations of many kinds of innovation. This was well recognized, for instance, in the classificatory analyses of Freeman and Perez during the 1980s and 1990s. In particular, the taxonomy they suggested in Freeman and Perez (1988, pp45–47) had four categories. Only two of these, incremental and radical innovations, were about individual innovations; the other two were about constellations of innovations – 'new technology systems' and 'techno-economic paradigms'.[6]

New technology systems were seen as *combinations of innovations* that had pervasive effects that went beyond a single industry to cut across several branches of the economy, as well as giving rise to new industries. An example was the constellation of innovations associated with synthetic materials between the 1920s and 1950s (various 'plastics', synthetic fibres, synthetic rubber and so forth). These involved a succession of chemical process innovations, together with the development of downstream processing machinery (e.g. for plastic moulding and extrusion), plus innumerable product application innovations, as well as organizational and managerial innovations.

Freeman and Perez (1988) identified changes in techno-economic paradigm (or what they also called 'technological revolutions') as much more *extensive constellations of innovations* that had yet wider-reaching impacts affecting not merely several industries but the entire economy. A new paradigm, they explained, 'not only leads to the emergence of a new range of products, services, systems, and industries in its own right; it also affects directly or indirectly almost every branch of the economy' (Freeman and Perez, 1988, p47). They identified five such changes in techno-economic paradigm. The first was the revolution in mechanization of production between the 1770s/1780s and 1830s/1840s, and the most recent was the revolution based on information and communication technologies (ICTs) that took off in the 1980s. But a central feature of these 'technological revolutions' was that they were not just technological. To an extent that was even greater than in the emergence of new technical systems, they involved fundamental changes in the organization of production, and they depended upon, as well as resulted in, pervasive change in the wider socio-institutional context of production.

A key further feature of these Freeman–Perez ideas about constellations of innovations, both new technological systems and new techno-economic paradigms, was that the innovations they involved were not only 'radical'. The authors argued that new technology systems 'were based on a combination of radical and incremental innovations' (Freeman and Perez, 1988, p46); and in the case of new techno-economic paradigms, they emphasized that: 'A change of this kind carries with it many clusters of radical and incremental innovations, and may eventually embody a number of new technology systems' (p47). Thus, for example, alongside the well-known radical innovations that underpinned the ICT-centred shift in techno-economic paradigms, there were numerous more incremental innovations.

Even the limited illustrative material already mentioned in this chapter can illustrate this point. Recall, for example, Christensen's (1997) description of the 'technologically straightforward {…} [and] off-the-shelf components' that contributed substantially to a series of 'disruptive' innovations in the computer disk-drive industry during the 1970s and 1980s. Recall also the description by Abernathy and colleagues (1983) of 630 innovations in the US automobile industry between 1893 and 1981. These were spread across seven categories of significance of their impact (transilience), and those towards both the 'incremental' and 'radical' ends of that spectrum occurred even during the period up to the mid-1950s, the more transformative phase of the industry's development contributing to the emergence of the Freeman–Perez 'mass production' techno-economic paradigm. All nine of the innovations in the level 7 category, including Ford's revolutionary assembly line, were generated during this period, but those in levels 1 to 3 (far out of sight of anything that might be described as radical) accounted for 74 per cent. Is it really the case that the combined impact of such innovations upon the emergence of the new mass production paradigm can be dismissed as insignificant?

Much the same picture is provided by the studies of chemical process innovations by Enos (1958, 1962) and Hollander (1965). Both of these examined patterns of innovation that occurred in industries that were at the heart of the emergence of the fourth Freeman–Perez techno-economic paradigm.

As noted earlier, Enos examined innovation in petroleum refining during the first half of the twentieth century when it contributed to a massive reduction in the cost of petroleum – a key input for the automobile-centred component of the emerging mass production paradigm.[7] In a striking analysis, he identified the relative contribution to these cost reductions made by his two categories of innovation: the Alpha phase initial introduction of technologically novel processes and the subsequent incremental improvements to existing processes. He demonstrated that the cumulated magnitudes of the gains from the Beta phase improvements were as significant as the gains from the Alpha phase novel processes.

Hollander's examination of innovation in rayon production by DuPont covered a period of about 30 years (late 1920s to 1950s) during the emergence of the new synthetic materials technology system. The development of this new technology system was another strand of multiple innovations that contributed to the emergence of the mass production paradigm, and the connection was particularly strong in the case of rayon because an important area of product innovation in the industry during the mid-1930s was the development of rayon cord for automobile tyres.

In Enos's terminology, the Alpha phase innovator of rayon was a European company, and DuPont initially entered the industry via licensing and international technology transfer from the original innovators. Hollander examined a long stream of subsequent innovations that DuPont introduced to improve the technology that the company had initially acquired. As noted earlier, he classified these improvements as 'major' and 'minor' changes depending upon whether they were easy or difficult to develop on the basis of the existing knowledge base available to the firm. These improvements were implemented, in part, by investment in a succession of five new plants and, in part, as improvements within existing

TABLE 2.2 The role of 'minor' innovations in cost reduction in DuPont rayon plants

A Du Pont plant (Acronym)	B Period examined	C Number of years	D Average annual reduction in unit costs	E Proportion of cost reduction due to technical change	F Proportion of technical change effect due to 'minor' changes
OH	1929–1951	22	4.9%	85%	79%
S I	1932–1950	18	4.5%	97%	80%
S II	1937–1951	14	2.3%	35%	100%
S III	1938–1952	14	4.9%	95%	46%
S IIA	1945–1952	7	3.7%	100%	83%

Source: Hollander, 1965

plants during their operating lifetimes. His results with respect to the economic gains from those two modes of implementing improvement are summarized in Table 2.2. Three points are striking:

1. Remarkably high rates of cost reduction were achieved by improvements to plants during their operating lifetimes – approaching 5 per cent per year in three of the plants (column D). One consequence was that, when new plants were constructed using the latest technology, their unit costs in some cases were not substantially lower than those of much older plants whose performance had been improved over time.
2. Very large proportions of those cost reductions (generally 85 to 100 per cent) resulted from technical changes (i.e. innovations) as opposed to other sources, such as scale economies (column E).
3. Very large proportions of those cost-reducing innovations (about 80 to 100 per cent in four of the plants) were 'minor' (i.e. largely derived from the existing internal and external stocks of knowledge available to the firm at the time).

Such paths of increasing efficiency in existing plants are commonly shrouded beneath terms such as 'learning curve' and 'experience curve'. These are frequently described as arising from a process of 'learning by doing', something that yields improved performance as a result of 'experience' – a kind of byproduct gain in productivity that arises automatically from simply undertaking production or implementing repeated investment projects. The Hollander analysis suggests that such productivity gains are not free lunches. They are more likely to arise from paths of deliberately developed and cost-incurring innovations with characteristics lying towards the 'incremental' end of the spectrum in terms of both the novelty of the knowledge base they draw upon and the impact they each achieve individually.[8]

Incremental and radical innovations: Complementarities and interactions

The examples in the previous section illustrate ways in which three kinds of technological innovation contribute to long-term paths of productivity growth and change in the structure of economic activity.[9]

- *'Improvements' made to existing production systems*. In these illustrations the innovations were primarily centred on improvements to plants and processes; but roughly the same pattern can be identified in terms of the improved price-performance characteristics of existing models of products; and, in any case, product and process innovations are commonly combined together. These kinds of 'intra-plant' (or intra-model) innovations typically seem to be 'incremental' in one or both of two senses: first, they involve limited technological 'novelty' in the sense that they draw on and incorporate elements from the existing stock of knowledge or marginal additions to it; and, second, the impact and consequences of the individual changes involve quite modest adjustments in performance, even if the cumulated effect of a stream of such innovations can be substantial.
- *'Improvements' incorporated within substantial new units of production capacity*. These are mainly 'new plants' in the preceding examples, but the category could no doubt also encompass the introduction of major new product models. These again, being 'improvements' on the past, are also incremental in the senses of drawing on limited novelty in the knowledge they incorporate and/or resulting in modest consequences relative to pre-existing practices.
- *Significantly novel production systems (or subsystems)*. These are 'radical' in the sense that, like Enos's Alpha phase innovations, they incorporate underlying knowledge that is 'radically different' from what has been applied before (even though it may have been known for some time); or, like some of Christensen's 'disruptive' innovations, they may involve radically different architectures consisting largely of elements of existing technology. The consequences of the initial commercial introduction of such novel technologies seem highly variable. In some cases they may be quite modest, requiring substantial subsequent 'improvement' innovations before realizing their full potential. But, occasionally, the impact of even such first steps on their own may be substantial.

It is important to note that each of these different types of change seem to run into diminishing returns. The 'learning curves' that reflect paths of 'intra-plant' (or intra-model) innovation typically peter out with very low or absent rates of improvement after a time. The potential for continuing to improve successive new vintages of plants (or model) also seems to become increasingly limited as it is progressively more difficult and costly to extract improvement from the existing body of knowledge and marginal advances in it. Similarly, as stressed by Freeman and Perez, existing techno-economic paradigms regularly appear to run into barriers for further development.

But the main point here is that these types of innovation do not stand in isolation. They are linked together as complementary components of long streams of change, and it is fairly pointless to debate whether one or the other type is more or less important than the others. Such streams of change may run within particular industries, as with Enos's Alpha and Beta phase innovations in petroleum refining. But they may also lead to the establishment of new industries as the streams of innovation branch out in different directions that later come to be seen as distinct industries. Some streams may be initiated by radical changes, but others may also involve a succession of radically novel steps – as with the sequence of Alpha phase innovations over 30 to 40 years in petroleum refining. Several such streams, linked by commonalities in their underlying knowledge bases, may also run along parallel and closely related trajectories – giving rise to the kinds of interconnected phenomena that Freeman and Perez described as 'new technology systems'. Such parallel and interacting streams may also diversify much more pervasively, becoming the heart of 'new techno-economic paradigms'.

The differences between such streams of interconnected types of innovation highlight the point that radical innovations differ widely in their consequences. Some may be radical in their technological origins but run quite narrowly in their subsequent streams of innovation; others may have similarly radical technological origins but with subsequent streams of change that run much more broadly in their consequences. The latter fall into the category that Mokyr (1990) described as 'macro-inventions' and which he distinguished from 'micro-inventions'. But he went further and identified an important interaction between them. 'Radical advances in the manipulation or understanding of physical processes are usually the beginning of a prolonged series of improvements and modifications.' The 'chief importance of radical inventions', he argued, is that they open up the potential for a sequence of subsequent improvements. Thus, he suggested, 'in periods of radical inventions, we observe an intensification of smaller inventions as well' (Mokyr, 1990, p352).

In other words, while incremental forms of innovation seem to be needed to exploit the full potential of radical advances, those kinds of change appear to run into diminishing returns, and more radical innovative steps are required to open up the potential for incremental improvement to yield more pervasive and deeper consequences.

Aspects of this pattern are illustrated in the next section by a case study that covers more recent experience than the examples drawn on already, while also being more explicitly connected to issues about the development and transfer of low-carbon technology – in this case involving technology to achieve rising energy efficiency in steel production, one of the most energy-intensive manufacturing industries.

Raising energy efficiency in the casting-rolling stage of steel production: A 50-year trajectory of innovation

The casting-rolling stage in steel-making transforms liquid steel produced by various kinds of furnace at the previous stage into a set of rolled products for further processing at later stages.[10] Until the 1950s, production of almost all the world's steel followed a route through this stage via the steps summarized in the path along the top of Figure 2.2. Molten steel was cast into ingots – very large blocks weighing between 10 tonnes and 50 tonnes, though often around 20 tonnes (typically with cross-sections between $500mm^2$ and $800mm^2$). These were then reheated and repeatedly rolled down to three kinds of smaller intermediate product:

- *Billets*, lengths of steel usually with square cross-sections between $50mm^2$ and $125mm^2$. These were subsequently further hot-rolled down to 'long' products such as small steel sections, reinforcing bar for concrete or rod for processing into wire.
- *Blooms*, lengths with larger, usually square, cross-sections ($150mm^2$ to $300mm^2$). These were typically then rolled down to beams and girders (I-shaped and other cross-sections) as structural products for use in construction (e.g. of buildings or ships).
- *Slabs*, lengths with rectangular cross-sections (mainly in the range of 50mm to 230mm thick x 600mm to 1500mm wide). These were hot-rolled down to steel strips that would later be processed into 'flat' products for use in producing such things as automobile bodies, consumer durables, roofing materials or tinplate for food and beverage cans.

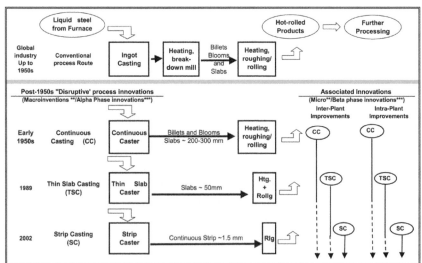

FIGURE 2.2 Innovation in the steel industry: The casting-rolling process stage (1950s to 2000s)

Source: Adapted from Luiten, 2001, Chapter 5; and Luiten and Blok, 2003

Notes: The terminology of Mokyr (1990) and Enos (1962).

In other words, very large amounts of equipment, energy and labour were used to do not a lot more than squash large lumps of solid steel down to smaller sizes that were more suitable for subsequent manufacturing processes.[11] At the same time, a considerable proportion of the initial volume of steel from the prior furnace stage was wasted when poor-quality sections were cut from the ends and sides of the rolled material at various points. This stage of steelmaking was undertaken by very large companies, colloquially known as Big Steel, in large-scale and capital-intensive plants that were usually concentrated in a small number of geographical locations within a relatively small number of the world's most industrialized economies.

However, over the next 50 years all of these characteristics of this section of the industry were dramatically transformed as a result of a series of innovations – primarily technological, but also organizational. The technological heart of this transformation is often presented as having involved the introduction of three disruptive innovations, each involving significant novelty in the knowledge and principles incorporated within the innovations – as per the bird's eye summary running down the left of Figure 2.2.

Step 1: The introduction of continuous casting

This new process bypassed the ingot casting and initial rolling steps of the conventional process of the 1950s by casting steel into continuous strands already with the relatively small cross-sectional dimensions of billets, and relatively small blooms. These only required hot rolling to pass on as semi-finished long products for further downstream processing. The process was first introduced during the early 1950s in Germany and the UK. It was taken up gradually by other firms through the later 1950s and 1960s, and then much more rapidly during the 1970s and 1980s.

An estimate made in the early 1980s suggested that continuous casting resulted in an 80 per cent reduction in specific energy consumption compared with the conventional ingot casting route (Mizoguchi et al., 1981, pp152–153). But the development of continuous casting led to a much wider set of changes in the industry. The much lower capital costs of the continuous casting technology, combined with developments in furnace technology at the preceding process stage (mainly associated with scrap-using electric arc furnaces) sharply reduced barriers to entry in the industry and a considerable number of smaller firms started up – the mini-mills. By the end of the period this had dramatically changed the structure of the overall steel industry as the mini-mills captured a large share of the long products market from integrated Big Steel companies (Barnett and Crandall, 1986; Crandall, 1996). At the same time, the smaller scale and lower barriers to entry transformed the geography of the industry. Mini-mill plants not only moved to a much wider range of locations in large advanced economies, but also to smaller developing countries such as Malaysia and Singapore long before conventional integrated plants could have been established.

For a number of years, however, technical limitations confined the application of continuous casting primarily to billets and blooms for the long and structural product segments of the industry. However, by the 1980s, a number of the large steel companies introduced developments of continuous casting technology to produce relatively thick slabs (around 220mm) to feed their existing flat-product rolling mills.

Step 2: Thin slab casting

In 1989, Nucor, a pioneering US mini-mill, introduced the novel technology of thin slab casting, following its development via a sizeable experimental plant by the German equipment company Siemens. The new process cast wide slabs (132mm) that were only 50mm thick, about one quarter of the thickness of slabs from conventional slab casters. As a result, compared with conventional thick slab casting, the process substantially reduced reheating and the initial stages of rolling flat steel products. In addition, a particularly striking feature was the more or less direct link between casting and rolling to form a single (and almost) continuous process. One set of estimates suggests that, compared with the thick slab casting route, these changes resulted in a reduction of 40 per cent in specific energy consumption, and of 30 per cent in CO_2 generation per tonne (Matsushita et al., 2009, p2).

This step change in the development of the technology not only achieved much lower energy costs, but also required substantially less equipment, considerably reducing capital costs. Nucor quickly expanded its use of the highly profitable technology, and other mini-mills followed, capturing a rapidly rising share of the flat products market that had been dominated for decades by the capital-intensive plants of the 'Big Steel' companies. In the US, this contributed to a striking further transformation in the structure of the industry. Integration among the leading mini-mill companies (the most successful of which, such as Nucor, were no longer very 'mini') led to a new form of industry concentration that replaced the prior dominance of the Big Steel companies.

Step 3: Strip casting

In 2002, based on technology that had been developed on an experimental plant by a Japanese equipment supplier (IHI) and an Australian steel producer (BHP), Nucor pioneered the commercial introduction of a further innovative step to reduce rolling requirements in the production of strip steel: the direct casting of strip at a thickness of only 1.6mm. Heating and rolling requirements for strip production were yet further reduced from those that had been achieved via thin slab casting.

It remains to be seen whether this step will have a wide-ranging transformative effect on the steel industry along the same lines as the previous two. By 2010, the only reported further commercial plant was a second commissioned by Nucor in 2009. However, several other companies and consortia were developing experimental plants. These included a German company that was reported to be planning to expand to an industrial scale its large experimental stainless steel strip caster at Krefeld.

It may be that this third step in the development of casting-rolling technology is passing through a phase which is similar to that experienced with continuous casting technology during the 1950s and 1960s – a pause in diffusion of the technology while technological and market constraints and uncertainties are reduced. If that is so, the *potential* for a further step in the transformation of the industry seems high. Once again, compared with conventional casting routes, the reduction in capital costs for strip casting seems to be substantial (Luiten and Blok, 2003, p1341). Similarly with energy intensity, industry estimates reported by Luiten and Blok (2003, pp1341–1342) suggest that, compared with the thin slab casting route, energy consumption in strip casting carbon steel is likely to be about 90 per cent lower. For stainless steel, data for the Krefeld plant suggest a reduction of about 87 per cent compared with conventional casting and rolling. Alternative estimates (Matsushita et al., 2009, p2) suggest slightly more modest reductions for carbon steel relative to thin slab casting – about 80 per cent for specific energy consumption and 70 per cent for CO_2 generation. But relative to thick slab casting, the reductions are about 90 and 80 per cent, respectively. A broadly similar perspective is provided by Schlichting et al., (2009).

This three-step overview of the 50-year technological development of this segment of the industry seems consistent with a view of techno-economic transformations arising from strikingly novel innovations towards the radical/disruptive end of the spectrum. But that is only part of the story because these three steps were intimately associated with, and interacted with, a host of other innovations at the incremental/minor end.[12] As illustrated down the right-hand side of Figure 2.2, these minor/incremental innovations were brought into use in two ways: first, by investment in new plants or substantial elements of new plant – new continuous casting billet/bloom plants (CC), then later new thin slab casting plants (TSC), but as yet only very few new or large experimental strip casting plants (SC); and, second, by changes that were incorporated within existing plants of the three types during their operational lifetimes.

Limited space here precludes the provision of even illustrative detail about these changes. However, the technical and industrial literature over these decades carries innumerable descriptions of such incremental/minor improvements – such things as changes in the overall structure of the main mould and roller system; modifications to the shape of the container into which the molten steel is initially poured (tundish); alterations to the mode of oscillation of the mould; improved coatings for the tundish; mould and rollers; changes to the mould and roller configuration in order to cast strands for I-shaped structural products nearer to their final shape; changes to the cooling system and to methods for controlling shell formation and other aspects of the solidifying strand; improved methods of controlling scale formation on the cast surfaces; and pervasive improvements to instrumentation and control systems.

As a consequence of all this, and much more, the same broad kinds of impact were being achieved as with the three more radical steps:

- *Extending the range of continuously cast products.* Just as the radical steps had opened up changes in the types of product that could be continuously cast (e.g. the shift from long products to flats), so also the range of continuously cast products was further stretched by minor/incremental improvements: for example, wider slabs and strips, improved levels and consistency of quality (reduced flaws and inclusions) to capture new markets, or changed metallurgical properties (such as those needed to produce flat products to meet the demanding requirements of the automobile industry).
- *Raising capital and labour productivity.* A common direction for incremental/ minor innovation was not only to raise the scale of new plants,[13] but also to increase the rate of production within them, and this often seems to have been achieved with little or no increase in labour input and a less than commensurate increase in capital expenditure, consequently raising the productivity of both capital and labour.[14]
- *Energy efficiency.* In much the same way as with the three more radical steps, incremental changes that cumulatively expanded the product range consequently raised the proportion of total steel production that was being processed via one of the much more energy-efficient continuous casting routes. But at the same time, such incremental changes also raised the energy efficiency of the continuous casting routes themselves. In addition, the gains in energy efficiency at this process stage were increased by incremental improvements that enhanced material yields. This reduced the extent to which energy used in the prior steel production stages was wasted as scrap during the casting-rolling stage.

In summary, then, the paths of technological change (in this case, history) encompass the same combination of innovation at both ends of the radical-incremental spectrum as in the more thorough analyses drawn on earlier in this chapter's third section. At the same time, that path of technological change seems to have contributed directly to dramatic changes in the global steel industry – changes in its organizational structure, in its intra-national and international geographical location, and in the productivity of all the resources it used, including a dramatic reduction in its energy efficiency and, consequently, its carbon-intensity.

A few further observations about this process should be noted. First, the technological changes noted here – changes in the casting-rolling stage of production – were embedded in a wider set of related innovations, especially at the upstream stage. In part, these were about the development of increasingly efficient furnaces that initially used scrap steel (electric arc furnaces). These were much smaller scale and less capital intensive than the equivalent process steps in conventional steelmaking, supporting the wider technological and organizational process of mini-mill development and structural change in the industry.[15]

Second, although one consequence of this path of innovation was a dramatic increase in energy efficiency, the technological bases for that trajectory could in no sense be described as advances in 'energy technologies'. Instead, a wide range of different areas of technology underpinning steel production were involved. Indeed, the achievement of higher energy efficiency was not even a major aim and driver of the innovation process,[16] no doubt a reflection of the structure of relative prices during the period. Instead, the primary goal was broader: to achieve and sustain efficient methods of steel production at this process stage. This was closely linked to entrepreneurial ambitions – among both steel producers and steel equipment suppliers – to open up opportunities for new businesses in the steel industry, followed by ambitions to sustain those profitable positions.

Third, it is very difficult to fit this history of innovation into the standard 'innovation chain' model with its sequence running from R&D via demonstration and deployment to diffusion. A rough fit between the two might be possible if the 50-year history had consisted only of the three discontinuous technological novelties, and if the model consisted only of the initial steps from R&D to the start of deployment. But the model also claims to cover what it identifies as the largely innovation-free stages of deployment (beyond its initial step) and diffusion. In practice, however, the history involved innovation-intensive activities running through these supposedly innovation-free stages – extending far beyond the initial deployment of the three technologically novel innovations. In addition, R&D was not something that happened exclusively at the start of this sequence. Obviously, substantial R&D *was* involved in the introduction of three discontinuous novelties. But it is evident from the technical and industrial literature that the subsequent streams of minor/incremental change also drew extensively on inputs from R&D that was undertaken by steel producers, by suppliers of equipment for casting and rolling facilities, by their suppliers of the myriad subsystems incorporated within those facilities, by research institutes, and by universities. In other words, R&D was undertaken not only at the start of the innovation chain, but also *in parallel with all* its stages.

Finally, it may be useful at this stage to forget that the case history is specifically about the steel industry. The purpose of presenting the history is not to claim anything in particular about the future development of this industry and its further evolution towards a lower carbon future. Instead, the aim is, in one sense, much broader: to suggest that the pattern of innovation towards lower carbon futures in many other industries – both those that 'supply' energy and those that 'use' it – is likely to be broadly similar: an interacting sequence of innovation steps that will embrace both those towards the radical/novel end of the spectrum and those towards the minor/incremental end. The relative importance of each will differ quite widely between industries, times and places; but one generality seems likely to be important: there will be very few circumstances where the interacting complementarity between the two is so imbalanced that those at the incremental/minor end of the spectrum can be dismissed as unimportant in achieving disruptive transformations leading to lower carbon futures. Indeed, in considering the wide spectrum of technological change that will be required to achieve those transformations, it may be useful to bear in mind Mokyr's (1990, p352) view that 'in periods of radical inventions, we observe an intensification of smaller inventions as well'.

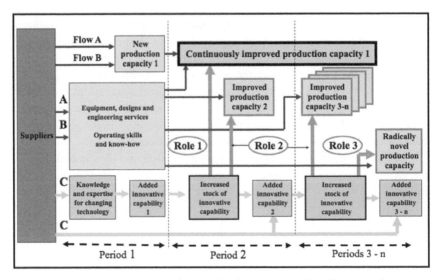

FIGURE 2.3 Transferring lower-carbon technologies: The roles of increased innovation capability

Conclusions and questions

Conclusions

The question posed at the end of the second section asked about the developmental purpose of trying to organize the international transfer of low-carbon technologies so that the qualitative content of transfer contributes to strengthening innovation capabilities in the technology-acquiring organizations and economies. What kinds of innovation are envisaged? Hence, what is included in the largely empty box on the right-hand side of Figure 2.1?

The third and fourth sections suggest at least the outline skeleton of an answer that differs substantially from those that seem to be commonly presumed. As illustrated in Figure 2.3, cumulatively strengthened innovation capabilities may play three roles:

Role 1. These capabilities may contribute to continuous improvement of initially installed production capacity that incorporates the imported technology – that is, by engaging actively with types of innovation (probably incremental) that raise performance during the operational lifetime of the production facility or system (production capacity 1).

Role 2. They may contribute to types of innovation (probably incremental) that are incorporated within subsequent new facilities and systems based on the same broad area of low-carbon technology (e.g. production capacity 2). This role may extend further in time by contributing to innovation embedded in a stream of new facilities or systems over a number of years (successive vintages of improved production capacity 3-n). These may be based on the same area of low-carbon technology, but they may also involve 'diversification' into related areas where generically applicable aspects of the cumulating innovation capability can be used effectively in implementing different kinds of incrementally innovative investment in new facilities and systems.

Role 3. They may contribute to innovation that incorporates distinct discontinuities within the particular area of low-carbon technology (radically novel production capacity).

Capabilities for contributing to, or engaging with, these three kinds of innovation are not only capabilities for directly undertaking innovative activities themselves. They are also capabilities for assessing, reviewing and taking decisions about the introduction of new technologies: which to use, what elements to modify, where to acquire them from, and how? Capabilities to address those kinds of question in connection with all three roles are likely to be all the more important the more complex the technologies involved, the greater the intensity and diversity of innovation within a particular area, the greater the 'radical' components within that diversity, and the more globally dispersed the innovation actors involved. In other words, significant depths of 'hands-on' innovative capability are likely to be particularly important for taking decisions about innovation in precisely the conditions that are usually involved in achieving significant transitions towards low-carbon development paths.

   The chapter's third and fourth sections also suggest that the carbon-using characteristics of technologies are inextricably entangled with their other dimensions, and improvements in one are likely to require and contribute to improvements in others. Hence, the kinds of innovation achieved by these three roles are likely not only to lower the carbon intensity of production activities, but also to change other characteristics with wider developmental implications (e.g. increasing the productivity of capital and labour or altering product quality standards and specifications).

   But insights from the wider literature about innovation capabilities in developing countries suggest that, via these three roles, they are also likely to contribute in other ways to broader developmentally positive outcomes. In particular, they are likely not only to increase efficiency in existing production activities, but also to accelerate entry into new production activities, hence contributing to faster change in the *structure* of the economy.[17] This is likely to happen in two ways.

   First, innovation capabilities provide an important basis for identifying opportunities to start up new production activities and for implementing the types of innovation needed to exploit them. As noted above, for example, this might arise via role 2 in the form of diversification steps in a succession of new investment projects associated with areas of technology acquired at an earlier stage.[18]

   Second, the capabilities to engage in innovation also provide a basis for opening up local sourcing of inputs to existing production activities (role 1), to new investment projects involving incremental advances on previous practice (role 2), or to projects involving more radical advances in technology (role 3).

Questions

Numerous further questions are raised by the conclusions above about the roles of innovation capabilities in areas of lower carbon technology, and about the wider developmental outcomes that might follow from strengthening capabilities to play those roles.

How pervasively relevant, and how variable, are the observations about the three kinds of innovation role? As noted earlier, the conclusions are derived largely from understanding about innovation processes in a range of manufacturing industries, especially some of the more materials- and energy-intensive industries that will have to be dramatically transformed or replaced in low-carbon development paths. But they are also relevant across energy 'supply' industries based on both old and new technologies, and the relative importance of the different roles of innovation capabilities is likely to vary quite widely across these. For example, the potential malleability and 'improvability' of the technology during the operating lifetimes of large-scale sub-sea marine power systems seem much more limited than in the case of biomass-conversion systems. But beyond such limits to our understanding about innovation in all these kinds of industry, there appears to be even less understanding about how international technology transfer to developing countries and local innovation within them interact further 'downstream' in the activities that use the outputs of these industries. For example, to continue downstream from the steel industry illustration in this chapter, we seem to know little about the transfer-innovation interaction in the context of the engineer/architect service industries whose innovative activities shape the intensity of steel use in the development of infrastructure and the built environment.

To what extent do the outcomes from the three kinds of innovation role depend upon **localized** *accumulation of the underlying innovation capabilities, or can the same outcomes be achieved by drawing on international suppliers of innovation-implementing goods and services?* The illustrations of innovation paths in this chapter gave little attention to the geographical location of the underlying innovation capabilities. But they were very diverse. For example, Nucor set up its pioneering strip casting plant in 2002 in a small town in central Indiana on the basis of radically novel technology that had been developed in collaboration between a Japanese equipment manufacturer and an Australian steel producer. But Nucor's own innovation capabilities were central to the intensive improvement of the subsequent performance of that plant, and they then appear to have played a significant role in shaping the technological configuration of its second plant at the end of the decade. However, what are the more general characteristics of those patterns of complementary interaction between 'local' and 'distant' innovation capabilities? In particular, what would be lost by not intensifying the development of 'local' innovation capabilities, and consequently relying more heavily on distant sources, for inputs to innovation in transitions towards lower-carbon income growth in developing countries?

What do those innovation capabilities consist of? Most discussions about strengthening innovation capabilities in developing countries jump immediately, and usually exclusively, to focus on capabilities for undertaking R&D. But those kinds of capability were only a part of the capabilities contributing to the paths of innovation described in the earlier sections of the chapter. Other kinds of capability were also critically important. In particular, most of the companies involved in those innovation histories drew heavily on various kinds of non-R&D engineering capability, together with accumulated operational experience; indeed, Nucor, the company that pioneered the introduction of thin slab casting and strip casting did not have any R&D function for a large part of that period while it engaged very significantly in innovation via roles 1 and 2. But there appears to be quite limited understanding about the details of such non-R&D capabilities for innovation, and about how their characteristics and significance vary across innovation in different kinds of technology.

In what kinds of organization are innovation capabilities most effectively accumulated, and are there merits in accumulating capabilities in any particular sequence? Most of the studies about successfully acquiring and accumulating innovation capabilities in developing countries (primarily in East Asia and some Latin American countries) have highlighted the central role of industrial enterprises as the key organizational location for doing so. Moreover, they have identified the process of building those capabilities as one that runs through a cumulative sequence from capabilities to play role 1, through those needed for role 2, and then later towards those for the more radical and disruptive innovations involved in role 3. There is also an extensive contrasting literature that describes the very limited innovation impact that is typically made by capabilities for R&D that have been created and accumulated in more centralized types of organization – primarily public research institutes and universities. There is consequently considerable discussion about a 'gap' or 'innovation chasm' between such central organizations and production enterprises. That seems to change as the firm-centred process of accumulation reaches stages when the deepening innovative capabilities of enterprises begin to direct towards such central organizations a demand for R&D-derived knowledge to use in their own innovative activities. But how general is the apparent effectiveness of that dispersed 'firm-first' kind of sequence? In what ways might it need to be different in a context of transitions towards lower carbon development paths that are likely to involve multiple trajectories of innovation that are relatively intensive in more radical/disruptive forms of innovation?

Above all, and central to the issues addressed at the start of this chapter, how might processes of international transfer be organized and managed in order to contribute much more significantly to the development of localized innovation capabilities? Even in the limited discussion around international technology transfer, innovation capabilities and climate change, there is a substantial omission: discussion centres almost entirely on the transfer dimension of the process – the arrows in Figure 2.1 (repeated in Figure 2.3) that appear to *deliver* Flow C-type content via one-way flows that run *from* suppliers *to* technology importers. That leaves out of account two other important parts of the process that have been illuminated by studies of international transfer and the strengthening of innovation capabilities in other areas of technology in developing countries. First, and fairly obviously, firms add substantially to their innovation capabilities by other mechanisms than international transfer. Second, the effectiveness with which they use international transfer to add to their innovation capabilities depends very heavily on the existing stocks of capability they have already accumulated, and on their organization and management of those capabilities in the process of transfer.[19] In other words, the key issues about the Flow C-type content of transfer are perhaps less about whether and how suppliers transfer it, and more about how firms in developing countries set about acquiring it. One recent study has begun to draw this kind of understanding into the examination of issues about international transfer and climate change: Doranova et al., (2011) in connection with technology transfer and the CDM. But this needs to be taken very much further in order to illuminate issues about policy for, and the management of, international technology transfer in the area of low-carbon technologies.

Notes

1 This is merely a transfer-centred focus on a more general perspective – as, for example, in the *World Development Report 2010* that sees technology-related capacities to tackle mitigation and adaptation as part of a basis for building strong and competitive economies (World Bank, 2010).
2 The sharp distinction between technologically creative innovation and technologically passive and replicative diffusion may also lead in to views that exclude most developing countries from engaging in innovation because it is deemed efficient for only a very few large and/or technologically advanced emerging economies to engage in innovation. All the others should focus on adopting and using available technologies, perhaps involving a certain amount of marginal improvement and adaptation for local contexts (e.g. McArthur and Sachs, 2002).
3 Though at times he seemed to identify such radical novelty in terms of whether the knowledge originated in formally organized R&D (Freeman, 1994, pp474–475).
4 This distinction might be thought of as a difference between, first, innovations that could be developed and introduced on the basis of the existing stock of knowledge available to the firm, either internally or from external sources; and, second, those that required the development of new knowledge via the firm's own R&D or by suppliers and other external actors.
5 For a much more elaborate treatment of the terminology problem in the management literature, see Garcia and Calantone (2002). Note also the further complication that has arisen from the increasing use of a framework developed in the Oslo Manual (OECD/Eurostat, 2005) to standardize the classification of different types of innovation in national surveys. This distinguishes between innovations that are new to the world, to the firm's own industry/market, and to the firm itself. This framework is helpful for many purposes. But, in contrast to some suggestions, it is not even roughly aligned with the radical/incremental distinction. In particular, new-to-the-world innovations (i.e. those meeting the criteria for international patenting) may not be very 'radical' in any sense of the term.
6 Over subsequent years the terminology and conceptual framework in this area varied across the joint and separate work of Freeman and Perez; but most of the basic principles remained very similar to those in the 1988 book chapter, as summarized below.
7 An interesting aspect of this massive cost reduction was that between 1913 and 1955, alongside rising capital and labour productivity and increasing yields from material inputs, there was also a 92 per cent improvement in specific fuel consumption in petroleum refining (Enos, 1962, Table 5, p318).

8 This critique of the conventional 'learning-by-doing' framework is developed more extensively in Scott-Kemmis and Bell (2010), with specific comment on its potentially misleading application in econometric modelling of growth and technological change as a basis for assessing policy options for accelerating innovation to reduce the costs of carbon abatement.

9 However, they illuminate little about various kinds of organizational innovation that also contribute to these long-term paths of change in the economy – and beyond. But it may be the case that these also fit into the tripartite distinction sketched below.

10 This excludes the route that runs from liquid steel to forged steel shapes and final products that are typically used in the electrical and mechanical engineering industries. Also, in some cases this stage results in simpler types of final rolled product.

11 This obviously oversimplifies things a bit because other things were also achieved – for instance, rolling changes the metallurgical properties of the steel.

12 It is also incomplete because it omits a number of important organizational innovations.

13 Crandall (1996) noted that, during the phase of expansion of continuous casting, the scale of mini-mill plants increased from less than 500,000 tonnes per year to around 2 million tonnes per year.

14 This was far from being only a matter of scaling up new plants. For example, the author's unpublished observations of performance improvement over selected periods during the lifetime of two Asian mini-mills (in Japan between 1977 and 1980 and 1989 and 1992, and in Malaysia between 1982 and 1985, and during 1997 and 1998) indicated that output increased by about one third and labour productivity nearly doubled in the former, while in the latter output almost trebled, as did labour productivity.

15 It is difficult to identify in this associated path of innovation any significant technological discontinuities (equivalent to the three more radical innovation steps in continuous casting) – though the development of the technology during the 1950s from its prior use in the small-scale 'special steels' segment of the industry might qualify as a more major/radical step. Thereafter, the story seems to be much more one about a steady cumulative innovations towards the minor/incremental end of the spectrum. However, the development of technology for the direct reduction of iron (DRI) was a distinctly novel innovation, and it was important in opening up a new source of materials for the mini-mills –reducing their dependence on steel scrap.

16 This point is well developed by Luiten and Blok (2003).

17 Over the last ten years or so, Hausmann, Rodrik and colleagues have elaborated upon the argument that changing the composition of production, and not only raising efficiency in existing production activities, is a centrally important component of the growth and development process, thereby reviving an earlier 'structural' tradition in the analysis of growth. See, for example, Hausmann and Rodrik (2003) and Hausmann et al., (2007).

18 Recall, for example, the relatively narrow diversification by DuPont from textile rayon products into tyre-cord products. Innovation capabilities for wider kinds of diversification steps were important – for instance, in the industrial development in Korea during the 1970s and 1980s, as emphasized by Amsden and Hikino (1994), who highlighted the role of what they called 'project execution capabilities' in enabling Korean firms to deal effectively with the technological aspects of entry into new lines of production.

19 Among other contributions in this area, the work of Linsu Kim on the transition of Korean firms from technology imitators to innovators highlights the significant of this point (e.g. Kim 1997, 1998). Dantas and Bell (2011) illustrate the same issue in connection with the emergence of the Brazilian oil company Petrobras, from technology imitation to innovation at the global frontier of offshore technology. In particular, as the company approached the international innovation frontier, its own substantial innovation capabilities were a necessary 'entry-ticket' to participate in the international consortia developing radically novel technologies for deep-water operations.

References

Abernathy, W. J., Clark, K. B. and Kantrow, A. M. (1983) *Industrial Renaissance: Producing a Competitive Future for America*, Basic Books, Inc., New York, NY

Amsden, A. H. and Hikino, T. (1994) 'Project execution capability, organisational know-how and conglomerate corporate growth in late industrialisation', *Industrial and Corporate Change*, vol 3, no 1, pp111–147

Barnett, D. F. and Crandall, R. W. (1986) *Up from the Ashes: The Rise of the Steel Minimal in the United States*, The Brookings Institution, Washington, DC

Bell, M. (1990) *Continuing Industrialisation, Climate Change and International Technology Transfer*, SPRU, University of Sussex, Brighton, UK

Crandall, R. W. (1996) 'From competitiveness to competition: The threat of minimills to large national steel companies', *Research Policy*, vol 22, no 1/2, pp107–118

Christensen, C. M. (1997) *The Innovator's Dilemma: When New Technologies Cause Great Firms to Fail*, Harvard Business School Press, Boston, MA

Dantas, E. and Bell, M. (2011) 'The co-evolution of firm-centered knowledge networks and capabilities in late industrializing countries: The case of Petrobras in the offshore oil innovation system in Brazil', *World Development*, vol 39, no 9, pp1570–1591

Dechezleprêtre, A., Glachant, M. and Menière, Y. (2008) 'The Clean Development Mechanism and the international diffusion of technologies: An empirical study', *Energy Policy*, vol 36, pp1273–1283

Doranova, A., Costa, I. and Duysters, G. (2011) Absorptive Capacity in Technological Learning in Clean Development Mechanism Projects, UNU-MERIT Working Paper Series No 2011-010, United Nations University–Maastricht Economic and Social Research and Training Centre in Innovation and Technology, Maastricht

Enos, J. L. (1958) 'A measure of the rate of technological progress in the petroleum refining industry', *Journal of Industrial Economics*, vol 6, no 3, pp187–194

Enos, J. L. (1962) 'Invention and innovation in the refining industry', in R. R. Nelson (ed) *The Rate and Direction of Inventive Activity: Economic and Social Factors*, Princeton University Press, Princeton, NJ

Freeman, C. (1992) *The Economics of Hope*, Pinter Publishers, London and New York

Freeman, C. (1994) 'The economics of technical change', *Cambridge Journal of Economics*, vol 18, pp463–514

Freeman, C. and Perez, C. (1988) 'Structural crises of adjustment: Business cycles and investment behaviour', in G. Dosi, C. Freeman, R. Nelson, G. Silverberg and L. Soete (eds) *Technical Change and Economic Theory*, Pinter Publishers, London and New York, pp38–66

Garcia, R. and Calantone, R. (2002) 'A critical look at technological innovation typology and innovativeness terminology: A literature review', *The Journal of Product Innovation Management*, vol 19, pp110–132

Geels, F. W. and Schot, J. (2007) 'Typology of sociotechnical transition pathways', *Research Policy*, vol 36, pp399–437

Hausmann, R. and Rodrik, D. (2003) 'Economic development as self discovery', *Journal of Development Economics*, vol 72, no 2, pp603–633

Hausmann, R., Hwang, J. and Rodrik, D. (2007) 'What you export matters', *Journal of Economic Growth*, vol 12, no 1, pp1381–4338

Hollander, S. (1965) *The Sources of Increased Efficiency: A Study of Du Pont Rayon Plants*, MIT Press, Cambridge, MA

Kim, L. (1997) *Imitation to Innovation: The Dynamics of Korea's Technological Learning*, Harvard Business School Press, Boston, MA

Kim, L. (1998) 'Crisis construction and organisational learning: Capability building in catching-up at Hyundai Motor', *Organization Science*, vol 9, pp506–521

Luiten, E. E. M. (2001) *Beyond Energy Efficiency: Actors, Networks and Government Intervention in the Development of Industrial Energy Process Technologies*, Ph.D. thesis, Utrecht University, Utrecht

Luiten, E. E. M. and Blok, K. (2003) 'Stimulating R&D of industrial energy-efficient technology: The effect of government intervention on the development of strip casting technology', *Energy Policy*, vol 31, pp1339–1356

Matsushita, T., Nakayama, K., Fukase, H. and Isada, S. (2009) 'Development and commercialization of twin roll strip caster', *IHI Engineering Review*, vol 42, no 1, pp1–9

McArthur, J. W. and Sachs, J. D. (2002) 'The Growth Competitiveness Index: Measuring technological advancement and the stages of development', in World Economic Forum (ed) *The Global Competitiveness Report 2001–2002*, Oxford University Press, New York and Oxford

Mitchell, C., Sawin, J., Pokharel, G. R., Kammen, D., Wang, Z., Fifita, S., Jaccard, M., Langniss, O., Lucas, H., Nadai, A., Trujillo Blanco, R., Usher, E., Verbruggen, A., Wüstenhagen, R. and Yamaguchi, K. (2011) 'Policy, financing and implementation', in O. Edenhofer, R. Pichs-Madruga, Y. Sokona, K. Seyboth, P. Matschoss, S. Kadner, T. Zwickel, P. Eickemeier, G. Hansen, S. Schlömer and C. von Stechow (eds) *IPCC Special Report on Renewable Energy Sources and Climate Change Mitigation*, Cambridge University Press, Cambridge and New York

Mizoguchi, S., Ohashi, T. and Saeki, T. (1981) 'Continuous casting of steel', *Annual Review of Materials Science*, vol 11, pp151–169

Mokyr, J. (1990) 'Punctuated equilibria and technological progress', *American Economic Review*, vol 80, no 2, pp350–354

Ockwell, D. G., Watsin, J., Mackerron, G., Pal, P. and Yamin, F. (2008) 'Key policy considerations for facilitating low carbon technology transfer to developing countries', *Energy Policy*, vol 36, pp4104–4115

Ockwell, D., Ely, A., Mallett, A., Johnson, O. and Watson, J. (2009) *Low Carbon Development: The Role of Local Innovative Capabilities*, STEPS Working Paper Series 31, STEPS Centre and Sussex Energy group, SPRU, University of Sussex, Brighton, UK

Ockwell, D. G., Haum, R., Mallett, A. and Watson, J. (2010) 'Intellectual property rights and low carbon technology transfer: Conflicting discourses of diffusion and development', *Global Environmental Change*, vol 20, pp729–738

OECD (Organisation for Economic Co-operation and Development)/Eurostat (2005) *Oslo Manual: Guidelines for Collecting and Interpreting Innovation Data*, OECD, Paris

Schlichting, M., Ondrovic, J., Woodberry, P. and Michael, D. (2009) 'Energy and environmental advantages with the Castrip process', *Stahl und Eisen*, vol 129, pp44–51

Schneider, M., Holzer, A. and Hoffmann, V. A. (2008) 'Understanding the CDM's contribution to technology transfer', *Energy Policy*, vol 36, pp2930–2938

Scott-Kemmis, D. and Bell, M. (2010) 'The mythology of learning-by-doing in World War II airframe and ship production', *International Journal of Technological Learning, Innovation and Development*, vol 3, no 1, pp1–35

Seres, S., Haites, E. and Murphy, K. (2009) 'Analysis of technology transfer in CDM projects: An update', *Energy Policy*, vol 37, pp4919–4926

Slocum, A. and Rubin, E. S. (2008) *Understanding Radical Technology Innovation and Its Application to CO_2 Capture R&D: Interim Report, Volume One – Literature Review*, Department of Engineering and Public Policy, Paper 66, Carnegie Mellon University, Pittsburgh, PA, http://repository.cmu.edu/epp/66

UNFCCC (United Nations Framework Convention on Climate Change) (2009) *Advance Report on Recommendations on Future Financing Options for Enhancing the Development, Deployment Diffusion and Transfer of Technologies under the Convention*, UNFCCC, FCCC/SB/2009/INF.2

World Bank (2010) *World Development Report 2010: Development and Climate Change*, International Bank for Reconstruction and Development/World Bank, Washington, DC

PART II

Learning from Technology and Country-Specific Analysis

3

CASE STUDY

Technology Transfer of Energy-efficient Technologies among Small- and Medium-sized Enterprises in India

Prosanto Pal and Girish Sethi

Background

The small- and medium-sized enterpriss (SME)[1] sector plays a vital role in the Indian economy, contributing around 45 per cent of manufacturing output and 40 per cent of exports, and employing an estimated 59.7 million people spread over 26.1 million enterprises according to recent estimates. SMEs today account for almost 90 per cent of the total number of industrial units in the country (Government of India, 2011). In order to promote small industries, a large number of items were reserved to be produced exclusively in small sector during the 1970s and 1980s. However, having functioned for five decades within an overly protective economic and industrial framework, a substantial proportion of Indian SMEs remain isolated from modern technological developments. Apart from certain new-age sectors such as information technology, biotechnology, pharmaceuticals, etc.), SMEs in traditional manufacturing sectors such as castings, forgings, glass and ceramics, food processing, textile processing and so on use obsolete, inefficient technologies to burn commercial fuels (coal, oil and gas), leading to wastage of fuel as well as the release of high volumes of greenhouse gases (GHGs) and particulate emissions that are harmful to health and damage the atmosphere. A large number of energy-intensive SME clusters (around 178 clusters manufacturing about 15 product categories) are energy intensive, with fuel costs making up 2 per cent to 50 per cent of the total cost of production (TERI, 2009). While individual SME units are relatively small in size, their sheer numbers, coupled with the fact that they depend on low-efficiency fuel burning technologies, make the SME sector a sizeable source of carbon emissions. Hence, there is a clear and urgent need for SMEs to adopt EE (energy-efficient) technologies that will help them reduce both fuel consumption and carbon emissions.

This case study summarizes two successful TT (technology transfer) projects undertaken by The Energy and Resources Institute (TERI) with the support of the Swiss Agency for Development and Cooperation (SDC), to promote energy efficient technologies in two energy-intensive SME sub-sectors: small-scale foundries and small-scale glass industries. The technology development and demonstrations were undertaken during the period of 1995 to 2000. A detailed account of the process adopted under the two projects is provided in the books published by TERI (Pal, 2006; Sethi and Ghosh, 2008). A summarized version of the account also appears in the UK–India collaborative study on the transfer of low carbon technology (SPRU and TERI, 2009).

It is important to note that TERI played two roles during this project:

- As primary collaborator of the improved technological know-how, through on-going capacity building of the TERI team by its overseas partners.
- As facilitator and technology service provider for the development, adaptation, demonstration and promotion of the improved technologies in the concerned SME sectors, through direct field-level interactions with entrepreneurs and other industry stakeholders.

Small-scale foundries

Profile

Foundries make iron castings from molten iron. Castings find diverse applications, such as in the manufacture of sanitary pipes and fittings, automotive parts and engineering equipment (e.g. pumps, compressors and electric motors). There are about 5000 small-scale foundry units in India, with a collective annual output of about 6 million tonnes of castings. While their output predominantly caters to domestic markets, a small percentage is exported. The foundry sub-sector is growing at 8 to 9 per cent annually and provides direct employment to an estimated 0.5 million people.

The Indian foundry industry had its roots in the 19th century, when industrialization and rapid expansion of railways provided an assured and growing market for castings. After Independence, the steel and coal-mining sectors largely remained under government control, hence the foundry industry was assured about both the availability and prices of its primary raw materials, pig iron and coke. In this scenario, energy efficiency was not a major concern for foundries. However, the situation changed following the opening up of the Indian economy during the early 1990s. Competition has forced the large integrated steel plants to reduce pig iron production, leading to a rise in the price of pig iron. In addition, with the removal of licences and permits to obtain coke, small-scale foundries are no longer assured of coke supplies at steady prices. Hence, energy efficiency has become of vital concern for small-scale foundries as a means of reducing fuel consumption and costs and thereby increaeing profitability and competitiveness.

Foundries are mainly located in clusters across the country. The clusters vary in size: some have less than 50 units, while others have over 500 units. Typically, each cluster specializes in producing castings for specific end-use markets. For instance, the Howrah cluster in eastern India has around 300 foundries that mainly produce low value-added castings such as manhole covers and pipes; while the Rajkot cluster in western India has around 500 foundries that mainly produce grey iron castings for the local diesel engine industry.

Mirroring the different products they make and the diverse markets to which they cater, foundry clusters differ from one another in terms of technology, operating practice, and commercial dealings. Within a cluster, foundry units usually operate in isolation; there is little sharing among them of information related to technology, operating practice, and so on. The units may form loose associations; but these are primarily for the purpose of obtaining fuel (coke) at favourable prices from suppliers and for other trade-related issues.

Technology

A foundry makes iron castings by melting a variety of iron-containing materials such as pig iron and cast iron scrap in a furnace called a cupola. The resulting molten iron is then poured into moulds to make castings of desired shapes. Usually, cupolas burn coke as fuel. Melting is by far the most energy-intensive stage of a foundry's operations.

Until the early 1990s, most Indian foundries were using the conventional 'cold blast' cupola. As described later, TERI partnered with the UK-based BCIRA (now known as BCIRA Cast Metals Technology Centre)[2] to identify, transfer and adapt a more energy-efficient melting technology for small-scale Indian foundries – the DBC (divided blast cupola).

Scope for reducing carbon emissions

The total coke consumption by the foundry sub-sector is estimated to be around 600,000 tonnes per year (equivalent to around 1,640,000 tonnes CO_2). There is considerable scope for saving 20 to 40 per cent fuel and reducing carbon emissions in this sector by switching to the energy-efficient DBC technology.

Small-scale glass industries

Profile

Almost the entire small-scale glass industry in India is located in a single cluster in Firozabad, about 4 km from Agra. According to a TERI estimate, each day, glass units in Firozabad produce around 2000 tonnes of glass products, including 50 million bangles, and provide direct employment to an estimated 150,000 people.

Besides having a near-monopoly in the production of bangles, the Firozabad glass cluster also produces popular low-value glass products (bowls, tumblers, lamp shades and so on). There is a steady demand for such products across India. However, with very little to distinguish the glass products made by one unit from another, and in the absence of direct linkages with consumers or retail markets, units sell their products at prices dictated by dealers and middlemen. As a result, competition is vicious among units, and profit margins are thin and unpredictable.

During the early 1990s, glass entrepreneurs had very few options to increase returns. There was no room to reduce manpower or wages. Fuel (coal) costs were beyond their control. The only way to reduce fuel costs was by increasing the energy efficiency of the glass melting furnace; but this task required technical knowledge and skills that neither the entrepreneur nor the traditional furnace builders (*mistrys*) possessed.

Technology

Glass is made by melting silica sand (which contains about 96 per cent, by weight, silicon dioxide – SiO_2) together with chemicals that reduce melting temperature and give strength and colour to the end-product. The molten glass is drawn from the furnace, blown or formed into desired shapes, and then annealed (heated and cooled in a controlled manner) to impart hardness to the glass. Depending upon their nature, the products are then subjected to various cutting and finishing operations. The glass industry is highly energy-intensive, with fuel cost accounting for over 40 per cent of product cost.

In Firozabad, three basic kinds of melting furnaces are used to make glass: tank furnaces, open-pot furnace, and closed-pot furnaces. Glass for making bangles is melted almost exclusively in open-pot furnaces. Until the early 1990s, almost all of these furnaces operated on coal. By 1996, most tank furnaces in the cluster (about 55 in number) had switched from coal firing to oil firing, but pot furnaces (about 100 in number) were still being fired by coal. As described later, TERI worked with British Glass and other partners to transfer know-how for the development and promotion of a more energy-efficient melting technology for small-scale glass melting units in Firozabad – the gas-fired recuperative pot furnace. The fuel switch from coal to natural gas was mandated by the Supreme Court of India in December 1996 with the Taj Trapezium Zone (TTZ), an area of 10,400 square kilometre^2 around the Taj Mahal, within which Firozabad is located.

Scope for reducing carbon emissions

According to a study by TERI (1995), the pot furnace units in the Firozabad glass cluster consumed around 100,000 tonnes of coal annually. As a result, there was considerable potential to reduce coal consumption and decrease carbon emissions in this sector by shifting to more energy-efficient technologies for producing molten glass.

Collaborative research, development, demonstration and diffusion (RDD&D)

In 1992, the SDC initiated a macro-level study by TERI of energy consumption patterns in the Indian SME sector. Based on this study, the SDC partnered with Indian non-governmental organizations (NGOs)/research institutions and international consultants to initiate a programme aimed at introducing clean energy-efficient technologies in four energy-intensive SME sub-sectors. Two of these sub-sectors – namely, the foundry and glass industries – are discussed in this case study. The other two sub-sectors identified were the small-scale brick industry and small/micro enterprises that burn biomass fuels. The overall goals of the SDC programme were to:

- help SMEs achieve energy savings and thereby improve profitability of operations'
- bring about reduction in CO_2 and other emissions and thereby address environmental concerns at both local and global levels.

Technology transfer process

In the case of both small-scale foundries and small-scale glass industries, energy-efficient technologies were identified, transferred and adapted to local conditions and requirements in the following broad stages:

- conducting energy audits to identify areas in which to improve energy efficiency (needs assessment);
- identification, design, development and adaptation of energy-efficient technological solutions in collaboration with international experts, industry associations and local experts;
- demonstration and fine-tuning of the improved technologies through unit-level demonstration/pilot projects;
- strengthening the knowledge and skills of local entrepreneurs and building their confidence and capabilities in the new/improved energy-efficient technologies through on-going capacity building programmes (thus preparing the ground for dissemination and mainstreaming of the demonstrated technologies).

During the technology transfer processes, TERI obtained strategic support and inputs from Sorane SA, Switzerland. Sorane SA provided advice in energy management and systems integration, helped to identify and coordinate activities with international energy and environmental consultants, and assisted by way of technical support.

Energy audits

During 1993 to 1994, TERI conducted detailed energy audits of representative units in the Agra foundry cluster. The audits revealed that the 'coke feed ratio', or CFR[3] (a measure of cupola efficiency), ranged between 31 and 19 per cent in the existing cupolas, compared with the best CFR levels achieved within India and abroad of about 10 per cent. This indicated large potential for improving energy efficiency by introducing optimally designed cupolas and adopting best operating practices (BOPs). TERI also conducted energy audits of representative glass units in the Firozabad glass cluster in 1994. These audits revealed that the energy efficiency of the coal-fired open pot furnace was as low as 10 per cent, with flue gases escaping from the furnace at a temperature of around 950°C. These findings indicated considerable potential to increase energy efficiency in pot furnaces through heat recovery from flue gases.

Identifying technologies for transfer

Having identified the areas in which to introduce energy-efficient technologies – namely, the cupola furnace in foundry units and coal-fired open pot furnace in glass units – TERI teamed up with international partners to select appropriate technologies which could be transferred and adapted to the local industry needs.

In each case, the first step was to evaluate existing technologies to identify those that could be adapted or modified to meet the standards set for better energy efficiency and environmental performance. Thereafter, from among the available options, the most appropriate one – that is, the one most suited to adaptation to meet local needs and conditions – was selected through a bottom-up participatory process and transferred for further development to meet the needs of the industry concerned.

Divided-blast cupola

In the case of foundries, the results of the Agra cluster energy audits were discussed and validated by experts from Cast Metals Development Limited, UK, a group company of BCIRA. The discussions also focused on finding ways to improve the efficiency of the melting furnace. Based on its consultations with the British partners, TERI chose the divided-blast cupola or DBC as the best option to improve energy efficiency in small-scale foundries at a modest investment. The DBC offered the following advantages over the conventional cupola:

- coke consumption reduced by about 25 per cent;
- tapping temperature increased by about 50° C; and
- melting rate increased.

Gas-fired recuperative pot furnace

The identification of EE technology options for the open pot furnace in Firoz-abad presented a unique challenge, primarily because a pot furnace is intrinsically inefficient in design. Pot furnaces had been used in countries such as the UK and Germany; but these burned better-quality coal to make very high-value products such as crystal ware, as a result of which the proportion of fuel cost in the product cost (i.e. the energy intensity) remained low, making operations profitable. In contrast, the pot furnaces in Firozabad burned medium-grade coal to make relatively low-value items hence the fuel cost made up a substantial portion of the product cost. The result: low profitability of operations.

Following extensive consultations between TERI, its British partners and Sorane SA, a new pot furnace design was evolved through research, development and demonstration in the Firozabad glass cluster: a gas-fired pot furnace with its burner mounted on the crown, and with a recuperator to recover and reuse waste heat from flue gases.

Technology development and demonstration

Divided blast cupola

TERI developed and demonstrated the DBC at a foundry unit in the Howrah cluster. In setting up the demonstration plant, the project brought together local and international experts in many disciplines – project management, foundry technology, energy management, cupola operation, and environmental technology.

In particular, Cast Metals Development Limited, UK, provided crucial support and expertise in transferring technical know-how related to the DBC, and at every stage during the design and commissioning of the demonstration plant. The British partner assisted the TERI team in conducting an energy audit of the existing cupola in the demonstration unit; in analysing the results of the audit so as to evolve design parameters for the new DBC; in ensuring that quality and design norms were adhered to during the fabrication of various components of the DBC; and in fine-tuning various sub-systems during the trial runs.

The demonstration DBC was successfully commissioned in mid-July 1998. The DBC showed a marked improvement in energy efficiency (CFR 8 per cent) compared with the existing cupola (CFR 13.3 per cent). In effect, the new plant yielded an energy saving of about 40 per cent compared to the earlier cupola. The DBC also yielded additional benefits in terms of an increase in metal temperature and a substantial reduction in silicon and manganese losses. On an average monthly melting of 430 tonnes, the demonstration DBC yielded an annual saving in coke of 270 tonnes. The payback period worked out to less than two years on the investment in the DBC alone.

Although the new DBC had proved itself to be far more energy-efficient than the existing cupola, proper operating practices had to be followed to reap the full benefits of its improved design. Hence, following the demonstration, TERI and its British partners worked for several weeks in training the furnace operators and maintenance personnel to follow BOPs in the day-to-day running of the plant.

Recuperative pot furnace

In designing, developing and demonstrating the new gas-fired pot furnace, TERI worked closely with a number of British partners whose key roles are summarized as follows.

- British Glass, UK, provided expertise in glass technology. Along with other partners, British Glass also finalized the conceptual and detailed designs of the new pot furnace.
- AIC (Abbeville Instrument Control Ltd), UK, helped in developing the concept and design of the new furnace, including its heat recovery unit (recuperator).
- Chapman and Brack, UK, provided guidance in constructing the crown of the furnace.
- TECO (Toledo Engineering Co Inc), UK, provided expertise in commissioning the recuperator for the furnace.
- NU-WAY, UK, supplied the burners for the furnace.

The gas-fired recuperative pot furnace was successfully commissioned in February 2000. Following the demonstration, TERI and its British partners trained furnace operators and other workers in monitoring and operating the new system. The British partners continued to provide support for a few years after demonstration in trouble-shooting the furnace system and in fine-tuning its performance parameters.

The energy consumption of the gas-fired recuperative furnace was measured at 16.5 gram calories per day (Gcal/day), against 39.4 Gcal/day in the traditional coal-fired pot furnace. This represented a 58 per cent reduction in energy consumption, of which around 28 per cent came from heat recovery alone. Estimates by the project team indicated that the recuperative furnace was also 34 to 38 per cent more energy efficient than the retrofitted gas-fired furnaces being used by other pot furnace units in the Firozabad cluster.[4] Because of its increased fuel efficiency, the recuperative furnace promised a payback within two years.

Technology dissemination and results

Following the successful demonstration of the two EE technologies – the DBC and the recuperative pot furnace – TERI focused its efforts on disseminating these technologies through:

- providing customized design solutions and installation/commissioning support to other entrepreneurs on the new energy-efficient technologies;
- awareness generation among industry stakeholders at both policy and cluster levels;

- capacity-building programmes involving entrepreneurs, fabricators, local consultants, masons and other stakeholders.

The Swiss and British partners continued to provide technical support to the project in order to facilitate replication of the EE technologies.

As of September 2010, around 95 TERI-design DBCs of different capacities (based on local requirements) have been adopted by foundry clusters across the country (Ahmedabad, Rajkot, Coimbatore, Nagpur, etc.).[5] The adoption of this energy-efficient technology has yielded an estimated cumulative energy savings of 33,000 tonnes of oil equivalent (toe) and a cumulative CO_2 savings of about 120,000 tonnes to date. The DBC technology has also been replicated in two foundry units in Bangladesh with technical support from TERI – an example of successful South–South technology transfer. Widespread adoption of the DBC will make it possible to save about 25 per cent of the coke consumed by the Indian foundry industry (i.e. 150,000 tonnes of coke annually). The overall CO_2 emissions from conventional cupolas used by the foundry industry are estimated at 2.5 million tonnes per annum. The CO_2 emissions could be reduced by around 0.6 million tonnes annually through the widespread adoption of DBC technology.

Similarly, with on-going technical support from TERI, 76 of the 100 odd operating pot furnace units in the Firozabad cluster have since switched over to the TERI-designed recuperative furnace, yielding an estimated cumulative energy savings equivalent to 79,000 toe. Most of the remaining pot furnace units are expected to follow suit in the next few years. These replications have brought about a cumulative reduction in carbon emissions of 245,000 tonnes CO_2. In addition to these 'direct' replications of the TERI-designed furnace, reportedly almost all pot furnace units in Firozabad have adopted the *concept* of heat recuperation from the TERI-designed furnace. This in itself indicates that the technology transfer process has succeeded; that the entrepreneurs have shed their traditional reluctance to consider changes in their technology, and are becoming increasingly confident in learning from the improved EE technologies and adapting them to suit their individual needs.

Intellectual property rights (IPRs)

Given the fact that SME entrepreneurs in developing countries are generally resource poor (as they operate on thin margins), it is difficult for them to raise the resources to invest in improved technology (which may have elements of IPR), even when they are willing to overcome their intrinsic reluctance to abandon their traditional technologies. Often, therefore, an improved technology that is introduced in the SME sector (which, in most cases, is likely to be costlier) is prone to be reverse-engineered by local manufacturers and sold in cheaper forms. The 'hardware' elements of technology (e.g. specific pieces of imported equipment such as burners, etc.) are particularly prone to be reverse-engineered. The 'software' elements (such as design drawings for a cupola furnace) are less prone to be engineered locally, primarily because they have an intrinsic element of knowledge which requires capacity-building of local stakeholders (from implementing agencies, local consultants, users, etc.). Besides, for each particular industry, the 'software' needs to be adapted to suit local conditions.

It is therefore vital for all project partners and collaborators – including both hardware and software suppliers – to be aware of these ground realities of the SME sector. However, it is important to note that the reverse-engineered models are usually not perfect and therefore perform far less efficiently than the original technologies. For instance, some pot furnace units in Firozabad are using a locally made gas burner that resembles the gas burner imported from NU-WAY, a leading manufacturer of energy-efficient burners in the UK, and costs much less. However, reports indicate that the locally made gas burner reportedly has a much shorter life than the NU-WAY burner, and its performance is unpredictable. Likewise, reports from the foundries in Howrah that have adopted DBCs without reference to the project indicate that their performance is suboptimal.

As a result, although the overall deployment of improved technologies (including original and self-replicated versions) is not hindered by the tendency in the SME sector to reverse-engineered technologies, the overall fuel and carbon savings achievable is reduced to some extent because the performance of these locally engineered technologies is usually suboptimal.

Policy implications

- Small-scale industries manufacturing energy-intensive products form the backbone of many developing country economies. There is an enormous potential to reduce CO_2 emissions at lower costs in this sector. This is important considering the fact that SMEs, in general, do not have either the inherent financial capacity or the technical capacity to undertake research or adaptation activities that would help them to improve their energy and environmental performance. It is therefore important to identify such SMEs in developing countries (preferably a group of enterprises having a similar technological base and similar operating practices), and then develop tailor-made RDD&D programmes for them (which may be industry/cluster specific), with support from multilateral/bilateral organizations, especially in the context of climate change.

- The RDD&D programmes should focus on cleaner production, which means conservation of resources and energy use in the production processes through improved technologies. The industry associations at the local/state levels can play an important role by identifying suitable locations for initiating such programmes, while multilateral agencies can help in facilitating inter-governmental partnerships and bringing international technical experts together in such programmes.

- The technology demonstrations could focus on 'incremental' technological improvements, which would be easily adopted by the SME sector. By their very nature, SME units find it hard to absorb rapid change; they are inhibited by factors such as lack of technical knowledge, resource constraints, low productivity, and so on. A step-wise *incremental* approach in TT allows them to adopt and absorb better technology/operating practices, and based on their own unique (unit-specific) requirements; all that they require is the necessary technical back-up support to be able to do so. The incremental approach also imparts a growing confidence in the entrepreneurs to experiment with, evolve and adopt their own cost-effective technological solutions.

- In order to ensure their sustainability in the long term, RDD&D programmes in the SME sector should have a strong partnership element and involve local actors right from the initial stages. Local actors can take over once the consultants and expert R&D organizations have helped the SMEs to identify and demonstrate the benefits of cleaner technologies.

- It is equally important for funding agencies to have a long-term commitment and a flexible approach in RDD&D programmes; change is invariably a slow process at the small-scale industry level. A prime factor that has contributed to the success of the TT project highlighted in this case study is the flexibility and long-term engagement with the project shown by the SDC, and the unique partnership arrangement that exists among the project partners at different levels: between the funding organization (SDC), implementing agency (TERI), local consultants, international consultants, industry associations at cluster level and grassroots-level agencies. There is a need to develop and replicate similar innovative partnership arrangements on a much larger scale for interventions in the SME sector.

- For a collaborative RDD&D venture in the small-scale sector to be effective in bringing about sustainable change, technology transfer should *not* take place directly between technology supplier(s) and end users. Rather, it should be routed through an intermediary institution (such as an R&D establishment or consultancy organization), which can act as a facilitator to disseminate the improved technology on a large scale. An example is the TT project described above, in which TERI as an intermediary institution was able to absorb two improved EE technologies and then transfer and adapt them for dissemination among a large number of SMEs across India.

- One of the financial mechanisms available for promoting low-carbon technologies under international climate change protocol is the Clean Development Mechanism (CDM). However, high transaction costs act as a barrier in initiating small CDM projects in the SME sector. For CDM projects to be workable in the SME sector, there is a need to revisit the CDM implementation cycle – specifically, the documentation and verification formalities. A simplified CDM project cycle for SMEs holds promise of reducing millions of tonnes of GHG emissions in diverse SME sectors in developing countries.

Notes

1 The manufacturing enterprises are defined in terms of investment in plant and machinery. They are classified as small scale if the investment does not exceed 50 million Indian rupees and medium scale if the investment is below 100 million Indian rupees. US$1 is about 45 Indian rupees.
2 BCIRA is the UK's leading organization involved in research and development within the field of cast metals.
3 The energy efficiency of a cupola is measured in terms of the amount of coke consumed per tonne of metal charged. Known as CFR, or coke feed ratio, this is usually denoted as a ratio or as a percentage. The lower the CFR, the more efficient is the cupola.
4 By December 2001, almost all pot furnace units in Firozabad had adopted a 'retrofitted' gas-fired furnace design, which provided some improvement in energy efficiency compared to coal firing. However, the recuperative furnace designed by TERI and its partners proved far superior in terms of energy efficiency, as described.
5 The number of replications needs to be viewed in light of the fact that they focused on a limited number of foundry clusters in India. An external review of the project conducted in July 2011 mentioned that a much larger number of cupolas have adopted certain innovative design features introduced under the project, if not the entire cupola. Hence, there are spin-offs of the innovation. Some factors which have inhibited quicker replication are the slow pace of change typical among small-scale industries, the higher capital cost, the lack of life-cycle costing by SMEs, and limited capacity for and availability of the DBC technology at the cluster level.

References

Government of India (2011) *Annual Report, 2010–2011*, Ministry of Micro, Small and Medium Enterprises, Government of India, New Delhi
Pal, P. (2006) 'Towards cleaner technologies: A process story in small-scale foundries', in G. Sethi, P. Jaboyedoff and V. Joshi (eds) *Towards Cleaner Technologies: A Process Story in Small-Scale Foundries*, The Energy and Resources Institute, New Delhi
Sethi, G. and Ghosh, A. M. (eds) (2008) *Towards Cleaner Technologies: A Process Story in the Firozabad Glass Industry Cluster*, The Energy and Resources Institute, New Delhi
SPRU and TERI (2009) *UK–India Collaborative Study on the Transfer of Low Carbon Technology: Phase II Final Report*, March, Prepared for the UK Department of Energy and Climate Change (DECC), London
TERI (The Energy and Resources Institute) (1995) *Energy Sector Study Phase 1*, Submitted to the Swiss Agency for Development and Cooperation (SDC), TERI, New Delhi
TERI (2009) *BEE SME Programme: Situation Analysis in 35 SME Clusters*, Prepared for Bureau of Energy Efficiency (BEE), Ministry of Power, Government of India, New Delhi

4

LOW-CARBON INNOVATION IN CHINA

The Role of International Technology Transfer

Jim Watson and Rob Byrne

Introduction

This chapter focuses on low-carbon innovation in China, the world's second largest economy and largest user of energy. It draws on the results of a UK–China collaborative research project that was funded by the UK government. Through the use of four case studies, the chapter examines the relative contributions of indigenous innovation and international technology transfer to China's low-carbon innovation journey so far. It uses this evidence to draw lessons for national and international policy, and to reflect on the on-going debate about technology assistance to developing countries under the United Nations Framework Convention on Climate Change (UNFCCC).

During the past two decades, China's economy has continued to grow rapidly, at an average rate of around 10 per cent per year (Wang and Watson, 2009). At the same time, this economic expansion has led to large increases in energy demand and carbon emissions. These increases have continued through the recent financial crisis, which has led to falling emissions in many of the Organisation for Economic Co-operation and Development (OECD) countries. China is now the world's largest emitter of carbon dioxide (CO_2). However, on a per capita basis, China's CO_2 emissions are lower than those of most industrialized countries. Per capita emissions reached the global average level of about 4.5 tonnes in 2006 (UNDP China, 2010, p29). China is particularly vulnerable to the expected impacts of climate change. Some of these impacts are already being observed, and they are projected to increase in the future (NDRC, 2007). They include shrinking glaciers in mountain regions (with impacts upon water availability), changes in rainfall patterns which have already caused droughts in the North, and sea-level rise which could have serious impacts upon coastal areas.

Coal continues to dominate China's energy system, accounting for two-thirds of primary energy. According to recent official statistics, total energy consumption rose to 3250 million tonnes of coal equivalent in 2010. The official Chinese news agency stated that China's power generation capacity reached approximately 960 gigawatts (GW) by the end of 2010 – an increase of 85GW over the figure a year earlier. Around three-quarters of total capacity is coal fired. This is the second largest generation capacity in the world, and is now close to that of the US (1010GW).[1] Imported oil is also increasing sharply to over 50 per cent of total oil consumption in 2009 – up from 29 per cent in 2000 as domestic output has matured. Demand for natural gas keeps growing, but plays a small role in overall primary energy supply – it accounted for approximately 3.8 per cent of primary energy consumption in 2008.

These trends lead to a number of pressing economic and environmental challenges. In its report to the 2011 National People's Congress, the National Development and Reform Commission (NDRC) stated that 'total energy and resource consumption is too large and is growing too quickly, and emissions of major pollutants are high … energy-intensive and highly polluting industries are still growing too fast. Consequently, we face mounting pressure to save energy, reduce emissions, and respond to climate change' (NDRC, 2011, p18).

A new target for carbon intensity reduction was announced in the run-up to the Copenhagen Conference of the Parties (COP) in 2009 and requires that carbon intensity should fall by 40 to 45 per cent by 2020 from 2005 levels. Related to this, the State Council ratified a target that 15 per cent of China's overall energy supply should come from non-fossil sources by 2020. Implementation is already under way within the context of successive *Five-Year Plans. The 11th Five-Year Plan* from 2006 to 2010 included a 20 per cent energy-intensity reduction goal. The NDRC claims that a 19 per cent reduction was achieved during this period, despite an increase in energy intensity in early 2010 which made the target more difficult to achieve (NDRC, 2011). Controversy has surrounded some of the methods used by provincial governments (such as power rationing) to meet their share of this target (Watts, 2010). Significant attention is now being paid to the implementation of the *12th Five-Year Plan*, which runs from 2011 to 2015. It includes targets for a 16 per cent reduction in energy intensity and a 17 per cent reduction in carbon emissions intensity (Wen Jiabao, 2011). It also states that non-fossil sources should account for 11.4 per cent of primary energy by 2015, up from 9 per cent in 2008 (Zhang, 2010).

To achieve these goals, there will be a central role for the development and deployment of low-carbon technologies in China, including technologies and measures to improve energy efficiency and low-carbon energy supply. This process is already under way, supported by China's own domestic policies (e.g. Lewis, 2007; Chatham House et al., 2010; Tan et al., 2010). Recent examples include research and development (R&D) support, incentives for technology deployment (e.g. renewables) and demonstration trials (e.g. of electric vehicles). Significant Chinese government funding for these technologies was included within the 4 trillion Yuan (approximately UK£400 billion) stimulus package implemented after the global financial crisis.

Of course, international policy frameworks also have a role to play in supporting such innovation processes. There is now some acknowledgement within China that international agreements on technology transfer need to take into account the concerns of leading international firms, including respect for intellectual property. This has led to a modified tone in speeches by senior officials. Rather than simply calling for subsidized access to intellectual property, officials often couple this with a general emphasis on intellectual property protection. For example, Zeng Peiyan, an influential former vice premier of China, stated in May 2010 that: 'regrettably, we haven't seen substantive progress in the sharing of these [low-carbon] technologies... There is a need to develop institutions and finance ... to transfer technologies on concessional terms whilst safeguarding intellectual property rights.'[2]

Our study approach

This chapter summarizes the results of a collaborative research project by the Sussex Energy Group (University of Sussex, UK) and the Laboratory of Low Carbon Energy (Tsinghua University, China) on low-carbon technology transfer to China (Watson et al., 2011). The project was carried out between February 2010 and April 2011, and was funded by the UK government's Department of Energy and Climate Change. The project was funded to analyse low-carbon innovation in China, and the particular role of technology transfer within this.

Much of the research was focused on four empirical case studies: energy efficiency in the cement industry; electric vehicles; offshore wind power; and efficient coal-fired power generation. The case studies were chosen in consultation with policy-makers in the UK and through a stakeholder workshop in China. The choice was designed to capture the diversity of low-carbon innovation in China. They encompass different stages of technological development, different markets (capital and consumer goods) and different parts of the energy system (electricity generation, industry and transport). The case studies include near-market technologies (improvements in efficiency in the cement industry and more efficient technologies for coal-fired power) but also focus on future technologies that are still in the process of being commercialized in China (electric vehicles and offshore wind). Some of these cases (e.g. offshore wind) have not been examined in depth within previous studies of low-carbon innovation. The research on each case examined the full range of factors that influence innovation and technology transfer, including technological capacity, access to intellectual property rights (IPRs) and the role of national and international policy frameworks. The research team was also asked to suggest policy implications, particularly for the international United Nations Framework Convention on Climate Change (UNFCCC) negotiations.

Whilst low-carbon technology transfer provided a useful entry point for the research, the study analysed such transfer within the context of broader processes of low-carbon innovation in China. Technology transfer from international firms and other organizations is only one source of such innovation. This cannot be analysed in isolation from other indigenous sources of innovation – or from the wider national and international policy contexts that will affect rates of low-carbon technology development and deployment. In addition to this, the research drew on theories of innovation that do not only emphasize the development of new technological hardware. They also place particular importance on the knowledge and skills required to acquire and further develop such hardware technologies (e.g. Bell, 1990; Ockwell et al., 2009). In doing so, the research added to an understanding of how technological capacity is developed in China. In the context of international discussions, it is important to gain a better insight into how such capacity contributes to both economic development and low-carbon technology deployment.

The remainder of this chapter comprises three further sections. The next section discusses the evidence from four project case study technologies and focuses on a number of key issues, including the development of technological capabilities and the role of national (Chinese) policies. This is followed by a discussion of the role of international policy frameworks with a particular emphasis on the Clean Development Mechanism (CDM). The final section summarizes the main conclusions and suggests some implications for policy.

Cases of low-carbon innovation in China

This section sets out our findings with respect to four case study technologies. For each case, the analysis focuses on two main aspects: the current status of technological capabilities within China; and the role of national policy frameworks in fostering innovation and deployment.

Energy efficiency in the cement industry

Globally, the production of cement contributed around 8 per cent of anthropogenic CO_2 emissions in 2006 (Müller and Harnisch, 2008). China has been the world leader in cement production for many years, reaching 1.65 billion tonnes in 2009, or more than 50 per cent of world production (CCA, 2010). The process of cement production generates CO_2 emissions in two important ways. Currently, the conversion of limestone into lime accounts for about 55 per cent of these emissions, and the combustion of energy carriers needed to drive this conversion process accounts for another 40 per cent (Müller and Harnisch, 2008, p2). There are many measures that can reduce primary energy consumption in the production of cement and therefore could be characterized as energy-efficiency improvements. These range from behavioural changes amongst staff to using the most efficient technological hardware and optimized processes (Worrell and Galitsky, 2008). In addition, the use of fossil energy carriers as the source of heat can be substituted with waste material or biomass, and waste heat itself can be recycled and/or recovered for power generation (Müller and Harnisch, 2008). All of these approaches have been implemented in China to some extent.

China's capabilities in cement production

The energy efficiency of cement production has steadily improved in China, although there remains a significant gap between international and Chinese average efficiencies. The rapid adoption of advanced new suspension pre-heater kiln (NSP) technology has been an important driver of improvements, with well over 1000 units now in operation in China. Some of the 'efficiency gap' can be explained by significant numbers of small and inefficient kilns that remain in operation (Müller and Harnisch, 2008). By 2007, the energy intensity of cement production in China had fallen to 158 kilograms coal equivalent per tonne (kgce/tonne) but was significantly higher than the international advanced level of 127 kgce/tonne (Ohshita and Price, 2011, p53). More recently, it was announced that the energy intensity of cement production fell 16 per cent during the *11th Five-Year Plan* period (2006 to 2010), suggesting that this gap has been closed further since 2007.[3]

In terms of manufacturing production equipment, we found that Chinese firms are able to make most of the equipment locally. Some have built cement production facilities abroad, beginning as early as 1992. The first plant exported was rated at 700 tonnes per day (tpd), but this had risen to 10,000 tpd by 2005. Moreover, there are now five firms able to construct 10,000 tpd production lines and more than 300 with the capability to construct 5000 tpd facilities. Having said this, some technologies and equipment are still imported, including vertical mills, grate coolers, precision weighing machines and x-ray diffraction instruments. In addition, some of our interviewees identified difficulties training staff in new and advanced energy-efficiency technologies, in integrating these technologies within production systems, and barriers to accessing IPRs.

In what is a familiar pattern across our case studies, Chinese firms have developed their technological capabilities through a sequence of increasingly sophisticated and larger-scale activities. With respect to NSP technology, this process started during the late 1970s with the installation of a 700 tpd line supplied by a Japanese firm. Later, a local firm was licensed to supply Japanese separator technology. By 1984, some Chinese firms were able to design and develop a 2000 tpd NSP kiln and source most of the equipment locally. During the early 1990s, the first joint ventures were established, beginning with a 4000 tpd NSP line in 1992 under Dalian Huaneng-Onoda Cement Company. Others followed, including Yantai Mitsubishi, Daewoo Sishui and Qinhuangdao Asano. However, the widespread diffusion of NSP cement kilns did not take place until the mid-2000s.

Through these activities, the Chinese innovation system for cement technologies has developed. A number of Chinese firms say that they undertake joint R&D with local research institutes and universities. Such local collaboration is favoured because the technologies are cheaper than foreign-made equipment, it is faster and more convenient to get after-sales service, it is easier to pursue continuous improvement strategies, and there is a wish to support the development of the local industrial system. One interviewee summed up this preference by stating that 'domestic techniques fit the Chinese cement development better'. In addition to this, there are organizations such as the China Cement Association that help to disseminate information and represent the industry at national level.

Policy frameworks for efficient cement production

The energy-intensity target within the *11th Five-Year Plan* has been one of the most important policy drivers of energy efficiency in recent years. According to the vice minister of the NDRC, Xie Zhenhua, the plan has led to a reduction in energy intensity by 16 per cent between 2006 and 2010.[4] The target was accompanied by the Top 1000 Energy Consuming Enterprises Programme, which was launched by the NDRC in 2006 (Wang and Watson, 2009). Its aim was to reduce energy intensity within firms accounting for 33 per cent of China's final energy consumption. Projected savings at the inception of the programme were 100 million tonnes of coal equivalent by 2010 (equivalent to 260 million tonnes of CO_2 compared to 'business as usual'). Cement is included within one of nine industrial sectors covered by the programme.

The programme included a number of elements (Price et al., 2008). Targets were agreed with individual provinces, which were then translated into agreements with individual firms. Performance evaluations of provincial officials were adjusted to take account of their relative success in meeting targets. Firms were required to develop goals and plans, and funding was made available for energy-efficiency projects specified in these plans. Funding for energy efficiency and pollution abatement from the Chinese central government was 23.5 billion Yuan (over UK£2 billion) in 2007 and 27 billion Yuan (UK£2.5 billion) in 2008 (Ohshita and Price, 2011). The 2008 figure included 4 billion Yuan (UK£400 million) for phasing out small inefficient plants – a policy that was backed up by surcharges on their electricity tariffs. Further financial incentives took the form of a reduction in export tax rebates for energy-intensive products. Our interviewees within the Chinese cement industry confirmed that incentives for cement plants to implement more efficient technologies and processes have been seen as significant. They had accessed grants from energy-saving project funds, subsidized loans and taken advantage of tax breaks. However, they also stated that more could be done – for example, to accelerate the uptake of more efficient technologies.

The evaluation of the programme has been difficult at times due to a lack of data (Price et al., 2008). With respect to cement, a team at the Lawrence Berkeley National Laboratory found that it was not possible to validate stated energy-efficiency savings because of the variety of cement technologies in use. With respect to the efforts to close smaller, less efficient plants producing steel, electricity, cement and other products, some progress has been made. Recent figures show that of the

250 million tonnes of cement capacity earmarked for closure within the *11th Five-Year Plan*, 140 million tonnes had been closed between 2006 and 2008 (Ohshita and Price, 2011). It is not clear whether further progress has been made since then – but one of our interviewees pointed out that national closure programmes are often only partially successful. Plants are sometimes kept in operation by local officials, despite being officially declared closed, for economic and employment reasons.

Electric vehicles

China was the third largest automobile producer and the second largest consumer in the world in 2008 (NBS, 2009). It became number one on both counts for the first time recently because of domestic subsidies for buyers and shrinking markets in Japan and the US because of the financial crisis (Xinhuanet, 2010). Vehicle sales rose to a record of 9.35 million in 2008 and motorcycle sales reached 25.5 million in 2007 (CATARC, 2008). However, vehicle ownership is still much lower than the world average. In 1990, road vehicles accounted for 54 per cent of transport energy demand in China, and grew to 65 per cent in 2005 (IEA, 2008). Under a business-as-usual scenario, this share is projected to become 77 per cent by 2030. Consequently, the Chinese government is keen to reduce transport oil demand through the promotion of 'new-energy' vehicles, including hybrid and electric vehicles (HEVs) (Ouyang, 2006; Wan, 2008). In principle, HEVs can have lower carbon emissions than internal combustion engine (ICE) vehicles, and could be three times more fuel efficient (Zhang et al., 2008). Clearly, this depends on the fuel used to generate electricity – a problematic issue for China where 75 per cent of electricity comes from coal.

China's capabilities in electric vehicles

Chinese firms have developed large-scale production capacity and significant technological capabilities in ICE vehicles. However, they are not yet able to serve the high-end market, which is still dominated by foreign firms. Reasons for this include inconsistent quality of production and risk aversion (UKTI and SMMT, 2010). It appears that the emphasis is on a cost-conscious mass market where quality matters less. This acts as an important constraint on their ability to adapt their capabilities to develop and commercialize advanced technologies such as HEVs. The design and manufacturing demands of HEVs appear to be much more complex – and costly – than those of traditional ICE vehicles.

Nevertheless, some Chinese firms have established capabilities in HEVs. In electric motors, large investments have been made to increase production capacity. Capacity in 2010 was estimated to be 272,000 electric motor sets – up from 73,000 a year earlier (Ouyang, 2010). With respect to batteries, most attention is focused on lithium-ion (Li-ion) technologies. One battery firm in Shenzhen, Build Your Own Dreams (BYD), became internationally famous by entering the automotive industry with plans for HEV models on an accelerated timescale – though these plans have not yet been fully realized. At present, while Chinese firms can manufacture batteries, there are parts of the process that they have not mastered. For example, there is still a need to import a critical membrane that is necessary to prevent overheating.

Capabilities are also weak in battery management systems, an area in which Chinese firms are still dependent on foreign suppliers. Similarly, some key components for electronic control systems for HEVs have to be imported into China. Finally, charging infrastructure is also important so that vehicles can be charged conveniently. The development of this infrastructure in China is at an early stage in some cities (UKTI and SMMT, 2010), though there are plans to increase coverage.

There is a significant amount of R&D being carried out by Chinese firms to build their capabilities in HEVs, but this has not yet had commercial impacts. This includes some joint R&D with foreign firms. Those firms have been attracted by the prospect of access to China's rapidly growing market for private vehicles. For example, the Shanghai Automotive Industries Corporation (SAIC) has now bought UK capabilities[5] and so has an R&D base in the UK. They have been using this over the past two years to help train Chinese engineers. Chang'An Automobile Company has also invested in an R&D base in the UK, recently entering into work with the University of Nottingham. Of course, the acquisition of firms and partnerships between firms/organizations in the OECD and developing countries is not unique to China. Previous research conducted by the Sussex Energy Group and The Energy and Resources Institute (TERI) on India also shows a similar trend in wind and solar power, as well as hybrid vehicles in India (Mallett et al., 2009).

The limited impacts of R&D, to date, are present across a range of important technologies for the commercialization of HEVs: in many aspects of batteries, advanced transmission, the integration and management of vehicle systems, and meeting the various regulations necessary for exporting to European and US markets, in particular. Where Chinese firms are enjoying more success in developing innovative capabilities appears to be in the domestic electric bicycle market. This may act as a platform from which they can build more complex capabilities over time that could see currently unexpected evolutions of HEVs.

Policy frameworks for electric vehicles

There are two specific ways in which public policy is providing incentives for innovation in electric vehicles in China. The first is in support for some of the R&D activities that have already been mentioned. There has been substantial Chinese government funding for R&D since at least 2002. Under the 863 programme, 860 million Yuan (UK£80 million) was spent between 2002 and 2006 on electric, hybrid and fuel cell vehicles. A follow-on programme which ran from 2006 has spent a further 1.1 billion Yuan (UK£105 million) on these technologies.

One of the significant results of this state-sponsored R&D drive has been China becoming the second most successful country for HEV patents. In such an R&D-intensive industry, IPRs can be seen as particularly important. Chinese firms have been successful in creating their own patents. But according to one analysis, this patent success may have become a source of paralysis in the local market (UKTI and SMMT, 2010). This argues that Chinese firms are reluctant to release their HEVs into the local market because they are fearful that their competitors will imitate them – the converse of the more common concern expressed by some international firms that Chinese firms might engage in reverse engineering.

The second area of support is through demonstration and deployment support programmes. There is a programme to deploy 60,000 'new energy vehicles' in 13 cities, and a plan that these vehicles should account for 5 per cent of total car sales in 2011. This would amount to more than 600,000 vehicles (total sales in China in 2010 were 13 million). Looking slightly further ahead, it is hoped that 0.5 to 1 million new energy vehicles will have been sold by 2015 (Levi et al., 2010).

Within this, the Chinese government intends to spend 20 billion Yuan (UK£1.9 billion) on the promotion, manufacture and sale of electric vehicles. This will underpin a new *Ten Cities, One Thousand Vehicles* plan which plans to demonstrate 1000 new electric vehicles (EVs) each year. A recent Accenture report quoted a higher figure for government support for EVs of 115 billion Yuan (UK£10.9 billion) between 2011 and 2020 (Accenture, 2011). This includes funding for R&D, commercialization, component manufacture and electricity infrastructure. The report notes that consumers could receive a subsidy of 50,000 Yuan (UK£4700) to purchase plug-in hybrid EVs, and slightly more for a pure EV. It also highlights electricity-charging infrastructure as a potential bottleneck, stating that as recently as 2009, 'there were only a handful of public charging stations located in a few cities, such as Shenzhen' (Accenture, 2011, p61).

Offshore wind power

Wind power is an important part of China's energy strategy as it looks to reduce the reliance on fossil fuels. It already has a highly active onshore wind industry that has grown from almost nothing at the end of the 1990s to become the largest in the world (Lewis, 2007; Levi et al., 2010). By the end of 2010, installed capacity had reached 42GW.[6] As a result of this explosive growth, the Chinese government raised its 2020 target for installed capacity from 30GW to 100GW.[7] It is now looking to exploit its offshore wind potential. However, unlike the onshore wind industry when China entered, the global offshore wind industry is still in its early stages.

Offshore wind presents a particularly difficult challenge. First, the turbines need to be designed specifically for the harsh marine environment. Second, there is a need to build in redundant systems in order to minimize the number of times a turbine needs to be accessed, and to avoid losing a turbine's output for the sake of a simple spare part. One of our interviewees described it as finding constant trade-offs between structural strength, flexibility, sophistication and cost. The blades will typically endure 20 to 50 times the duty cycle of an aircraft wing, while being longer and heavier, but at 5 per cent of the cost.

Chinese capabilities in offshore wind

China has become a world leader in production capacity for onshore wind power. With its long coastline, there is obvious potential to move offshore. There are specific challenges associated with this in China. It experiences frequent typhoons and the environment in areas where wind turbines are cited is harsher than in Europe, particularly in the silt base of the intertidal zone. The first offshore wind farm in China became operational in 2010. A young Chinese firm, Sinovel, won the contract for the project. It designed the 3 megawatt (MW) turbines in collaboration with Wintec (an Austrian firm), and manufactured them in China.

By the end of 2009, there were about 80 wind turbine manufacturers in China, although only 30 of these had actually sold turbines.[8] The top three manufacturers have a combined production capacity of about 8GW per year, supplying to a domestic market of 13.8GW per year. For offshore wind, the government introduced an access standard, which means that only those manufacturers that can produce turbines of 2.5MW or greater will be eligible for selection for offshore wind projects. This has spurred the larger firms such as Sinovel to develop prototype machines to meet this criterion.

The firm Goldwind (55 per cent state owned) serves as an exemplar of the catching-up strategies in the Chinese wind power industry. Established in 1998, it grew to become one of the world's top five wind turbine manufacturers in 2009 (GWEC, 2010, p10). It started by purchasing a licence to manufacture 600 kilowatt (kW) turbines from a second-tier German manufacturer, Jacobs, and subsequently bought licences for other turbines, increasing in size each time (Lewis, 2007). By 2009, Goldwind was able to manufacture turbines of 2.5MW with an annual production of 2.2GW. Similarly, Sinovel began by licensing to manufacture turbines, purchasing from the Austrian firm Windtec. In time, Chinese firms such as these have sent their employees overseas to learn from experienced firms or to study, and they have entered into joint ventures with foreign firms. Sinovel, for instance, worked with Windtec to design the 3MW turbines for the Shanghai offshore project, and Mingyang and Aerodyn of Germany co-designed the Super Compact Drive 3MW turbine. More recently, Chinese firms have acquired foreign capabilities by buying foreign firms. For instance, Goldwind bought Vensys in Germany in order to strengthen its R&D capabilities.

In general, however, while the production capacity for wind turbines is high in China, there are problems with certain aspects. The manufacture of blades, gearboxes, converters and spindle bearings is not yet fully indigenized. Discussions with interviewees indicate that poor capabilities for materials processing seems to be the problem rather than design skills, and there are certainly firms manufacturing these components. With respect to offshore wind, some Chinese representatives of industry and government are cautious about Chinese capabilities – and stress that power companies have little knowledge so far, and that it will take time for the technology to mature (Prideaux and Qi, 2010).

Policy frameworks for offshore wind

The Chinese government's policies and incentives for wind power have been well documented (e.g. Barton, 2007; Lewis, 2007; Levi et al., 2010). Legislation such as the 2005 Renewable Energy Law and incentives such as concessions and mandates have led to rapid deployment during the past five years. Targets for onshore wind power have been revised upwards as rapid growth has unfolded.

Despite this progress, there have been misgivings about the Chinese policy approach. For example, there have been criticisms that incentives aimed at encouraging wind power have focused on the construction of capacity rather than maximizing output at the best wind sites. There has also been an on-going problem of connection to the electricity grid. Around one third of wind farms have not been able to generate and sell their electricity due to bottlenecks in electricity transmission capacity. Recent reforms have sought to tackle this issue, with a greater emphasis on enforcing priority access for wind plants.

With respect to offshore wind, developments are relatively recent. The potential resource has been estimated by a number of official assessments. The Chinese Meteorological Association estimates this to be 750GW in water depths of less than 20m. However, some other assessments have provided lower estimates – and have led some officials to urge caution with respect to offshore wind (Prideaux and Qi, 2010). An initial 100MW demonstration plant was constructed by Sinovel for the World Expo in Shanghai in 2010. Coastal provinces are now required to develop plans for offshore wind, and specific targets have been agreed in some cases. The current 'wind base' programme includes a target of 7GW of offshore wind capacity off the coast of Jiangsu Province by 2020.

In order to support a first tranche of offshore wind capacity, a concession process was launched in 2010 to build 1GW of capacity in Jiangsu Province. Under this concession policy – which is well established for onshore wind – local grid operators are required to sign a long-term power purchase agreement with winning bidders. With respect to onshore wind, these agreements typically last for 25 years, with the price paid being fixed for the first ten years. At present, the rules for offshore wind stipulate that projects should be developed by Chinese firms or by international joint ventures in which the Chinese partner has a controlling share.

As in the case of EVs, the Chinese wind turbine industry has benefited from the government's 863 R&D programme. Support has also been provided under the companion 973 programme, which focuses on basic research. As Xiaomei Tan has explained, early efforts by the Chinese government to fund joint ventures between Chinese and international firms had limited success (Tan, 2010). She argues that this led to direct funding of Chinese firms' R&D centres under the 863 and 973 programmes. For example, Goldwind (one of the leading Chinese wind power firms) received grants under these programmes to scale up its wind turbines – and develop independent capabilities in turbines of up to 1.5MW. The firm also received further R&D support and tax concessions from the government of the Xinjiang Autonomous Region. Goldwind is now one of ten firms that have been officially accredited by the Chinese government to build offshore wind projects.

Improved efficiency in coal-fired power generation

China derives about 80 per cent of its power from coal. The rapid expansion of coal-fired power during the last decade has been a significant factor in China's increasing carbon emissions. Despite the implementation of policies designed to increase the contribution of non-fossil energy in China, coal's role is likely to remain significant for the next few decades. The average efficiency of China's coal-fired power generation stock has been improving over the past two decades, from 28.8 per cent in 1990 to 35.6 per cent in 2008 (IEEPS, 2009). One driver for this has been a shift towards more efficient coal-fired power station technologies. As in many other countries, attention has mainly focused on supercritical technology, with some activity to invest in integrated gasification combined cycle (IGCC) technologies.

Chinese capabilities in more efficient coal-fired power

The interest from China in gaining supercritical (SC) and ultra-supercritical (USC) coal-fired power plants is longstanding and the desire to localize the technology was designated a Key National Programme during the 1990s (Tan, 2010). As of 2008, there were 93 SC and USC units in operation, and by 2009 there were more than 100 USC units on order from Chinese power companies. In 2010, it was expected that SC and USC power plants would account for over 40 per cent of new thermal units in China (Chen and Xu, 2010).

As a result, there are Chinese firms that can now build SC and USC plants. However, there is still a significant gap in their capabilities compared to the international advanced level. For example, manufacturing companies such as Shanghai, Harbin and Dongfang have not mastered the core design software. The normal practice is for these manufacturers to collaborate with regional design institutions within China on power plant designs. However, they often need to collaborate with leading foreign companies such as Siemens, Hitachi and Alstom when they design new plants. There are also difficulties in manufacturing the high temperature components locally. The special steel materials all need to be imported.

With respect to IGCC technology, the capabilities picture is different. Chinese firms have a long history of coal gasification, through its application for chemicals and fertilizer production rather than for power generation. Since the 1990s, there has been a strategy of acquiring licences from leading international firms such as Shell (Watson et al., 1998). Chinese gasification technology has now developed to the stage where it has been specified for an IGCC plant in the US. The Thermal Power Research Institute in Xian has developed a design which is being used in China's first full-scale IGCC plant (Greengen). This plant eventually plans to fit carbon capture and storage. The gasification technology for the Greengen plant has been specified for the planned Good Spring IGCC in the US.

The gasifier is, however, only one component of an IGCC plant. Another critical technology is the advanced industrial gas turbine which burns the syngas produced by the gasifier. With respect to this component of IGCC plants, Chinese capabilities are considerably weaker. There are only a handful of leading suppliers of advanced industrial gas turbines worldwide – with GE, Siemens and Mitsubishi as market leaders. Chinese turbine companies have formed collaborations with these suppliers, but there is a long way to go before the Chinese partners have independent capabilities (Liu et al., 2008). The terms of these collaborations mean that cutting-edge technologies and knowledge embodied in high-tech parts (such as the first stage turbine blades) are not shared. This controlled approach to knowledge sharing in return for market access has been standard practice among the leading international gas turbine manufacturers for many years (Watson, 1997).

Policy frameworks for efficient coal-fired power

As noted above, policy incentives have aimed to improve the efficiency of coal-fired power generation for many years. The power sector was covered by the Top 1000 Energy Consuming Enterprises Programme under the *11th Five-Year Plan*. Wen Jiabao's recent report to the National People's Congress states that 72GW of small plant capacity had been closed by the end of the *Five-Year Plan* period (Wen Jiabao, 2011). The share of coal-fired power generation capacity with unit sizes of over 300MW rose from 47 per cent in 2005 to 69 per cent in 2010.[9]

In parallel with this closure programme, the Chinese government has also placed more emphasis on economic incentives for improved power plant efficiency. It reduced the prices paid to power plants with capacities of less than 50MW, and some plants of 100MW to 200MW (Andrews-Speed, 2009). New dispatching rules were trialled to reinforce the incentive for the most efficient plants to operate. However, the government has been slow to remove controls on final electricity prices until very recently (IEA, 2006). Historically, prices to end consumers have been kept artificially low, with consequent impacts upon power company finances. According to China's National Energy Administration, 43 per cent of China's coal-fired power plants operated at a loss in 2010.[10]

It is clear that Chinese government policy has played a strategic role in directing acquisition, innovation and deployment of more efficient technologies. For super-critical technology, the acquisition process started with the operation of China's first supercritical units in 1992. These were sourced from leading international firms – ABB for boilers and General Electric for steam turbines (Tan, 2010). The government managed and funded an iterative process of assessment, collaborative R&D and reverse engineering so that Chinese firms developed independent capabilities in this technology. The first Chinese manufactured 600MW supercritical unit entered service in Henan Province in 2004 (Chen and Xu, 2010).

The acquisition of more efficient USC technology followed in 2000 with the support of the 863 and 973 R&D programmes. It resulted in China's first ultra-supercritical unit at Yuhuan which entered service in 2007 and was part funded by the Shanghai government (Tan, 2010; Tan et al., 2010). This plant, which entered service in 2006 to 2007, included collaborations between Chinese and international firms for the main components. The boilers were co-supplied by Mitsubishi Heavy Industries and the Harbin Boiler Company, and the turbines were manufactured by Shanghai Electric and Siemens (to a Siemens design). This plant has since been followed by many more USC projects.

With respect to coal gasification – a key element of IGCC plants – the 863 programme played a particularly important role. It supported coal gasification research at the Thermal Power Research Institute (TPRI) in Xian (Osnos, 2009). The TPRIs coal gasification technology is being used in the first full-scale IGCC plant in China (GreenGen), which is currently under construction. Perhaps more remarkably, as noted earlier, it has also been licensed for use in the Good Spring IGCC plant being planned in the US. According to some reports, this was chosen over competing technologies from Shell and GE due to its higher efficiency (Osnos, 2009). However, it is important to note that other IGCC plant components (most importantly, the gas turbines) are not being sourced or licensed from China.

International finance and policy

International financial flows are clearly important for low-carbon innovation in China and other developing countries. According to the International Energy Agency, US$10.2 trillion of finance will be needed globally by 2030 to provide a reasonable chance of limiting average temperature increases to 2°C (IEA, 2009). While many analyses show that private sources will provide the majority of finance, public funding has a key role to play in leveraging private-sector investment (DECC, 2010). Furthermore, international policy frameworks to support low-carbon technology transfer and deployment in developing countries will be an important complement to national policy frameworks within these countries.

The Cancun Agreements that were agreed at the UNFCCC's 16th Conference of the Parties in December 2010 formalized financial commitments by industrialized countries to support mitigation and adaptation in developing countries. The text of the agreements included a goal of 'mobilising jointly USD 100 billion dollars a year by 2020 to address the needs of developing countries' (UNFCCC, 2010, p15). While this level of funding would represent a large increase from current levels, it would also build on existing international initiatives that have already had an impact upon low-carbon innovation in China.

With respect to our case study technologies, bilateral and multilateral initiatives have already made some contributions to the innovation process. One example of this is the Global Environment Facility (GEF), which is the official financial mechanism of the UNFCCC. Since it was created in the early 1990s, the GEF has provided modest funding to projects in China – with a total value of less than US$500 million.[11] This includes the China energy-efficient boiler project, which arguably had some success in subsidising licences to Chinese firms. It was a difficult project that suffered from delays, and only resulted in licences from 'second-tier' international suppliers (Birner and Martinot, 2005). These 'second-tier' firms believed that they would gain more from selling licences than they would by operating directly in the Chinese market, whereas leading international boiler firms held the opposite view. This demonstrated how difficult it can be to offer licencing terms that are attractive to leading players.

Another example is the support for supercritical and IGCC technology in China by the Asian Development Bank (ADB) (Watson et al., 1998). The ADB financed an early supercritical plant during the late 1990s. It was also involved in funding feasibility studies for an IGCC plant in Yantai during the same period. Plans for this plant suffered from repeated delays, and it has not been constructed. More recently, the ADB provided a loan of US$135 million to the GreenGen IGCC plant,[12] which is under construction in Tianjin.

Bilateral agreements for wind power have also been important sources of finance and other assistance. Two examples illustrate this: the Danish–Chinese Wind Energy Development Programme[13] (with Denmark as partner) and the China Wind Power Research and Training Project[14] (with Germany as partner). Both agreements have led to the transfer of onshore wind technologies from firms in Europe (Vestas and RE-Power) to their counterparts in China. Similarly, a number of international collaborative activities are under way on electric vehicles in China. One of the most prominent is a US–China cooperative programme that aims to develop standards and to implement joint demonstration projects in a number of cities and a technology roadmap (White House, 2009). To complement this, there are partnerships being built between Chinese and US firms (Levi et al., 2010). According to our interviews, some Chinese partners argue that the partnerships lack depth. At the same time, some international firms are wary about potential loss of their technological leadership as a result of such partnerships (Levi et al., 2010).

By far the most important international financial mechanism for the deployment of low-carbon technologies in developing countries is the Clean Development Mechanism (CDM). Created as part of the Kyoto Protocol, the CDM aims to reduce carbon dioxide emissions and to contribute to economic growth and development within developing countries. The mechanism combines these objectives by allowing approved emissions reduction projects to generate Certified Emissions Reductions (CERs) for each tonne of greenhouse gas emissions that they abate. In addition, CDM projects are also encouraged to contribute to sustainable development and to incorporate technology transfer. While the CDM started as a project-based system, it has become increasingly important amongst some beneficiaries in supporting specific national policies and programmes (Schroeder, 2009).

China has received more CDM project investment than any other developing country – with total investment of more than US$50 billion.[15] As of March 2011, 2923 CDM projects had been approved by the official Executive Board, 1282 of which were in China.[16] The total emissions reductions attributed to China's projects amounted to 287 million tonnes of CO_2 equivalent per year. Approximately 200 projects in China claim that they include a direct form of technology transfer, which is usually in the form of hardware transfer.

The CDM has played a fundamental role in the development of at least two of our case study sectors: wind energy (Yang et al., 2010) and energy efficiency in large cement companies (Yan et al., 2009). Based on the available data (Stua, 2011), we can see that other sectors and technologies within China have also benefited from the CDM. These include developments in hydroelectric power, coal methane recycling, the substitution of coal with natural gas in power plants, and energy-efficiency projects in the iron and steel industries.

The CDM has had a particularly important impact upon Chinese wind power development. It is clear that the Chinese government has made strategic use of the CDM to support the rapid expansion of wind power. More recently, China's first offshore wind farm in Shanghai has also been part-financed through the CDM.

A recent analysis by Joanna Lewis shows that a large number of Chinese wind power projects have been registered as CDM projects and have requested CERs (Lewis, 2010). More than half of wind projects built in 2007 and more than one third of those developed in 2008 were registered. As Joanna Lewis notes, there are questions to be asked about the extent to which CDM financing was necessary to make many of these projects financially viable (i.e. whether they are leading to 'additional' emissions reductions). According to He and Morse (2010), apart from one exception, the baselines used to determine the additionality of wind projects in China for the CDM are not benchmarked against coal-fired plants – something they suggest is problematic in a country that is so coal dependent for their electricity. Furthermore, they state that China's NDRC 'determines power tariffs in a propri-etary, non-market-based manner – as is their right in making sovereign decisions about energy policy – [and so] there is no real way to know what is business as usual and what constitutes gaming of the CDM' (He and Morse, 2010, p3).

With respect to the cement industry, the CDM has also been strategically impor-tant. Almost 50 per cent of China's CDM projects that focus on energy efficiency (43 out of 88 projects) involve cement plants. While many of these projects do not claim technology transfer – and deploy established Chinese technologies – a number of the project design documents refer to the Japanese origin of the hardware employed. More efficient coal-fired generation (at utility power-plant scale) has not been a focus for Chinese CDM projects. However, one project has recently been approved by the CDM Executive Board in December 2010. This is for an ultra-supercritical coal-fired power plant at Waigaoqiao, with a capacity of 2000MW.

A key question for this chapter is the extent to which CDM projects lead to technology transfer. A number of authors have found significant emphasis on technology transfer within CDM project documentation (e.g. Haites et al., 2006). It is important to interpret such claims with care since it is not possible *ex ante* to know the extent to which particular projects will lead to technology transfer in practice – and hence the improvement of innovative capabilities in recipient firms. Furthermore, it is not always clear to what extent the meaning of technology trans-fer within CDM project documentation is purely focused on hardware transfer – or whether it also includes knowledge transfer.

Conclusions and implications

The analysis in this chapter and the longer report upon which it is based (Watson et al., 2011) lead to six main conclusions.

First, the analysis of low-carbon innovation in China reveals important differences between low-carbon technologies. This confirms the findings from the previous UK–India studies that were led by our research group (e.g. Ockwell et al., 2008; Ockwell et al., 2009). The extent to which Chinese firms are 'catching up' with the international frontier varies widely. As might be expected, Chinese technological capabilities are stronger in more near-market technologies such as supercritical coal-fired plants and onshore wind. With respect to more early-stage technologies such as electric vehicles (EVs) and possibly offshore wind, significant gaps in capabilities are more apparent.

Second, the case of China is unique and should not be used as a proxy for developing countries in general. China is now the second largest economy in the world, the number one consumer of energy, and the largest emitter of carbon dioxide. While China still faces many development challenges, and many of its citizens remain on very low incomes, rapid economic development means that resources are available to support low-carbon innovation and deployment. The strong role of the Chinese government is evident in our case studies – in directing technology acquisition, providing R&D support, implementing policy frameworks and in taking advantage of the CDM.

Third, a range of policy mechanisms has been used within China to promote low-carbon technology development and deployment. The strong government role means that targets and regulations have often been favoured. Examples include the *11th Five-Year Plan*'s energy intensity target and regulations to mandate the closure of inefficient industrial capacity. While some of these policies are far from perfect, they have been extremely important. To complement this, we found examples of the use of economic incentives, such as reforms to coal-fired electricity tariffs which have started to encourage the use of the most efficient plants. While the policies required to support low-carbon innovation will differ between technologies and sectors, we support the NDRC's wish for a greater role for market-based mechanisms to help meet the new *Five-Year Plan* targets (NDRC, 2011).

Fourth, Chinese firms and institutions are developing their capabilities in our case study technologies rapidly. As noted above, domestic policy interventions have been important in supporting the acquisition and assimilation of many low-carbon technologies. A combination of market support, regulations and R&D support has been crucial in many cases. However, there are some limitations – for example, in access to advanced component technologies and associated knowledge, and in engineering and design skills. Some of these reflect the early stage of a technology (such as electric vehicles), while others reflect a significant competitive disadvantage (e.g. advanced gas turbines for IGCC plants).

Fifth, access to intellectual property rights (IPRs) is not a fundamental barrier to the development of low-carbon innovation capabilities in China. This does not mean that IPR issues are unimportant. We found that the resources required to identify, acquire and assimilate low-carbon technologies slowed the development of capabilities and/or diffusion of some of these technologies. Sometimes, the impression is given that Chinese firms have a wholly independent capability in some low-carbon technologies; but this can be misleading. In some cases of more advanced technologies (e.g. EVs and gas turbines for IGCC power plants), barriers to entry for Chinese firms remain high due to a lack of affordable access to IPRs and/or gaps in knowledge and capabilities. With respect to low-carbon technologies that are more mature, the capabilities of Chinese firms tend to be greater. However, some of these firms are still partly dependent on licences from international firms (e.g. for some wind turbines for offshore use and in supercritical boiler technologies). While such licences are clearly affordable, their continuing presence indicates that independent innovation by the Chinese firms concerned is some way off.

Sixth, international institutions and policy frameworks have also played an important role in our case study technologies. The Clean Development Mechanism has been used strategically by the Chinese government – a unique approach that helps to explain why such a large proportion of CDM projects are located in China. China's considerable institutional capacity has enabled this, and has included an alignment between the CDM and national policy frameworks. The result has been particularly significant support from the CDM for onshore wind power and cement industry energy efficiency. Other multilateral institutions have also been important in China, including multilateral development banks (though they have been criticized for not focusing enough on low-carbon technologies). There are also a large number of bilateral arrangements, some of which have yielded tangible results for Chinese technological capabilities – for example, in wind power technology.

These conclusions lead to a number of implications for the UNFCCC negotiations. They are particularly relevant to the debate about the form and functions of the new Climate Technology Centre and Network (CTCN) that was agreed in the Cancun talks in 2010 (UNEP, 2010; UNFCCC, 2010). The broad aim of the CTCN is to assist developing countries in the development and deployment of low-carbon technologies. At the time of writing their scope and remit remains the subject of significant debate.

Most importantly, they suggest that low-carbon innovation within the UNFCCC should not be approached from a 'one size fits all' perspective. The level of international assistance required (and its type) will differ by country and by technology. It is therefore important that the CTCN is implemented in a way that takes account of national, sectoral and market differences that characterize a complex low-carbon innovation landscape.

For developing countries such as China that have significant resources and capabilities, the CTCN could include existing institutions. The experience of China with respect to the CDM suggests that international mechanisms and institutions can be particularly effective when integrated within existing national strategies and policy priorities. This does not mean that international mechanisms are a prerequisite for low-carbon innovation. But if such mechanisms are to add value, it is important that they recognize and work with national policy frameworks and capabilities.

There are many potential functions that the CTCN might perform. The evidence from China suggests that a broad range of functions are likely to be important, including investment (e.g. in R&D), diffusion support (e.g. in wind power concessions), a focus on 'soft technologies' (e.g. for coal gasification), collaborative research, development and demonstration (RD&D) (e.g. for EVs) and national plans (e.g. for industrial energy efficiency). China's experience shows that a minimal version of the CTCN that is mainly concerned with sharing information and best practice would be less likely to make a significant contribution to innovation.

Finally, an important element of the implementation of the CTCN is evaluation of past programmes. Such evaluations will promote learning from existing institutions that run programmes designed to fulfil some of the functions foreseen for the CTCN. Examples include the World Bank Climate Investment Funds (which aim to achieve transformational investments in low-carbon technology diffusion) and smaller pilot innovation centres (such as those being run in India and Kenya by Infodev).

Acknowledgements

We would like to thank our colleagues who contributed to the original report that formed the basis for this chapter: Michele Stua, David Ockwell and Gordon MacKerron at SPRU, University of Sussex, UK; Alex Mallett of Carleton University, Canada; and Zhang Xiliang, Zhang Da, Zhang Tianhou, Zhang Xiaofeng and Ou Xunmin at Tsinghua University, China. We would also like to thank all those who agreed to be interviewed for our research and to provide comments on drafts of the final report. The research was made possible through funding from the UK Department of Energy and Climate Change.

Notes

1 Data from the US Energy Information Administration: http://www.eia.doe.gov.
2 Speech to International Cooperative Conference on Green Economy and Climate Change, Beijing, China, 9 May 2010.
3 Xie Zhenhua, speech to 2011 International Conference on Low Carbon Energy and Climate Change, Tsinghua University, 24 March 2011.
4 Xie Zhenhua, speech to 2011 International Conference on Low Carbon Energy and Climate Change, Tsinghua University, 24 March 2011.
5 For example, Shanghai Automotive Industries Corporation, who had worked with Ricardo in the past, now own MG Motor in the UK: http://www.insideline.com/saic/mg-motor-opens-uk-design-studio.html.
6 'China has world's most installed wind-power capacity', Xinhua News Agency, 13 January 2011, http://www.china.org.cn/business/2011-01/13/content_21733120.htm.
7 Speech by Zhang Guobao, chair of the National Energy Administration to the International Cooperative Conference on Green Economy and Climate Change, Beijing, China, 9 May 2010.

8 Levi et al., (2010, p88, endnote 144) report the presence of about 100 turbine manufacturers and that the Chinese government intends to set guidelines for the industry that will see only 'twenty to thirty survivors'.

9 Xie Zhenhua, speech to 2011 International Conference on Low Carbon Energy and Climate Change, Tsinghua University, 24 March 2011.

10 Reported at http://www.prlog.org/11349430-chinas-power-plants-operate-at-loss-cost-of-coal-and-energy-efficiency-blamed.html.

11 This figure includes on-going or completed projects documented on the GEF website, including those not directly connected to energy and/or mitigation activities, since 1994 up to June 2009.

12 See http://www.adb.org/Documents/News/PRCM/prcm201006.asp.

13 See http://www.dwed.org.cn/.

14 See http://www.cwpc.cn/cwpc/en/cwpp.

15 Data obtained from project design documents for electricity production CDM projects: http://cdm.unfccc.int/Projects/projsearch.html.

16 See 'CDM in numbers' at: http://cdm.unfccc.int/Statistics/index.html.

References

Accenture (2011) *The United States and China: The Race to Disruptive Transport Technologies Implications of a changing fuel mix on country competitiveness*, Accenture, US

Andrews-Speed, P. (2009) 'China's ongoing energy efficiency drive: Origins, progress and prospects', *Energy Policy*, vol 37, no 4, pp1331–1344

Barton, J. (2007) *Intellectual Property and Access to Clean Technologies in Developing Countries: An Analysis of Solar Photovoltaic, Biofuel and Wind Technologies*, Issue Paper no 2, International Centre for Trade and Sustainable Development (ICTSD), Geneva

Bell, M. (1990) *Continuing Industrialisation, Climate Change and International Technology Transfer*, Report prepared in collaboration with the Resource Policy Group, Oslo, Norway, Science Policy Research Unit, University of Sussex, Brighton, UK, December

Birner, S. and Martinot, E. (2005) 'Promoting energy-efficient products: GEF experience and lessons for market transformation in developing countries', *Energy Policy*, vol 33, no 14, pp1765–1779

CATARC (2008) *China Automotive Industry Yearbooks, 1991–2008* (in Chinese), China Automotive Technology and Research Centre and Chinese Automotive Manufacturers Association, China

CCA (China Cement Association) (2010) http://www.ccement.com

Chatham House, Chinese Academy of Social Sciences et al., (2010) *Low Carbon Development Roadmap for Jilin City*, Chatham House, London

Chen, W. and Xu, R. (2010) 'Clean coal technology development in China', *Energy Policy*, vol 38, no 5, pp2123–2130

DECC (Department of Energy and Climate Change) (2010) *Beyond Copenhagen: The UK Government's International Climate Change Action Plan*, DECC, The Stationery Office, London

GWEC (Global Wind Energy Council) (2010) *Global Wind 2009 Report*, GWEC, Brussels

Haites, E., Duan, M. et al., (2006) *Technology Transfer by CDM Projects*, Margaree Consultants Inc, . Toronto, Canada

He, G. and Morse, R. K. (2010) *Making Carbon Offsets Work in the Developing World: Lessons from the Chinese Wind Controversy*, Stanford University Program on Energy and Sustainable Development Working Paper no 90, March

IEA (International Energy Agency) (2006) *China's Power Sector Reforms: Where to Next?*, OECD/IEA, Paris.

IEA (2008) *World Energy Outlook 2008*, OECD/IEA, Paris

IEA (2009) *World Energy Outlook 2009*, OECD/IEA, Paris

IEEPS (International Energy and Electric Power Statistics) (2009) *International Energy and Electric Power Statistics: State Grid Corporation of China*, US

Levi, M., Economy, E. et al., (2010) *Energy Innovation: Driving Technology Competition and Co-operation among the US, China, India and Brazil*, Council on Foreign Relations, New York and Washington, DC

Lewis, J. I. (2007) 'Technology acquisition and innovation in the developing world: Wind turbine development in China and India', *Studies in Comparative International Development*, vol 42, pp208–232

Lewis, J. I. (2010) 'The evolving role of carbon finance in promoting renewable energy development in China', *Energy Policy*, vol 38, pp2875–2886

Liu, H., Ni, W. et al., (2008) 'Strategic thinking on IGCC development in China', *Energy Policy*, vol 36, pp1–11

Mallett, A., Ockwell, D., Pal, P., Kumar, A., Abbi, Y., Watson, J., Haum, R., MacKerron, G. and Sethi, G. (2009) *UK–India Collaborative Study: Barriers to the Transfer of Low Carbon Energy Technology: Phase II*, Report for UK Department for Energy and Climate Change (DECC), London, March

Müller, N. and Harnisch, J. (2008) *A Blueprint for a Climate Friendly Cement Industry*, Report prepared for the WWF-Lafarge Conservation Partnership, WWF International, Switzerland

NBS (National Bureau of Statistics) (2009) *2008 National Economic and Social Development Statistical Report*, National Bureau of Statistics of China (in Chinese), http://www.stats.gov.cn/tjgb/ndtjgb/qgndtjgb/t20090226_402540710.htm

NDRC (National Development and Reform Commission) (2007) *China's National Climate Change Programme*, NDRC, Beijing

NDRC (2011) *Report on the Implementation of the 2010 Plan for National Economic and Social Development and on the 2011 Draft Plan for National Economic and Social Development*, Fourth Session of the Eleventh National People's Congress, NDRC, Beijing, 5 March 2011

Ockwell, D., Watson, J., MacKerron, G., Pal, P. and Yamin, F. (2008) 'Key policy considerations for facilitating low carbon technology transfer to developing countries', *Energy Policy*, vol 36, pp4104–4115

Ockwell, D., Ely, A., Mallett, A., Johnson, O. and Watson, J. (2009) *Low Carbon Development: The Role of Local Innovative Capabilities*, STEPS Working Paper 31, STEPS Centre and Sussex Energy Group, SPRU, University of Sussex, Brighton, UK

Ohshita, S. and Price, L. (2011) 'Lessons for industrial energy efficiency cooperation with China', *China Environment Series*, vol 11, pp49–88

Osnos, E. (2009) 'Green Giant: Beijing's crash program for clean energy', *The New Yorker*, 21 December

Ouyang, M. (2006) 'Chinese energy-saving and new energy vehicle development strategy and actions', *Automotive Engineering*, vol 28, pp317–322 (in Chinese)

Ouyang, M. (2010) Presentation, Electric Vehicle Workshop, Shanghai, 5 August 2010

Price, L., Wang, X. et al., (2008) *China's Top-1000 Energy-Consuming Enterprises Program: Reducing Energy Consumption of the 1000 Largest Industrial Enterprises in China*, Report for the Energy Foundation, Lawrence Berkeley National Lab, Berkeley, CA

Prideaux, E. and Qi, W. (2010) 'Chinese wind sector takes on the world', *Windpower Monthly*, December

Schroeder, M. (2009) 'Utilizing the clean development mechanism for the deployment of renewable energies in China', *Applied Energy*, vol 86, pp237–242

Stua, M. (2011) *A Transition Management Approach to the Low-Carbon Development of the Chinese Electric Power Production System*, Ph.D. thesis, Sant'Anna School of Advanced Studies, Pisa, Italy

Tan, X. (2010) 'Clean technology R&D and innovation in emerging countries: Experience from China', *Energy Policy*, vol 38, no 6, pp2916–2926

Tan, X., Seligsohn, D. et al., (2010) *Scaling up Low Carbon Technology Deployment: Lessons from China*, World Resources Institute, Washington, DC

UKTI and SMMT (UK Trade and Investment and Society of Motor Manufacturers and Traders) (2010) *China: New Vehicles, New Market, New Opportunities*, UKTI and SMMT, London, February

UNDP China (United Nations Development Programme China) (2010) *China Human Development Report 2009/10: China and a Sustainable Future: Towards a Low Carbon Economy and Society*, UNDP China and Renmin University, Beijing, China

UNEP (United Nations Environment Programme) (2010) *An Exploration of Options and Functions of Climate Technology Centres and Networks*, Discussion Paper prepared by ECN, NREL and UNEP, Nairobi

UNFCCC (United Nations Framework Convention on Climate Change) (2010) *Outcome of the Work of the Ad Hoc Working Group on Long-Term Cooperative Action under the Convention: Advanced Unedited Version*, UNFCCC

Wan, G. (2008) 'Thinking on Chinese energy-saving and new energy vehicle development mode and the initiatives', *Traffic and Transportation*, vol 2, pp1–3 (in Chinese)

Wang, T. and Watson, J. (2009) *China's Energy Transition: Pathways to Low Carbon Development*, The Tyndall Centre for Climate Change Research, Sussex Energy Group, SPRU, University of Sussex, Brighton, UK

Watson, J. (1997) Constructing Success in the Electric Power Industry: Combined Cycle Gas Turbines and Fluidised Beds, D.Phil. thesis, SPRU, University of Sussex, Brighton, UK

Watson, J., Oldham, G. et al., (1998) *The Transfer of Clean Coal Technologies to China: A UK Perspective*, Report no COAL R196, DTI/Pub URN 00/977, Department of Trade and Industry, London

Watson, J., Byrne, R. et al., (2011) *UK–China Collaborative Study on Low Carbon Technology Transfer: Final Report*, SPRU, University of Sussex, Brighton, UK

Watts, J. (2010) 'China resorts to blackouts in pursuit of energy efficiency', *Guardian*, 20 September

Wen Jiabao (2011) *Report on the Work of the Government*, Delivered at the fourth session of the Eleventh National People's Congress, Beijing, 5 March 2011

White House (2009) *Fact Sheet: U.S.–China Electric Vehicles Initiative*, White House Press Release, Washington, DC

Worrell, E. and Galitsky, C. (2008) *Energy Efficiency Improvement and Cost Saving Opportunities for Cement Making*, LBNL-54036-Revision, Ernest Orlando Lawrence Berkeley National Laboratory, University of California, Berkeley, CA, March

Xinhuanet (2010) *China Has Reached to Number 1 in Vehicle Production and Sales in 2009* (in Chinese), http://news.xinhuanet.com/auto/2010-01/11/content_12791236.htm

Yan, Q., Zhou, C., Qu, P. and Zhang, R. (2009) 'The promotion of clean development mechanism to cement industry capturing waste heat for power generation in China', *Mitigation and Adaptation Strategies for Global Change*, vol 14, pp793–804

Yang, M., Nguyen, F., De T'Seracles, P. and Buchner, B. (2010) 'Wind farm investment risks under uncertain CDM benefit in China', *Energy Policy*, vol 38, pp1436–1447

Zhang, A., Shen, W., Han, W. and Chai, Q. (2008) *Life Cycle Analysis of Automotive Alternative Energy*, Tsinghua University Press, Beijing (in Chinese)

Zhang, X. (2010) 'China in the transition to a low-carbon economy', *Energy Policy*, vol 38, no 11, pp6638–6653

PART III

Intellectual Property Rights (IPRs)

5

THE UNEP–EPO–ICTSD STUDY ON PATENTS AND CLEAN ENERGY

Key Findings and Policy Implications

Ahmed Abdel Latif

Introduction

The widespread diffusion of clean energy technologies (CETs) plays a key role in global efforts to combat climate change and to reduce greenhouse gas (GHG) emissions. For this reason, enhancing the transfer of these technologies, particularly to developing countries, has been one of the main pillars of the United Nations Framework Convention on Climate Change (UNFCCC) since its inception. In 2007, the *Bali Action Plan* reaffirmed the centrality of technology transfer, making it one of the priority areas to be addressed in discussions aiming at the 'full, effective and sustained implementation of the Convention through long-term cooperative action, now, up to and beyond 2012'.[1] After three years of discussions, agreement was reached at the Cancun Conference in 2010 to create a Technology Mechanism under the UNFCCC to enhance action for technology development and transfer, particularly to developing countries, in support of climate change mitigation and adaptation.

In these discussions, the role of intellectual property rights (IPRs) has been actively debated, particularly since the Bali Conference in 2007. On the one hand, developing counties have pointed that IPRs could be a possible barrier to the transfer of climate friendly technologies while developed countries have underlined their essential role in encouraging innovation in clean energy technologies and in facilitating technology transfer. However, the debate on this divisive issue is often confined to generalizations and lacks a solid empirical basis (see Ockwell et al., 2010, for a more detailed discussion).

Against this background, the United Nations Environment Programme (UNEP), the European Patent Office (EPO) and the International Centre for Trade and

Sustainable Development (ICTSD) joined forces in April 2009 to undertake a project on the role of patents in the transfer of clean energy technologies in the field of energy generation.

The project involved mining patent data to track the diffusion of clean energy technologies. While there have been previous studies that have undertaken mining of patent data, this is the first study in which the retrieval of patent data is carried out by one of the major patent offices in the world – the EPO – with a specialized expertise in this area. The study also goes further than previous ones by enriching the existing body of knowledge in this area with findings from the first global survey of licensing practices in clean energy technologies.

In addition, the project resulted in the development by the EPO of a new ground-breaking patent classification scheme and a searchable database for CETs that considerably facilitates patent searches in this area. Thus, the project made a tangible contribution towards achieving greater transparency and accessibility of patent information in this area, which in itself is one important element, among many others, in fostering an environment that is more conducive to technology diffusion and transfer. The final report was released in September 2010 (UNEP, EPO and ICTSD, 2010).

The objective of this chapter is to present the background, rationale and key findings of the UNEP–EPO–ICTSD study on patents and clean energy and its efforts towards bridging the gap between evidence and policy in order to contribute towards a more informed discussion in climate change negotiations on the role of IPRs in the transfer of climate-friendly technologies.

Intellectual property rights: A contentious issue in climate change discussions on technology transfer

Technology transfer has been a key pillar of the United Nations Framework Convention on Climate Change (UNFCCC). Article 4.5 of the convention requires developed countries to 'take all practicable steps to promote, facilitate and finance, as appropriate, the transfer of, or access to, environmentally sound technologies and know-how to other parties, particularly developing country parties to enable them to implement the provisions of the Convention'.[2]

Furthermore, Article 4.7 establishes a clear link between the extent to which developing countries will implement their commitments under the UNFCCC and the effective implementation by developed countries of their commitments relating to financial resources and the transfer of technology.[3]

For many years, developing countries have been demanding concrete steps and measures to operationalize these provisions. Developed countries, for their part, have pointed to the lack of enabling environments and limited absorptive capacities in developing countries as the main barriers to technology transfer. Difficulties in reaching a common definition of technology transfer and disagreements over the role of IPRs were recurrent issues in UNFCCC discussions.

In 2007, the *Bali Action Plan*, agreed at the 13th Conference of the Parties (COP) of the UNFCCC, reaffirmed the centrality of technology development and transfer and called for:

> Enhanced action on technology development and transfer to support action on mitigation and adaptation, including, inter alia, consideration of: (i) Effective mechanisms and enhanced means for the removal of obstacles to, and provision of financial and other incentives for, scaling up of the development and transfer of technology to developing country Parties in order to promote access to affordable environmentally sound technologies.[4]

After three years of discussions, agreement was reached at the Cancun Conference in 2010 on the creation of a Technology Mechanism under the UNFCCC to enhance action for technology development and transfer, particularly to developing countries, in support of climate change mitigation and adaptation.

During this period, IPRs were – and continue to be – one of the most divisive issues in the climate technology negotiations, if not the most divisive one. Developing countries have argued that IPRs are one of the possible obstacles to achieving large-scale affordable access to climate-friendly technologies that is needed to effectively address climate change and reduce GHG emissions. Accordingly, developing countries have presented a number of options to address the role of IPRs, ranging from exclusion of climate-friendly technologies from patentability in developing countries and least developed countries (LDCs), to full and expanded use of existing flexibilities under the World Trade Organization (WTO) Agreement on Trade Related Aspects of Intellectual Property Rights (TRIPS). However, developed countries have opposed such a view, given the essential role they consider that intellectual property (IP) protection plays in providing incentives for innovation in clean technologies and subsequent technology transfer. A 'polarized' debate followed in which there was little chance for meaningful discussion based on evidence rather than rhetoric. As a result, there was no agreement on any mention of IPRs in the post-Bali negotiating texts; thus all the language relating to IPRs remained in brackets as it is the practice where there is no agreement on an issue or its wording in multilateral negotiations. Ultimately, there was no reference to IPRs in the final text of the Cancun Agreements.

Since the Cancun Conference in 2010, several developing countries have indicated the need to 'bring back' IPRs into the climate change discussions. After a meeting of BASIC countries – Brazil, South Africa, India and China – in New Delhi, in February 2011, the Indian Minister of Environment mentioned that 'there were a number of issues in the Bali Road Map that had not been presented in the Cancun agreements, in particular the issue of equity, intellectual property rights and trade which are all very important to BASIC countries'. 'We will make every effort to bring these issues back to the mainstream discussion,' the Indian minister added (*Xinhua News*, 2011). In April 2011 at a UNFCCC workshop on the operationalization of the Technology Mechanism, China, Ecuador, Bolivia and Bangladesh specifically mentioned that the Mechanism should – in one way or another – address the role of IPRs in the transfer of climate-friendly technologies.[5] Since then, India has made a formal proposal to add intellectual property, amongst several other issues, to the agenda of the Durban UNFCCC climate conference (India, 2011). However, IPRs did not ultimately feature in the outcome of the Durban Conference.

The UNEP–EPO–ICTSD project on patents and clean energy: Background

Empirical evidence on the role of IPRs in the transfer of climate-friendly technologies has been the subject of increased attention during the past five years (e.g. Barton, 2007; Lewis, 2007; Ockwell et al., 2008; Dechezleprêtre et al., 2009; Mallet et al., 2009). Thus, it is only relatively recently that empirical research has begun to appear on the issue for a variety of reasons, including difficulties in obtaining reliable data and tracking key technologies.

Seeking to build on these efforts and to complement them, the United Nations Environment Programme (UNEP), the European Patent Office (EPO) and the International Centre for Trade and Sustainable Development (ICTSD) joined forces to carry out a project on the role of patents in the transfer of clean energy technologies, particularly in the field of energy generation. The project aimed at providing sound and objective empirical evidence to better inform the policy debate on this issue in the climate change discussions. The partnership between the three organizations was announced on 26 April 2009 on the occasion of the world intellectual property day (UNEP, EPO and ICTSD, 2009).

The project consisted of three main parts: a technology mapping study of key clean energy technologies, a patent landscape based on the identified CETs and a survey of licensing practices. In addition, the project resulted in the development by EPO of a new ground-breaking patent classification scheme and a searchable database for CETs.

The premise of the collaboration between the three organizations was that a wide partnership, combining their respective perspectives and skills, could better succeed in undertaking the challenging task of gathering and analysing patent and licensing trends in relation to clean energy technologies on a worldwide level.

Each of the organizations involved in the project had developed, during a concomitant time period, a similar interest in the issue from the perspective of its own mandate, work programme and activities.

UNEP, as the voice for the environment in the United Nations system, brought to the project a comprehensive global perspective on environmental matters. UNEP had launched, in 2008, the Green Economy Initiative (GEI) 'whose collective overall objective is to provide the analysis and policy support for investing in green sectors and in greening environmental unfriendly sectors'.[6] Since early on, the GEI aimed to identify the enabling conditions and policy options for making a shift towards the green economy. This included a better understanding of the factors impacting upon the transfer and diffusion of clean energy technologies.

The European Patent Office contributed to the project with its in-depth expertise on matters relating to the examination of patent applications and patent searches.

The EPO is one of the largest intellectual property organizations in the world, with its seat in Munich, and offices in The Hague, Berlin, Vienna and Brussels It is one of the two bodies of the European Patent Organization, an intergovernmental organization that was set up on 7 October 1977 on the basis of the European Patent Convention (EPC) signed in Munich in 1973, and which comprises 38 member states. In 2006, the EPO received more than 200,000 applications, and granted close to 63,000 patents. It has a staff of over 6000 individuals (EPO, 2007).

The work of the EPO involves searches and substantive examinations of European patent applications and international applications filed under the Patent Cooperation Treaty (PCT). To perform its duties in these areas, the EPO needs to consider worldwide prior art, including patent documentation, in order to assess claims of novelty. For this purpose, it has developed what are considered the most comprehensive and thorough patent databases and search tools available worldwide.

This is of particular importance in the area of patent classification, the system of codes that groups inventions according to technical area in order to make patent searching easier. Most patent offices worldwide use the International Patent Classification (IPC), which has approximately 70,000 different codes for different technical areas. While the IPC is the most widely used classification system, it is not the most extensive, as patent offices often define further subdivisions internally. In this regard, the European Classification System (ECLA) developed by the EPO builds on the IPC and includes approximately 135,000 subdivisions (almost double the number of IPC entries).[7] Thus, patent data mining studies which mainly rely on the IPC have limitations.

In terms of patent data, the EPO has a collection of over 70 million patent documents from all over the world that is available to the public via the free esp@cenet service on the internet. The EPODOC (EPO DOCumentation) database contains references to patent documents which make up the systematically classified search documentation of the EPO. The documents consist of published applications, granted patents and classified non-patent literature. The EPODOC database essentially corresponds to the DOCDB database, which is the internal EPO master file used for management of the search documentation. The bibliographic data (i.e. the publication, application and priority numbers and dates, the IPC classes, the inventor and applicant data and the title) are available for patent documents of most countries or other patent authorities.[8]

Thus, using EPO classification schemes (such as the ECLA) and patent documentation (EPODOC) enables much deeper and accurate worldwide patent searches.

In addition to its unique technical expertise, the EPO had also been developing an interest in the broader challenges facing the patent system, including in relation to climate change.

In 2005, it undertook a project which sought to achieve a better understanding of how intellectual property and patenting might evolve over the next 15 to 20 years. Four scenarios were developed and the results were published, in 2007, as a compendium entitled *Scenarios for the Future* (EPO, 2007). The Blue Skies scenario envisaged a world 'where complex new technologies based on a highly cumulative innovation process are seen as the key to solving systemic problems such as climate change, and diffusion of technology in these fields is of paramount importance' (EPO, 2007).

In 2008, at the European Patent Forum in Ljubljana,[9] the Executive Secretary of the UNFCCC, Yvo de Beer, called for input in seeking some answers to the questions raised by intellectual property in the climate change negotiations. 'The process needs clarity on where IPRs are a barrier and where not. If they are a barrier, how can that barrier be overcome? How can IPRs be handled in the international climate change context?', he asked (de Beer, 2008).

For its part, the ICTSD brought to the project the perspective of a civil society organization that has been actively involved in global debates on IPRs and sustainable development for more than a decade through its multi-stakeholder dialogues and policy-oriented research. Achieving a better understanding of the relationship between intellectual property and technology transfer to developing countries had been one of the main priorities of its research agenda throughout this time period. In 2007, the ICTSD also began to be increasingly involved in discussions on trade and climate change.

In this context, the ICTSD published, in December 2007, a study by John Barton on *Intellectual Property and Access to Clean Technologies in Developing Countries* (Barton, 2007). The study, which was among the first to use patent data, looked into three renewable energy sectors (wind, solar and biofuels) in three different emerging economies (China, Brazil and India). It came to the conclusion that while intellectual property did not represent a significant barrier to access by these countries to wind, solar and biofuel technologies, the situation required close monitoring and further evidence-based research.

Thus, UNEP, EPO and ICTSD came to a common realization, during the same time period, that a major data-gathering effort was needed to provide a more comprehensive and objective empirical basis for discussions on the role of IPRs in the transfer of climate-friendly technologies. Furthermore, they recognized that the synergies between the work of the three organizations could significantly facilitate such effort.

The methodology

Since the inception of the project, UNEP, EPO and ICTSD were acutely aware of the need to provide a sound basis for their data-gathering effort. They also rapidly came to the view that this effort should not be limited to a patent landscape of selected clean energy-generation technologies for a number of reasons. First, beyond patents, other IPRs – such as copyright, trade secrets and industrial designs – play an important role in technology transfer. In addition, it was not only the ownership of patent rights that had implications for technology transfer, but also how these rights were exercised through licensing practices. Therefore, to complement the analysis of patenting trends, a licensing survey was conducted among private and public organizations in order to obtain further insights into how these organizations licensed CETs.

In addition, at each stage of the project, the involvement of relevant organizations with pertinent expertise was actively sought. For instance, once the EPO had extracted the patent data regarding the selected clean energy technologies, it was shared with the Environment Directorate of the Organisation for Economic Co-operation and Development (OECD) to check the quality of the information retrieved and to undertake a statistical analysis in view of its previous experience in this area. In carrying out the licensing survey, the help of leading business organizations – such as the International Chamber of Commerce (ICC), the World Business Council for Sustainable Development (WBCSD), the Licensing Executives Society (LESI) and the Fraunhofer Gesellschaft of Germany – was solicited. These organizations facilitated the dissemination of the survey to their member companies in order to achieve the biggest response rate possible.

An important methodological issue that arose at the outset of the project related to terminology and definitions as a number of terms are commonly used to describe technologies that hold the potential for reducing waste and emissions, including GHG emissions, such as 'environmentally sound', 'environmentally friendly', 'green', 'clean' and 'eco-friendly'.

The UNFCCC documents generally use the term 'environmentally sound technologies', coined in Chapter 34 of Agenda 21 (1992), when referring to technologies that protect the environment; are less polluting; use resources in a more sustainable manner; recycle more of their wastes and products; or handle residual wastes in a more acceptable manner than the technologies they replace. They also often refer to technologies for climate change mitigation and adaptation.

However, technologies, particularly in the energy generation field, do not always fall into simple categories. Although a technology may have a significant potential to reduce CO_2 emissions (compared with a given baseline), it may not be universally accepted as a genuine climate change mitigating technology. For example, some 'clean coal' technologies reduce CO_2 emissions when compared with traditional coal combustion, but still contribute to GHG (UNEP, EPO and ICTSD, 2010).

In view of the large number of technologies existing in the field, the study was limited to analysing patenting trends for selected energy-generation technologies. As part of the joint project, the EPO developed a specific taxonomy based on the technical attributes of technologies that have been loosely referred to as CETs. For the purpose of the project, CETs were defined as energy-generation technologies which have the potential for reducing GHG emissions.

The following CETs were examined in the context of the study: solar photovoltaic (PV), solar thermal, wind, geothermal, hydro/marine, biofuels, carbon capture and storage (CCS), and integrated gasification combined cycle (IGCC).

Key findings of the patent landscape[10]

Before going ahead with the patent landscape, it was first necessary to carry out a technology mapping of the selected CETs in the marketplace or under development in order to avoid one of the pitfalls faced by patent landscaping studies: the limited ability of patent classification systems to correlate accurately and comprehensively with CETs.

To avoid this, ICTSD, with UNEP's support, commissioned a study undertaken by experts at the Energy Research Centre of The Netherlands (ECN) to map both mature and emerging CETs (Lako, 2008). The study drew upon the technology categories within the energy supply sector identified by the Intergovernmental Panel on Climate Change (IPCC) providing an assessment of existing and potential technologies for mitigating climate change (IPCC, 2007). This mapping study was then further peer reviewed by the lead authors of the IPCC Working Group III report and to a number of other experts from relevant international organizations, academia and the private sector.

The study identified several renewable energy technologies which are commercially available or have strong prospects of commercialization in the near to medium term.

Based on the findings of the technology mapping, the EPO developed a list of approximately 50 technical fields related to CETs, which includes technology and application sectors as well as appropriate apparatuses and components (such as turbine blades, rotors, etc.).

Using this new classification, the EPO reviewed 60 million patent documents and reclassified patents according to 50 technical categories related to CETs, such as solar photovoltaic and geothermal. Some 400,000 patents matching these criteria were retrieved worldwide. For the final data extraction and grouping according to the defined indicators, the EPO/OECD Worldwide Patent Statistics Database (PATSTAT) was used.[11] The time period covered goes from 1978 to 2006.

The concept of 'claimed priorities' was used by the OECD Environment Directorate for counting patent numbers and identifying the relevant patent applications from the data pool provided by the EPO. Claimed priorities refer to patent applications that have been filed in other countries based on the first filed patent for a particular invention.

Key findings of the patent landscape point to the fact that patenting rates (patent applications and granted patents) in the selected CETs have increased at roughly 20 per cent per annum since 1997, following a period of stagnation, even of relative decline, until the mid-1990s. During that period, patenting in CETs has outpaced the traditional energy sources of fossil fuels and nuclear energy.

Figure 5.1 shows the growth rate of claimed priorities patenting for selected CETs.

The surge of patenting activity in CETs coincided with the adoption of the Kyoto Protocol in 1997, which provides a strong message that clear policy signals from climate negotiations can be effective in stimulating the development of CETs.

Breaking the data down further for each of the CETs examined, patenting rates in solar PV, wind and carbon capture and storage have shown the most activity in

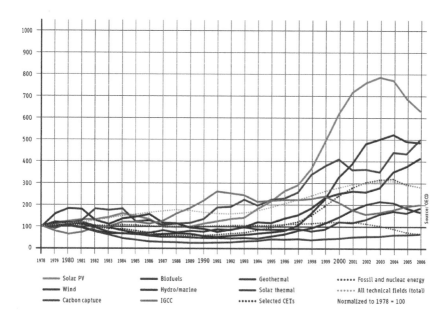

FIGURE 5.1 Relative growth rate for selected clean energy technologies
Source: UNEP, EPO and ICTSD, 2010

the past ten years, followed by hydro/marine and biofuels.

Patenting in the selected CET fields is currently dominated by OECD countries. Japan is far ahead, with the most patenting activity based on claimed priorities. The US and Germany are close together in second place, with the Republic of Korea showing a considerable increase in recent patenting. The UK and France complete the top six patenting countries in the selected CETs. Japan has almost twice as many patents for all selected CETs as the US.

Concentration of patenting activity in these countries reflects patenting trends in other technology sectors. Aside from geothermal, concentration in all CETs is relatively high. Notably, the top six countries account for almost 80 per cent of all patent applications in the CETs reviewed, each showing leadership in different sectors.

In terms of patent filing trends between countries, most activity is currently taking place in the patent offices of the top six patenting countries. Outside this group, inventors from Japan and the US have the largest numbers of claimed priority patents filed in China. Inventors from Germany, the UK, France and the Republic of Korea are the next largest patent filers in China. This indicates that China is considered an important market, but also a potential competitor.

In general, the patent landscape confirms the trends identified in previous studies (including Dechezleprêtre et al., 2009; Lee et al., 2009; Johnstone et al., 2010) that a few OECD countries dominate the field of CET patenting, as is the case in most technological fields.

However, a number of other countries emerge as significant actors in selected fields when CET patent data is benchmarked against total patenting activity (all technology sectors) in a given country. For instance, such an analysis reveals that India features within the top five countries for solar PV, while Brazil and Mexico share the top two positions in hydro/marine.

While patent numbers offer an indicator of innovation in a given technological field, reviewing a cross-section of patents owned by leading applicants in the different CET sectors would be a more precise method of assessing innovation and the relevance of such technologies. The analysis would look at the claims of the various patents held by entities to identify whether there is a depth of innovation in the field. Such information would help to identify patent quality issues.

In terms of technology transfer, some previous data mining studies (notably Dechezleprêtre et al., 2009) tend to consider patent filings in other countries as an indicator of technology transfer. However, such an approach has its limitations. The use of patent filings in other countries as an indicator of technology transfer is at best a crude measurement for several reasons. Patent filings are often made in another country for defensive purposes to prevent competitors from using the technology in a particular market or to be able to license it under certain conditions. Moreover, patents do not always fully disclose technology in such a way that it can be practised, developed or improved locally. Disclosure of a patent locally does not make the technology immediately accessible unless licensing is involved. In other words, the internationalization of the patent system and the resultant increase in patenting generally does not automatically mean an increase in technology transfer (UNEP, EPO and ICTSD, 2010). Thus, the study of licensing practices is an important element in order to have a clearer picture about the transfer of CETs.

Key findings of the licensing survey[12]

In addition to the patent landscape, the UNEP–EPO–ICTSD project carried out the first large-scale CET licensing survey amongst technology holders to better understand their licensing activities.

Obtaining information on licensing practices is challenging as they are often an integral part of the business strategy of firms. In addition, they can vary considerably according to products and geographical scope. Thus, as mentioned above, the collaboration of major business associations was important in reaching out to technology holders in this area. In addition, the results of the survey were based on an aggregate analysis of all respondents without reference to individual replies. This was done to preserve confidentiality and in view of the commercial sensitivity of the information.

The licensing survey was structured in three parts. First, it addressed different elements of the respondents' licensing practices and activities. This included questions about the importance of CET out-licensing (where the owner of the technology licenses it out for a financial return) and in-licensing (where an organization seeks access to a proprietary technology for its own purposes and activities);

whether there had been a shift in the organization's business strategy towards CET licensing within the past three years; and identifying activities based on additional collaborative IP mechanisms (patent pools, cross-licensing, joint ventures, strategic alliances, etc.). Second, the survey focused on aspects of CET licensing towards developing countries. Third, it included questions about whether the respondent was a private company, academic institution, governmental body, national laboratory, consortium, etc.; the location of the organization's headquarters; the size of the organization (i.e. multinational, large but focusing on domestic markets, small- and medium-sized enterprises (SMEs), non-profit, etc.); the type of CET it dealt with (i.e. wind, biomass, biofuels, solar, ocean, wave, waste, etc.) and the intensity of its research and development (R&D) activities.

The survey was carried out with the assistance of industry and business associations representing technology owners. The response rate amounted to 30 per cent of the nearly 500 organizations which were approached (160 organizations responded).

Private companies were the main respondents, with 66 per cent of the replies. Among private companies, 47 per cent were multinationals and 7 per cent were large companies, mostly focused on domestic markets. SMEs with fewer than ten employees made up 24 per cent of the private company respondents. Academic institutions and governmental bodies (including national research institutes) added up to 34 per cent of the total respondents. Respondent organizations with headquarters in Germany, the US, Japan, France and the UK amounted to 74 per cent of the total respondents.

The majority of respondents were active in the area of CETs, with 63 per cent focusing on biomass/biofuels, 46 per cent on waste-to-energy, 45 per cent on solar PV, 33 per cent on wind, 25 per cent on others, 25 per cent on solar thermal, 15 per cent on hydro/marine, 13 per cent on ocean/wave, and 12 per cent on geothermal.

The largest number of respondents (42 per cent) considered themselves engaged in full-scale R&D activities (i.e. from the early stages of research up to the final stages of development, including the ability to introduce new innovative products into the market). Approximately, one third of the respondents (32 per cent) saw themselves as having significant R&D capabilities, though mostly concentrating on the early and middle phases of the process. The remainder of the respondents categorized themselves as having limited R&D capabilities (18 per cent), focusing on improving existing technologies, or having low R&D capabilities (8 per cent) in that their business models were not focused on R&D.

Although not perfect in terms of a representative sample, the nature and type of organizations that did respond provide a useful cross-section for analysis.

Almost half (48 per cent) of the respondent organizations said CET-related patents constituted either a substantial or a significant part of their overall patent portfolio. Organizations reporting a low share of CET-related patents amounted to 37 per cent. The remainder of the respondents (15 per cent) said that CET-related patents constituted a negligible share of their overall portfolio.

Regarding collaborative IP-based mechanisms (categorized mainly as patent pools and cross-licensing) and cooperative R&D efforts (categorized as strategic

partnerships), the survey showed that CET-intensive organizations used collaborative IP-based mechanisms slightly more than other respondent organizations, with
41 per cent indicating that they were employed occasionally or frequently. This
trend repeated itself in relation to cooperative R&D efforts, with the vast majority
(83 per cent) of CET-intensive respondents stating they occasionally or frequently
used the process.

The findings of the survey concerning licensing of CETs to developing
countries were most revealing. For the purpose of the survey, developing countries
were defined as countries not members of the OECD.

In this regard, the survey showed that, overall, there is little CET out-licensing
activity towards developing countries among the survey participants. The majority
(58 per cent) of survey respondents indicated that in the past three years they had
not entered into licensing agreements with entities based in developing countries
(see Figure 5.2).

However, this trend has to be viewed in the broader context of reality in the
field of technology out-licensing. Findings from other industries indicate that there
are several hurdles to overcome in out-licensing due to a number of factors, such
as transaction costs, the challenges in identifying a suitable partner and mutually
agreed licensing conditions (i.e. pricing and the geographical or exclusive scope of
the agreement). Often, the willingness to out-license does not tend to reflect the
level of licensing (Pluvia Zuniga and Guellec, 2009).

For the time being, where licensing agreements have been entered into, the
main beneficiaries are entities based in China, India, Brazil and Russia.

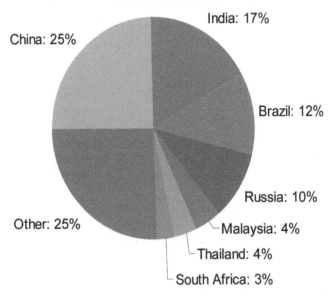

FIGURE 5.2 With which countries has your organization been most involved in
licensing or other commercialization activities of intellectual property in the field
of CETs?

Source: UNEP, EPO and ICTSD, 2010

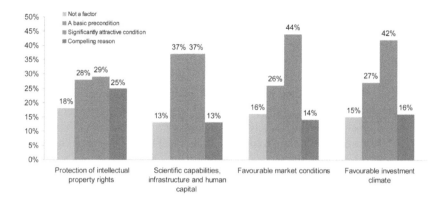

FIGURE 5.3 Factors affecting licensing or cooperation with developing countries
Note: When your organization is making a decision whether or not to enter into a licensing or cooperative development agreement with a party in a developing country, to what extent would the following factors positively affect your assessment?
Source: UNEP, EPO and ICTSD, 2010

The survey results also provide some useful insights as to the perceptions of technology holders in undertaking out-licensing activity.

Protection of IP in the recipient country is of importance to respondents when considering whether to enter into licensing agreements. It is considered an important factor by 82 per cent of organizations, with 54 per cent stating that it was either a significantly attractive condition or a compelling reason for an agreement. However, the protection of IP alone was not the only important factor in deciding whether to license to developing country entities. In line with findings in literature as well as empirical studies, scientific capabilities, infrastructure and human capital, favourable market conditions and investment climate were actually considered slightly more important, with between 85 and 87 per cent of respondents stating so (see Figure 5.3).

A separate analysis was conducted with respect to 'licensing-intensive' respondents (i.e respondents that had during the last three years occasionally or frequently entered into licensing agreements which involved recipients based in developing countries).

If the general response is compared with responses from licensing-intensive organizations, it appears that 89 per cent of licensing-intensive respondents attached greater importance to IP protection compared with 82 per cent of the general respondents. IP protection also carries slightly greater weight amongst licensing-intensive organizations than the other factors. For example, 87 per cent of licensing-intensive respondents saw scientific capabilities, infrastructure and human capital as important. It would seem that IP protection is a more important factor for organizations which have previously engaged in licensing agreements in which a proprietary technology is ready to be licensed to a developing country.

One significant lesson to draw from these findings is that it might be important for countries negotiating in the UNFCCC context to look beyond single factors as the cause for the lack of technology transfer and to search for collaborative solutions, but at the same time not to exclude or downgrade the importance of any one factor.

Finally, technology holders were asked whether they would be more willing to provide more flexible licensing terms in such circumstances where the country had limited financial capabilities. Overall, the majority of the sample (70 per cent) indicated they would be willing to provide more flexible licensing terms to recipients from developing countries. Respondents stating that they would be willing to offer 'substantially' more accommodating terms amounted to 5 per cent, with 15 per cent prepared to offer 'much more' accommodating terms. Comparing the general responses with licensing-intensive organizations, the latter group would be prepared to offer more flexible terms to licensees from developing countries (78 per cent). Notably, academic institutions and public bodies were slightly more willing than private enterprises to provide accommodating licensing terms to developing-country recipients. Small- and medium-sized enterprises (SMEs) were slightly more likely than multinationals to offer more flexible terms. However, it remains to be seen if such flexibility would be applicable to the latest technologies. More specific surveys might be needed to better understand licensing behaviour and prospects for improved forms of technology transfer.

While licensing in CETs is no lower than in other industries, the overall difficulty with markets for licensing may create particular challenges in the case of CETs, where rapid diffusion is needed.

In this regard, and building on the findings of the licensing survey, several measures could be envisaged in the context of climate change negotiations in order to improve market conditions and encourage licensing of CETs with a view to enhancing technology transfer to developing countries.

Such measures could include the development of models and platforms to assist companies to signal their licensing needs and preferences, including entities from developing countries. This would increase market transparency, help potential licensors and licensees to match supply and demand, and reduce transaction costs.

In addition, the elaboration of 'guidelines' to encourage licensing of climate technologies on 'fair and reasonable terms' to developing countries could also be envisaged, capitalising on the survey finding relating to the willingness by a majority of respondents to provide more flexible licensing terms to recipients from developing countries. This would be particularly relevant for the diffusion of the results of publicly funded research.

Finally, strengthening the capabilities of developing countries in the area of negotiating technology licensing agreements in order to maximize the benefits for their technological development is another area which could receive greater attention.

A new classification for clean energy generation patents[13]

As mentioned earlier, patent offices systematically classify patent documents and also non-patent literature in order to assist with administration and patent searching. Patent classification systems are arranged in a hierarchical structure and provide different technologies with different alphanumeric codes.

A major difficulty for research in the field of CET patenting is that current patent classification schemes often do not correlate with the type of information sought by the policy-makers in climate change discussions.

This makes it challenging for policy-makers and others to retrieve the patent information and produce patent technology landscapes without expending considerable resources and expertise. Moreover, even if the necessary resources and expertise are available, the data retrieved in such analyses may be of limited value as it reflects only the current snapshot of dynamically changing landscapes. The prevailing trends of today would not necessarily reflect tomorrow's realities, particularly in the CET field.

In addition, clean energy extends over vast technological sectors, which makes the task of establishing a new classification very complex. In order to deliver accurate patent data, a patent classification system must go beyond the level of the industrial sector (in classification language, the subclass level) to also consider the applications of a technology, such as apparatuses and even components and subcomponents (including hardware and software).

Relying only on the European classification collection to retrieve the relevant patents was also not considered sufficient in the context of CETs. An additional effort (e.g. using a combination of the IPC with keywords, even if no ECLA code exists) made it possible to capture documents from the Republic of Korea or Japan as well without having a family member already captured by the European classification.

Ultimately, close to 700,000 patents (not necessarily all in force) were retrieved and tagged within the two sectors of energy generation and carbon capture. This involved hundreds of new categories to be defined in a formal technical vocabulary and introduced into the patent examiner's workflow. While EPO had the previous experience of establishing a new classification for nanotechnology, it entailed the creation of only six new categories. The reclassification undertaken by EPO for CETs was unprecedented on this scale.

The new classification developed by EPO in the course of the project in the areas of energy generation and carbon capture is now available on the EPO's public patent information service (esp@cenet).

The classification presents numerous advantages: it includes more than 200 new categories relating to the clean energy generation technologies examined under the project; it provides a continuous and public flow of information; its scope encompasses worldwide coverage of all available patent data; it is regularly updated with the latest patent publications; it presents a high level of expertise as patent searches are carried out by EPO examiners; and, finally, the classification can possibly be extended in the future to other relevant mitigation and adaptation sectors.

Areas for future research

The study pointed to the need to explore further areas of research in order to guide future action at the international level.

One question that is still open is whether further climate change technology sectors (buildings, transport, industry, agriculture, waste management) can be addressed in the same manner as energy generation. This would require collaboration with experts from the UNFCCC, IPCC and key stakeholders to make the necessary fine-tuning to the technology mappings undertaken by the ICTSD with the support of UNEP in the remaining sectors (buildings and transport).

Another area where more information is needed is the demand side of the debate. Most studies have focused on the supply-side perspective. A survey capturing the views of entities in the developing world would be essential for a broader understanding of the challenges facing these entities in accessing clean energy technologies.

Future work and refinements should also be done on landscapes which identify patented inventions that have been commercialized in the marketplace. This would give a better idea of which technologies are working and inducing technological change. Furthermore, a study of patenting by publicly funded institutions and universities would be important in helping to understand the source of new technologies and the role of government funding in their development.

Conclusion

Bridging the gap between evidence and policy-making was the main *raison d'être* of the UNEP–EPO–ICTSD project. In this context, three key findings are of particular importance: the increase in patenting of CETs coincides with the adoption of the Kyoto Protocol, which shows that policy signals in climate change negotiations can play an important role in spurring development of CETs; second, providing accurate and publicly available information is urgently needed on existing and emerging CETs, including IP and licensing aspects; and, finally, options to facilitate licensing of CETs to developing countries should be considered. These findings and the further areas for research, identified in the study, could possibly be addressed in the future work programme of the Technology Mechanism and its bodies, taking into consideration persistent disagreements about whether and how to examine IP within the scope of the mechanism's current mandate.

Another important lesson from the project points to the fact that gathering, analysing and providing access to information on clean energy technologies, including IPRs and licensing aspects, is a costly and complex task. It involves a wide and diverse set of actors, such as governments, IP authorities and the private sector. It requires a collaborative effort on the part of all these stakeholders. Thus, technology information platforms should be an essential component of the emerging new technology transfer architecture under the Technology Mechanism.

As previously mentioned, reliable and accurate patent and technology data is not an end in itself. However, such information is an important component – among others – of an enabling environment for innovation and technology transfer.

Similarly, the considerable untapped licensing potential towards developing countries identified by the licensing survey, and the possible measures to address it, could provide a meeting point for collaboration between developed and developing countries and other stakeholders (international organizations, the private sector, etc.).

In conclusion, greater and better availability of technological information and facilitating licensing of clean energy technologies to developing countries appear to be concrete and practical measures that could find a more immediate echo in the climate change discussions.

Notes

1 UNFCCC (2007) *Bali Action Plan*.
2 See http://unfccc.int/resource/docs/convkp/conveng.pdf.
3 See http://unfccc.int/resource/docs/convkp/conveng.pdf.
4 UNFCCC (2007) *Bali Action Plan*, paragraph 1(d).
5 Presentations are available online at http://unfccc.int/meetings/awg/items/5928.php.
6 See http://www.unep.org/greeneconomy/AboutGEI/WhatisGEI/tabid/29784/Default.aspx.
7 See http://www.epo.org/searching/essentials/classification.html.
8 More detailed information can be found under www.epo.org/data.
9 European Patent Forum, 2008: a conference on IP and climate change organized by the European Patent Office, the European Commission, the Slovenian government and the Slovenian Intellectual Property Office in Ljubljana in May 2008. For summaries of the conference results, see http://www.epo.org/aboutus/events/archive/2008/epf2008/forum-1.html.
10 This section draws on Chapter 3 of UNEP, EPO and ICTSD (2010).
11 PATSTAT is a snapshot of the EPO master documentation database (DOCDB) with worldwide coverage, containing 20 tables, including bibliographic data, citations and family links. This database is designed to be used for statistical research.
12 This section draws significantly on Chapter 4 of UNEP, EPO and ICTSD (2010).
13 This section draws significantly on Chapter 5 of UNEP, EPO and ICTSD (2010).

References

Barton, J. (2007) *Intellectual Property and Access to Clean Energy Technologies in Developing Countries: An Analysis of Solar Photovoltaic, Biofuel and Wind Technologies*, ICTSD Programme on Trade and Environment, Trade and Sustainable Energy Series, Geneva, Switzerland, http://ictsd.org/i/publications/3354/, accessed 12 October 2011

de Beer, Y. (2008) *Address to the European Patent Forum*, Lubljana, Slovenia, 7 May 2008, http://unfccc.int/files/press/news_room/statements/application/txt/080507 speech_lubljana.pdf, accessed 12 October 2011

Dechezleprêtre, A., Glachant, M., Hăščič, I., Johnstone, N. and Ménière, Y. (2009) *Invention and Transfer of Climate Change Mitigation Technologies on a Global Scale: A Study Drawing on Patent Data*, Cerna, Mines Paris Tech and Agence Française de Dévelopement, Paris, France, http://personal.lse.ac.uk/dechezle/Innovation_diffusion_climate_techs_AFD.pdf, accessed 12 October 2011

EPO (European Patent Office) (2007) *Scenarios for the Future*, Munich, Germany, http://documents.epo.org/projects/babylon/eponet.nsf/0/63A726D28B589B5BC12572DB00597683/$File/EPO_scenarios_bookmarked.pdf, accessed 12 October 2011

India (2011) *Proposals for Inclusion of Additional Agenda Items in the Provisional Agenda of the Seventeenth Session of the Conference of the Parties – Addendum*, FCCC/CP/2011/INF.2/

Add.1, http://unfccc.int/resource/docs/2011/cop17/eng/inf02a01.pdf, accessed 12 October 2011

IPCC (Intergovernmental Panel on Climate Change) (2007) *Mitigation of Climate Change*, IPCC Fourth Assessment Report, http://www.ipcc.ch/pdf/assessment-report/ar4/syr/ar4_syr.pdf, accessed 12 October 2011

Johnstone, N., Hăščič, I. and Popp, D. (2010) 'Renewable energy policies and technological innovation: Evidence based on patent counts', *Environmental and Resource Economics*, vol 45, issue 1, pp133–155

Lako, P. (2008) *Mapping Climate Mitigation Technologies/ Goods within the Energy Supply Sector*, Energy Research Centre of the Netherlands for the International Centre for Trade and Sustainable Development, Geneva, Switzerland, http://ictsd. org/?p=67954&preview=true, accessed 12 October 2011

Lee, B., Iliev, I. and Preston, F. (2009) *Who Owns Our Low Carbon Future? Intellectual Property and Energy Technologies*, Chatham House Report, London, http://www.chathamhouse. org/sites/default/files/public/Research/Energy,%20Environment%20and%20 Development/r0909_lowcarbonfuture.pdf, accessed 12 October 2011

Lewis, J. I. (2007) *A Comparison of Wind Power Industry Development Strategies in Spain, India and China*, Center for Resource Solutions, San Francisco, CA

Mallett, A., Ockwell, D., Pal, P., Kumar, A., Abbi, Y., Haum, R., MacKerron, G., Watson, J. and Sethi, G. (2009) *UK–India Collaborative Study on the Transfer of Low Carbon Technology, Phase II Final Report*, University of Sussex, SPRU (Science and Technology Policy Research), Institute of Development Studies and the Energy Resources Institute, www.sussex.ac.uk/.../documents/decc-uk-india-carbon-technology-web.pdf, accessed 12 October 2011

Ockwell, D., Watson, J., MacKerron, M., Pal, P. and Yamin, F. (2008) 'Key policy considerations for facilitating low carbon technology transfer to developing countries', *Energy Policy*, vol 36, no 11, pp4104–4115

Ockwell, D., Mallett, A., Haum, R. and Watson, J. (2010) 'Intellectual property rights and low carbon technology transfer: the two polarities of diffusion and development', *Global Environmental Change*, vol 20, pp729–738

Pluvia Zuniga, M. and Guellec, D. (2009) *Who Licenses Out Patents and Why? Lessons from a Business Survey*, OECD STI Working Paper 2009/5

UNEP, EPO and ICTSD (United Nations Environment Programme, European Patent Organization and International Centre for Trade and Sustainable Development) (2009) 'UNEP, EPO and ICTSD join forces to look at the role of patents in the development and transfer of technologies to address climate change', *Press Communiqué*, 24 April 2009, http://ictsd.org/downloads/2009/07/press-communique.pdf, accessed 12 October 2011

UNEP, EPO and ICTSD (2010) *Patents and Clean Energy: Bridging the Gap between Evidence and Policy*, UNEP, EPO and ICTSD, http://ictsd.org/i/publications/85887/, accessed 12 October 2011

UNFCCC (United Nations Framework Convention on Climate Change) (2007) *Bali Action Plan*, UNFCCC, FCCC/CP/2007/6/Add.1, http://unfccc.int/resource/docs/2007/cop13/eng/06a01.pdf, accessed 12 October 2011

Xinhua News (2011) 'India holds post-Cancun BASIC ministerial meeting on Climate Change', *Xinhua News*, 28 February 2011, http://new s.xinhuanet.com, accessed 12 October 2011

6

TECHNOLOGY TRANSFER, IPRS AND CLIMATE CHANGE

Krishna Ravi Srinivas

Introduction

The role of intellectual property rights (IPRs) in technology transfer (TT) in the context of climate change has been a controversial one. While some state that IPRs do constitute a barrier, it is argued by others that IPRs are not a barrier and are needed for incentivizing innovation. While a case-by-case approach has been suggested by some, solutions that use the patent system to develop commons and other similar arrangements for sharing technology have also been suggested. Patents are the most preferred options for intellectual property (IP) protection when it comes to technologies; hence, a focus on patents is inevitable. There is thus a wide variety of views in this issue and the debate is nuanced despite extreme positions.[1]

This chapter takes these views into account and points out that there is an urgent need to go beyond the dichotomy in the perspectives to arrive at workable solutions, within the patent system and outside the patent system. The role of IPRs in facilitating TT in climate change is an issue for open debate, and the literature indicates the diversity in approaches and solutions.[2]

The patent system cannot be divorced from the larger innovation system.[3] Technology transfer enables the recipient firm/country to use the technology, study it and absorb it and engage in learning-by-doing so that the innovative capacity is increased. Flow of knowledge and technology transfer has significant impacts in the innovation system.[4]

Hence, if the patent system becomes a constraint in TT, solutions to overcome this have to be found, and this may include novel solutions that promote sharing and incentivising, enabling diffusion rather than exercising monopoly rights in a manner that inhibits diffusion. Thus, the patent system can be considered as part of the problem as well as part of the solution, and the need to go beyond the patent system in some circumstances cannot be ignored. In other words, a pragmatic solution-oriented approach is needed without ignoring the controversial aspects of IPRs and TT and the North–South divide in light of the role of IPRs in technology and technology transfer under Multilateral Environmental Agreements (MEAs).

While TT has been defined in many ways, this chapter uses the definition given by the Intergovernmental Panel on Climate Change (IPCC), which defines technology transfer 'as a broad set of processes covering the flows of know-how, experience, and equipment for mitigating and adapting to climate change amongst different stakeholders such as governments, private sector entities, financial institutions, non-governmental organizations, and research/educational institutions'.[5] In this chapter, the focus is more on new ideas and initiatives in TT than on debating older issues such as the role of the Trade Related Aspects of Intellectual Property Rights agreement (TRIPS) or using options such as compulsory licensing.

Technology transfer and IPRs: Overview of the debate

Do stronger IP regimes stimulate more transfer of technology and should developing nations opt for stronger IP regimes so that TT is facilitated? The views on this diverge, and there is no consensus that stronger IP regimes will always result in more technology transfer. Similarly, there is no consensus that IP regimes have no impact upon TT.[6] According to Keith Maskus, an expert in this topic:

> ... how IPR and ITT [international technology transfer] interact in these areas [is] highly context specific and broad claims are not particularly helpful. Secondly, economists have barely begun the task of analysing the task of linkages between public-goods externalities and ITT. Finally ... it is possible that transparent and enforced IPR could reduce the cost of ITT.[7]

The North–South divide in this issue is an important element in the discussions on the role of IPRs in TT, particularly in the case of MEAS such as the United Nations Framework Convention on Climate Change (UNFCCC). This should be understood in the context of the debate on technology flow from North to South.[8] The South and North have had divergent perspectives and efforts, such as the Code of Conduct in Technology Transfer, that did not succeed.

After reviewing the processes and debates, Krishnachar concludes:

> In general the international legal regime has failed in providing for effective transfer of technologies for resource poor nations. And the market factors dictating the technology[9] trade are furthering the technological gap between nations.[10]

In the context of climate change, TT and IPRs, there is a marked divergence in views. At the risk of simplification, they can be classified as:

- IPRs are no barrier to TT, and IPRs are necessary for innovation. While there are issues in TT, IPRs are not the constraining factor.[11] For example, this view is expressed by industry organizations such as ICC and Copenhagen Economics and IPR Co (2009).[12]
- IPRs are a barrier to TT and there is a North–South divide in terms of ownership of patents, innovative capacity and the reluctance to transfer technology.[13] For example, Srinivas (2009) articulates this view.
- IPRs are not a barrier to many climate change technologies because most of the technology is either old or is in public domain. Besides this, developing nations are becoming innovators.[14]
- Even if IPRs are a problem, they need not be barriers as there are solutions and options available for governments to facilitate TT. The private sector is also keen to transfer technology for large developing countries such as India and China.[15] For example, Peter Drahos argues: 'Probably the best strategy here is to keep intellectual property rights out of climate change negotiations and deal with specific issues as they rise on a case to case basis.'[16] Maskus (2010) points out the various options available and argues that TRIPS is flexible enough and cautions against moves that will undermine the use of IPRs as an incentive for innovation.

It is better to approach this issue on a case-by-case basis than to assume that IPRs are/are not barriers. In some instances, factors such as capacity to absorb technology and size of the market are more important in TT than IPRs. Thus, the solution as to whether IPRs are a barrier also has to be seen on a case-by-case basis. For example, studies conducted by a team at the Sussex Energy Group (SEG), University of Sussex, UK, in collaboration with researchers in India and China are a prominent exponent of this view.[17] A similar view is expressed by Ueno (2010), who argues that IPRs are not a barrier in TT to China. These studies argue that the significant barrier is not IPRs; rather, access to technology and using technology is determined by many factors, which vary from industry to industry. Based on these studies, some suggestions have been made for facilitating diffusion of clean energy technologies. The studies take a holistic view; but as they are based on case studies, their strengths and limitations are obvious.

Over the last few years the debate on this issue has been enriched by empirical findings, case studies, new ideas and analysis of data on patents, patent filings and ownership and geographical spread of patents in different technologies. Hence, today the debate is more nuanced and many new ideas have been put forth. This should be viewed in the context of suggestions such as using open innovation, 'open source' and patent pools to facilitate more innovation and further TT, as well as using patent pools for facilitating transfer and access to technology.[18]

'Open source' is a model of production originating in software development in which source code is developed through collaborative mechanisms and the software is shared under specific licences. Open innovation is a mode of innovation in which companies/teams collaborate, sharing their resources and knowledge to develop new products and services; this involves agreeing to share the IP and outcomes of the joint effort. Patent pools are a mechanism for sharing technology covered by patents. Here, the relevant patents are pooled together to form a patent pool and a mechanism is developed to share them through licensing/cross-licensing.

While each of the five views mentioned above have some merit, it is difficult to arrive at one single grand truth. Thus, even if IPRs constitute a barrier, the challenge lies in finding appropriate solutions so that IPRs do not remain a barrier. In the case of patents on climate change technologies, the data shows that firms, universities, research institutions and others from a handful of developed countries own a lion's share of the patents.[19] While the share of developing countries owning patents is increasing, their share is not significant in many technologies. While numbers provide only one part of the story, the relative importance of technology covered by patents is crucial in TT because, ultimately, what counts is not patents *per se* but the technology protected and disclosed by them.

We can illustrate this through a study conducted on wind power patents, which classifies them as low, medium, medium/high and high in terms of their relevance to the industry. For example, according to this study, only 2 per cent of the patents issued are of high importance to the entire industry, while 89 per cent of the patents provide only defensive protection.[20] Such patent landscaping exercises will be useful in identifying trends in technologies covered by patents, their relevance for the industry and ownership patterns. Such studies go beyond the numbers and help us to understand the innovation dynamics captured in patents.

As a result, we need more studies that go beyond the numbers and give us a fairly accurate picture about the ownership of patents that are critical for the industry, or patents that indicate significant breakthroughs in technology and the issues in TT, such as second- and third-generation biofuels.[21] In the context of innovation and climate change, in sectors such as agriculture, IPRs may be important in some applications but not in all; here also we need more studies on patenting trends, technologies covered, patent trolls and transfer of technology.[22] A recent study by the Economic Commission for Latin America and Caribbean indicates that in Latin America, except for Brazil, other countries in 'Latin America account for a very small share of patents relating to the biofuel production chain'.[23] A patent landscaping exercise may reveal the relative importance of these patents in the production chain.

Many companies that own the patents are not into manufacture and use licensing to earn revenue. Hence, they fiercely protect their IP, and litigation in clean energy patents is on the rise.[24] While most of these cases are fought and settled in the US and Europe, their global dimension cannot be ignored because companies in the US increasingly resort to using the Federal Trade Commission (FTC) as a forum and try to block imports of the products that consist of alleged infringed technology.

These facts do not negate the importance of case-by-case approaches or the argument that there are many options and that IPRs need not be a barrier. Rather, they point out the various facets of the issue; hence, more than one approach may be needed to facilitate TT. The approaches can be complementary. For example, in global negotiations, developing nations raise the issue of IPRs as a barrier to remind the North that no meaningful TT is possible if IPRs are viewed from a narrow perspective of using monopoly rights to extract rent than to promote sharing and using technology on reasonable terms.

Developing nations are promoting innovations and have harmonized their laws with TRIPS, and thus understand the role of IP protection.[25] At the same time, they point out the global dimension of climate change to argue that a business-as-usual approach is not the right one. Thus, cautioning against IPRs becoming a barrier does not mean an anti-IP or anti-patents perspective. In fact, many developing countries are part of various bilateral and multilateral initiatives for development and diffusion of technologies in which they work with governments and industry from developed nations.[26]

From an academic point of view, the debate is proceeding in the right direction, with studies and empirical research giving new insights and thereby stimulating new thinking on these issues. For example, studies on the relevance of open innovation in development and TT, with the Consortium of International Agricultural Research Centres (CGIAR) as a model for development and TT, indicate that IPR issues can be handled in many ways.[27] The idea of patent commons has been adopted to develop a commons for technologies that can be relevant in adaptation and mitigation, while the Creative Commons Model has been used to develop a similar commons in which companies decide what rights they will hold, what rights they will forego and what uses they will permit without giving up patents.

Nevertheless, we need more studies on TT to least developed countries (LDCs) in specific sectors such as renewable energy (e.g. biofuels) to find out whether these countries are able to access the relevant technologies and whether IPRs are a barrier in this. Similarly, we need to know more about the patterns in TT, licensing practices and how firms in developing countries overcome the hurdles in TT and their access to patented technologies. While there are studies on India and China, and the literature on litigation related to clean technology patents is increasing, there are many gaps in our understanding in terms of law and practice, particularly on infringement and validity of patents.[28]

At one level, TT in clean technologies is more like a business transaction; at another level, understanding the overall impact of these transactions in reducing CO_2 emissions and increasing energy efficiency and sustainable development is necessary for innovators, business firms and policy-makers so that diffusion of clean technology is possible.

Thus, the debate today is not just about the old questions; it is also about new ideas and new perspectives even when old viewpoints persist. Thus, even if one is of the view that IPRs do constitute a barrier, the relevance of various options to overcome this and the proliferation of new approaches and new solutions cannot be dismissed. Rather, the position one can take is that while IPRs may be a barrier to TT, proposed solutions deserve closer scrutiny. In fact, the broader question of developing and transferring relevant technologies can be approached and answered from various viewpoints, taking into account innovation, public policy, institutional aspects, human rights and access to advances in science and technology, etc.[29]

The linkages between IPRs and some of the above factors, such as human rights, and public policy deserve more attention: explorations into such linkages in the context of access to medicines for HIV/AIDS have had major ramifications for enabling access, and the Doha Declaration of 2001 and its paragraph 6 solutions are the outcomes of campaigns that used the findings from such explorations. I hope that the broader and wider debate on IPRs and TT in the context of climate change will also result in better understanding of the linkages and lead to informed decision-making, besides helping the global community to encourage new solutions and new approaches.

Technology transfer, IPRs and the UNFCCC process

Technology transfer has been one of the important objectives of the UNFCCC (e.g. Articles 4.5 and 4.7) and, in general, the emphasis has been on technology transfer from developed to developing countries. In the discussions on implementing technology transfer under the UNFCCC process, the North–South divide on the role of IPRs is too well known to be elaborated upon here. During the recent negotiations that started with the *Bali Action Plan*, although IPRs have been a contentious issue, in the final text of the Cancun Agreements there was no reference to IPRs even though the establishment of a Technology Mechanism was one of the key outcomes of the Cancun Conference.[30] TT and IPRs, in particular, have been one of the contentious issues in the negotiations right from the beginning, particularly after the *Bali Action Plan*, which identified technology as a key element of future action. While an analysis of the negotiating positions and proposals by various countries is beyond the remit of this chapter,[31] it is important to emphasize that the absence of reference to IPRs in the Cancun Agreement does not mean that the North–South divide on this issue has vanished. The BASIC group, for example, consisting of Brazil, South Africa, India and China, has stated that some of the issues, such as equity, intellectual property rights and trade, were important to BASIC countries and efforts would be made to bring them back within mainstream discussions.[32]

It is important to note that the proposed Climate Technology Centre and Network will also facilitate South–South cooperation, thereby going beyond the traditional North–South approach in technology transfer. Currently, the Technology Mechanism is not functional, although it is mandated that the mechanism will be operationalized by 2012. In the Expert Workshop on the Technology Mechanism, held in April 2011 in Bangkok, many suggestions were made regarding operationalizing the Technology Mechanism in 2012. Some of the participants were of the view that organs of the Technology Mechanism (i.e. the Climate Technology Centre and Network or the Technology Executive Committee) should address issues related to IPRs, while other participants did not envisage any role for them regarding IPR issues; some participants came up with a wish list of what these bodies can do in terms of TT and IPRs.[33]

It is important for the Technology Mechanism to deal with IP issues in technology transfer as this will help developing countries and LDCs. Moreover, much work has been done under the UNFCCC in assessing technology needs and identifying technologies. The Clearing House under the UNFCCC has little experience in handling IPR issues.[34] In fact, although many United Nations agencies are involved in TT or related activities, none of them, except the World Intellectual Property Organization (WIPO), have expertise in handling IP issues or have worked at the interface between TT and IPRs.[35] Similarly, although TT figures in many international agreements, these agreements have not been major avenues for TT.[36] These limitations of UN agencies, international treaties and conventions in facilitating TT must therefore be borne in mind.

It has been suggested that the Technology Mechanism should work closely with national governments in the plans for transitions to a low-carbon economy and to accelerate diffusion of technology and enhance energy efficiency. This would be possible only if the mechanism and the country share the same views on TT and IPRs; if a country does not want the mechanism to deal with IP issues, the mechanism will have no role to play in IP and TT issues. Thus, whether the Technology Mechanism will deal with TT and IPRs is not clear at this stage. It is likely that while it may be asked to deal with them, this may not form part of its core mandate. Much depends upon the structure of the mechanism, its mandate and the financial flows through the mechanism for TT-related activities.

An important factor is that much of the technology transfer occurs outside the purview of the UNFCCC; during recent years, the tendency has been to develop regional/bilateral networks/agreements for technology transfer and development. Thus, unless there is a radical change in the mandate of the Technology Mechanism, IP issues may not be covered in operationalizing the Climate Technology Centre and Network, and these issues will be left for individual partners or cooperating agencies to resolve amongst themselves.

As developing nations are also becoming important innovators at least in some technologies, it is likely that they will also give focus to IP rights in technology transfer agreements.[37] It can therefore be concluded that the North–South divide in the role of IPRs is likely to persist in negotiations, although the Technology Mechanism may not have the mandate to deal with IP issues in TT. Thus, even as the global community affirms its faith in the UNFCCC by agreeing to provide funding for the Technology Mechanism and to make the mechanism the main force in technology development and TT under the UNFCCC, IPRs may not be covered by the mechanism, but may continue as a contentious issue in further deliberations in fora such as the Conference of the Parties (COPs).

Patents, commons, patent pools, climate change, IPRs and TT

What will happen if companies donate patents and create a commons consisting of such patents, where access to which is available to all, subject to some conditions and restrictions. The idea of creating a commons with patents is not new. In 2005, the Patent Commons was created by Open Source Development Labs (OSDL) for the benefit of the open source development community and industry. Information about software patents pledged in support of open source software and patent pledges is available through this commons.[38] This enables the developers and users to know the boundaries of the commons; industry majors such as IBM, SUN Microsystems, Novell, Computer Associates and Red Hat either support this and/ or contribute to it. The idea is to promote the development and use of open source software and to ensure that both the industry and developers and users know their obligations, commitments and responsibilities to each other and to the public, for the use of patents made available through this commons.

In 2009, a similar initiative, the Eco-Patent Commons, was launched by the World Business Council for Sustainable Development (WBCSD), based in Geneva, in order to share and collaborate in eco-innovation through the use of patents.[39] Since then, many companies have contributed; currently, approximately 100 patents are available through this commons. The initiative received wide attention and has been extensively discussed.[40] While largely welcomed, Hall and Helmers (2010) point out that there are still some issues with the available patents in the commons.[41]

From the perspective of TT, the effectiveness of this commons is limited by many factors. For example, while access to these patents is available, this itself may not be sufficient to commercialize a technology because access to many patents, most of which may be protected, will be required. Tacit knowledge − the capacity to understand and operationalize the knowledge embodied in the patent − can hinder effective utilization of such patents by others. Thus, TT may not take place, but the user will be able to access the patents and through them may be able to innovate further, or put them to use for the listed purposes.

Another issue here is that patents made available through the commons are valid in their respective jurisdictions only. For example, a patent granted in Japan is valid in Japan, but it might not have been patented in India or China. A potential user may thus not have the freedom to operate beyond the jurisdictions in which they have been patented, and the availability of the patents need not result in effective TT in the absence of such a freedom to operate. The patents are made available under certain conditions, and rules are complicated by the defensive termination option. As the commons is not a public domain, access is not unconditional.

The rules appear to be clear, and the boundaries in terms of obligations and rights are also evident; but there is uncertainty about the right to use and freedom to operate: patent owners, for example, exercise some rights, such as the option of defensive termination. Nancy Cronin has made some suggestions for enhancing the effectiveness of the commons.[42]

Thus, while the idea of developing and maintaining a commons for furthering innovation and use of patents in environmental protection and promoting sustainable development is laudable, the effectiveness of this commons is not known. Given the wide range of technologies and applications, the number of patents in the commons is too small to make a significant difference. Another factor that limits use in TT is the lack of a bundle of relevant patents that can be applied together for specific applications without relying on patents not in the commons. It is hoped that based on this experience, the commons will be revamped and significant changes will be made.

Nations can pick up this idea and promote such commons at the national and regional level for facilitating better access and diffusion of relevant technologies. They can also supplement the commons by buying patents and donating them. Such commons can be application specific or sector specific (e.g. biofuel patents relevant for use of Jatropha or patents relevant for enhancing energy efficiency in the cement industry). Unless such initiatives are taken up at different scales, such patent commons will have a limited role in facilitating TT to developing countries and LDCs.

GreenXchange is another initiative modelled after the Creative Commons for promoting sustainable innovations by using intellectual assets in such a way that owners grant some rights and reserve other rights for them.[43] Creative Commons facilitates the use of different types of licences to encourage sharing and innovation as an alternative to the traditional copyright 'all rights reserved' mode. Under these licences, the creator can decide what rights (s)he will exercise and what uses will be permitted and for what purposes. For example, a creator can allow non-commercial use without prior permission from the creator, subject to some conditions. The various types of licences under the Creative Commons enable the creator to pick and choose and mix and match his rights and what (s)he is willing to share with (or offer) others.

In this there are three types of pledges and licences: research non-exempt, standard and standard plus. Intellectual assets are available, with owners permitting some actions and forbidding others, just as in the case of the Creative Commons licence, which allows sharing and use subject to some rights being reserved by the owner. The licensing mechanism is at the core of this initiative. This mechanism enables the owners of the patent to choose what licensing terms they wish to offer to whom and under what conditions. For instance, a patent may be offered under a licence to an academic institution for academic research or for working on it to develop new innovations, and the same may be offered to a firm under licence for commercial use. An important difference between the Eco-Patent Commons and GreenXchange is that in the latter the patent owner has not donated the patents, but makes them available, with flexibility in their use and licensing. The application of the Creative Commons principle to patents in clean technologies is interesting; but whether it will facilitate TT is yet to be seen – it may, for example, further innovation and research in academic institutions. Currently, Nike is the major donor/pledger in this new initiative, and most of its partners/collaborators are based in the US.

While both of these initiatives indicate the potential for creative use of the commons and licensing practices, they suffer from many flaws as they are not patent pools that facilitate sharing and cross-licensing. In other words, what is missing is the participation of major players in patenting clean technology, such as GE or Toyota. From a user's perspective, such players would enable better and easier access to patents and perhaps also enable non-commercial research; but from a TT perspective, they have miles to go to enable TT in different sectors. On the other hand, such initiatives can be used in combination with other efforts to facilitate technology transfer; and in the case of GreenXchange, patents for innovation in applications may be used where the patent holder is not interested or does not see any commercial value. Nevertheless, understanding and using the Commons and GreenXchange is not an easy task because their practices are complicated when compared to commercial licensing. Thus, users need expert advice to make the best use of them.

Another important option, Patent Pools, which has been widely used elsewhere, is yet to be applied or tested in the context of TT in climate change. By now there is sufficient literature on the scope and limitations of Patent Pools; yet there is not much literature on the use of patent pools as a tool for technology transfer and sharing in terms of climate change. In fact, the patent pool launched by UNITAID has become functional within a short time. Hence, there is also scope for launching a patent pool in clean technologies. The main difference, unlike with pharmaceuticals and electronics, is that the number of players and the range of technologies is too vast, and this constrains developing a patent pool with limited players. Moreover, the availability of substitute/alternative technologies (with most of the basic technology too old or in the public domain) restricts the need to have many patent pools for effective utilization of patents held by companies who are in the same sector. In some sectors, such as wind energy, the number of companies with significant technology patents is limited, and it is the numerous users who generate electricity who have access through licensing. Here, too, availability of substitutes or alternatives limits the need for patent pools. Hence, one can conclude that the need for a patent pool has not been experienced either by users or by holders of technology. Nevertheless, the idea may become a reality if it is initiated by governments or United Nations agencies as in the case of a patent pool for drugs to treat HIV/AIDS.

Conclusion

Complex issues demand innovative solutions and often call for new approaches to old problems. The question of TT of environmentally sound technologies (ESTs) for sustainable development is an old one, and in the case of climate change, it has been a much debated issue for more than two decades. Recently, however, it acquired a new urgency in the wake of the IPCC report published in 2007. Inevitably, the role of IPRs became a controversial issue and the controversy persists. This was evident in the negotiations under the UNFCCC, and although the Cancun Agreement is silent on this, one cannot assume that the issue has been resolved. In the debate on the role of IPRs in TT, many views were proposed. By now, the most common view is that while IPRs do play an important role in TT, they need not be considered as a barrier in all cases, although the North–South divide still exists. One important outcome of the debates is that while IPRs were identified as a factor, many proposals were made to find innovative solutions for accelerating TT. This chapter has highlighted some of these solutions, pointing out their potentials and limitations. At the same time, it should be highlighted that even with all of these proposals and initiatives, the problem will not be solved because there is no global effort, as was the case of the Montreal Protocol to facilitate TT.[44]

In case of the proposed Technology Mechanism, it is not clear as to whether it will deal with IPRs in TT. We are not witnessing initiatives similar to the Manhattan Project or the Green Revolution in climate change technologies, nor is there a global initiative to acquire patents and technologies to accelerate their diffusion.

Instead, what we are witnessing is a combination of many efforts, important in themselves, but not sufficient enough to make a significant impact in terms of TT. While open innovation models sound promising, unless the private sector and governments take the initiative, they may again end up as small scale initiatives with limited focus and reach. The big issue here is the question of TT, and the role of IPRs is just a part of it. Whether the proposed Technology Mechanism and other initiatives will result in more and better technology transfer is a question that is beyond the scope of this chapter. Nevertheless, what is clear is that while IPRs are not insurmountable problem, current efforts to deal with them may be insufficient to meet the challenges of technology transfer.

Notes

1 See the next section for an elaboration of this.
2 For example, see Reichman et al., (2008); Adam (2009); Lee et al., (2009); Hall and Helmers (2010); Hăščič et al., (2010); Maskus (2010); Maskus and Okediji (2010); Sommer et al., (2010); and Sarnoff (2011).
3 See Mathews (2010) and WIPO (2011) for an overview.
4 For an overview, see http://www.leydesdorff.net/th9/Paper%20O-109%20JOFRE.pdf.
5 IPCC (2000, p3). The report acknowledges that this definition is broader than specified in the UNFCCC. A useful overview of the literature on issues in TT in environmental technology can be found in Johnson and Lybecker (2009), while WIPO (2011) gives an excellent overview of patent systems and TT. For an overview of the concept of TT and TT in the context of MEAs, see Andersen et al., (2007, Chapter 2). See also Shepherd (2007).
6 See Hoekman and Javorcik (2006).
7 See Maskus (2009, p136).
8 For an overview, see Kariyawasam (2007) and Ockwell et al., (2010).
9 See Krishnachar (2006).
10 See Krishnachar (2006, p30).
11 Similar views were expressed in the context of access to medicine and TRIPS. See Abbott (2010).
12 See also Brandi et al., (2010); but see Parthasarathy (2010) for a different view in the context of IP in geoengineering technology.
13 This view is similar to the views expressed by G77 and developing countries in many international fora. In UNFCCC negotiations, such views have been expressed right

from the beginning and the UNFCCC recognizes the importance of North–South TT, although it does not take any position on the role of IPRs in this.

14 See Barton (2007). World Bank (2008) and UNESCO (2010) acknowledge that developing countries such as China, Korea, Brazil and India are more innovative now than a decade ago. This is evident in the number of publications, patents filed abroad and other indicators, and the remarkable progress made by China is well acknowledged in the literature. See also Fu and Soete (2010) for essays on the technological capabilities of developing countries.

15 Of these solutions, 'compulsory licensing' is the most controversial.

16 See Drahos (2009, p132).

17 See Ockwell et al., (2007); Mallett et al., (2009); Ockwell et al., (2010); and Watson et al., (2011).

18 See Iliev and Neuhoff (2009); Rattray (2009); Petrusson et al., (2010); and van Overwalle (2010).

19 See de la Tour et al., (2010); Dechezleprêtre et al., (2009); Dechezleprêtre et al., (2010); OECD (2010); UNEP, EPO and ICTSD (2010); and Veugelers (2010).

20 Torato, Philip Current and Future Trends in Wind Turbine Technology (July 2011); see http://www.totaro-associates.com/windpatentwatch.

21 See Juma and Bell (2009); Wolek (2011).

22 See Thomson and Webster (2010); see also Cahoy and Glenna (2009) in the context of biofuels and the application of private ordering mechanisms for commercialization.

23 See http://www.eclac.org/cgi-bin/getProd.asp?xml=/prensa/noticias/comunicados/8/42938/P42938.xml&xsl=/prensa/tpl-i/p6f.xsl&base=/tpl-i/top-bottom.xsl.

24 For example, *Hydro-Quebec versus A 123 Systems*: http://www.gizmodo.com.au/2011/05/knock-down-drag-out-fight-over-next-generation-batteries; *Osram versus Samsung and LG* (on LEDD patents); *Philips versus Seoul Semiconductor*. After six years of legal battle that extended from US courts to FTC, Toyota and Paice settled their patent suits in August 2010.

25 See Deere (2008), who points out that many countries have gone beyond TRIPS by adhering to TRIPS Plus norms in their national laws.

26 See Levi et al., (2010).

27 See Correa (2009).

28 The recent controversy over the validity of wind energy patents in India is an example of using the legal system to contest the validity of important patents. As more such cases emerge from developing countries, we will have a better idea about enforcement, patent validity and granting of patents in emerging economies. See http://iiprd.wordpress.com/2011/06/18/enercon-india-ltd-eil-vs-enercon-gmbh-eg/?like=1.

29 See, for example, CIEL (2009) and Brown et al., (2010).

30 See ICTSD (2011) and Romano and Burleson (2011).

31 See Gerstetter et al., (2010) and Rimmer (2009).

32 See http://www.twnside.org.sg/title2/climate/info.service/2011/climate20110504.htm.

33 See Expert Workshop on the Technology Mechanism in Conjunction with the Fourteenth Session of the Ad Hoc Working Group on Long-Term Cooperative Action under the Convention, Report by the Chair of the Workshop, FCCC/AWGLCA/2011/INF.2, pp12–13.

34 For more information on the UNFCCC and TT, see the presentations available at http://unfccc.int/ttclear/jsp/TrnDetails.jsp?EN=TNAWshpBonn. See also http://www.iisd.ca/vol12/enb12501e.html.

35 See UNDESA/UNIDO (2010).

36 See UNCTAD (2001) for a list of such agreements and provisions on TT.

37 The study by UNEP, EPO and ICTSD (2010) demonstrates that the North–South gap in patenting clean technologies is significant even as it points out that some developing countries are patenting more than ever before.

38 See http://www.patentcommons.org/.

39 See http://www.wbcsd.org/templates/TemplateWBCSD5/layout.asp?type=p&MenuId
    =MTQ3NQ&doOpen=1&ClickMenu=LeftMenu.
40 For example, Boynton (2011).
41 See Hall and Helmers (2011).
42 See   http://www.matternetwork.com/2008/5/invention-disclosures-could-improve-
    eco.cfm. See also Christopher (2011) for an analysis of the interface between tax credits
    and donations to the commons.
43 See http://greenxchange.cc/.
44 See Andersen et al., (2007) for details.

References

Abbott, F. M. (2010) *Innovation and Technology Transfer to Address Climate Change: Lessons from the Global Debate on Intellectual Property and Public Health*, ICTSD, Geneva

Adam, A. (2009) 'Technology transfer to combat climate change: Opportunities and obligations under TRIPS and Koyoto', *Journal of High Technology Law*, vol 9, no 1, pp1–20

Andersen, S. O. et al., (2007) *Technology Transfer for the Ozone Layer: Lessons for Climate Change*, Earthscan, London

Barton, J. H. (2007) *Intellectual Property and Access to Clean Energy Technologies in Developing Countries: An Analysis of Solar Photovoltaic, Biofuel and Wind Technologies*, ICTSD, Geneva

Boynton, A. (2011) 'Eco-Patent Commons', *William & Mary Environmental Law and Policy Review*, vol 35, issue 2, pp659–685

Brandi, C. et al., (2010) *Intellectual Property Rights as a Challenge to Providing Public Goods*, DIE, Bonn

Brown, A. et al., (2010) *Towards Holistic Approach to Technology and Climate Change*, University of Edinburgh, Edinburgh

Cahoy, D. R. and Glenna, L. (2009) 'Private ordering and public energy innovation policy', *Florida State University Law Review*, vol 36, p415

Christopher, K. (2011) *Reclaiming Our Technological Posterity at the Intersection of Intellectual Property and Taxation*, American University Intellectual Property Brief, pp31-43

CIEL (2009) *Technology Transfer in the UNFCCC and Other International Legal Regimes: The Challenge of Systemic Integration*, CIEL and International Council on Human Rights Policy, Geneva

Copenhagen Economics and IPR Co (2009) *Are IPRs a Barrier to the Transfer of Climate Change Technology?*, Copenhagen Economics, Copenhagen

Correa, C. (2009) *Fostering the Development and Diffusion of Technologies for Climate Change: Lessons from the CGIAR Model*, ICTSD, Geneva

Corvagila, M. A. (2011) *South–South Technology Transfer Addressing Climate Change*, WTI (NCR Trade Regulation), Berne

de la Tour, A. et al., (2010) *Innovation and International Technology Transfer: The Case of China Photovoltaic Industry*, CERNA, Paris

Dechezleprêtre, A. et al., (2009) *Invention and Transfer of Climate Change Mitigation Technologies on a Global Scale: A Study Drawing on Patent Data*, CERNA, Paris

Dechezleprêtre, A. et al., (2010) *What Drives International Transfer of Climate Change Mitigation Technologies*, CERNA, Paris

Deere, C. (2008) *The Implementation Game: The TRIPS Agreement and the Global Politics of Intellectual Property Reform in Developing Countries*, Oxford University Press, Oxford

Drahos, P. (2009) 'The China–US relationship on climate change, intellectual property and CCs', *WIPO Journal*, vol 1, no 1, pp125–132

Fu, X. and Soete, L. (2010) (ed) *The Rise of Technological Power in the South*, Palgrave, New York, NY

Gerstetter, C. et al., (2010) 'Technology transfer in the international climate negotiations – the state of play and suggestions for the way forward', *CCLR*, vol 1, no 1, pp3–12

Hall, B. H. and Helmers, C. (2010) *The Role of Patent Protection in (Clean/Green) Technology Transfer 2010-046*, UNU-MERIT, Maastricht

Hall, B. H. and Helmers, C. (2011) *Innovation and Diffusion of Clean/Green Technology: Can Patent Commons Help?*, NBER, Cambridge, MA

Håščič, I. et al., (2010) *Climate Policy and Technological Innovation and Transfer*, OECD, Paris

Hoekman, B. and Javorcik, B. S. (ed) (2006) *Global Integration and Technology Transfer*, Palgrave/World Bank, New York, NY

ICC (International Chamber of Commerce) (2009) *Climate Change and Intellectual Property*, ICC, Paris

ICTSD (International Centre for Trade and Sustainable Development) (2011) *The Climate Technology Mechanism*, ICTSD, Geneva

Iliev, I. and Neuhoff, K. (2009) *Intellectual Property: Cross-Licensing, Patent Pools and Cooperative Standards as Channel for Climate Change Technology Cooperation*, Climate Strategies, Cambridge

IPCC (Intergovernmental Panel on Climate Change) (2000) *Methodological and Technological Issues in Technology Transfer: Summary for Policy Makers*, WMO/UNEP, Geneva

Johnson, D. K. N. and Lybecker, K. M. (2009) *Challenges to Technology Transfer*, Colorado College, Colorado Springs, CO

Juma, C. and Bell, B. J. (2009) 'Advanced biofuels and developing countries: Intellectual property scenarios and policy implications', in *The Biofuels Market: Current Situation and Alternative Scenarios*, United Nations, New York, NY

Kariyawasam, R. (2007) 'Technology transfer', in S. D. Anderman (eds) *The Interface between Intellectual Property Rights and Competition Policy*, Cambridge University Press, Cambridge, pp466–504

Krishnachar, N. (2006) *Impediments to International Transfer of Technology: A Developing Country Perspective*, NYU School of Law, New York, NY

Lane, E. L. (2011) *Clean Tech Intellectual Property*, Oxford University Press, New York, NY

Lee, B. et al., (2009) *Who Owns Our Low Carbon Future? Intellectual Property and Energy Technologies*, Chatham House, London

Levi, M. et al., (2010) *Energy Innovation, Council on Foreign Relations*, Washington, DC

Mallett, A., Ockwell, D., Pal, P., Kumar, A., Abbi, Y., Watson, J., Haum, R., MacKerron, G. and Sethi, G. (2009) *UK–India Collaborative Study: Barriers to the Transfer of Low Carbon Energy Technology: Phase II*, Report for UK Department for Energy and Climate Change (DECC), London, March

Maskus, K. E. (2009) 'Intellectual property and the transfer of green technologies: An essay on economic perspectives', *WIPO Journal*, vol 1, no 1, pp133–137

Maskus, K. (2010) 'Differentiated intellectual property regimes for environmental and climate technologies', in *OECD Environment Working Papers*, no 17, OECD, Paris

Maskus, K. E. and Okediji, R. L. (2010) *Intellectual Property Rights and International Technology Transfer To Address Climate Change*, ICTSD, Geneva

Matthews, D. (2010) *Patents in the Global Economy*, Intellectual Property Office, London

Morey, J. et al., (2011) *Moving Climate Innovation into the 21st Century*, Clean Energy Group, Montpelier, VT

Ockwell, D. et al., (2008) *UK–India Collaboration to Identify Barriers to the Transfer of Low Carbon Energy Technology*, SPRU/TERI/DFID, Sussex/New Delhi/London

Ockwell, D. G., Mallett, A., Haum, R. and Watson, J. (2010) 'Intellectual property rights and low carbon technology transfer: the two polarities of diffusion and development', *Global Environmental Change*, vol 20, pp729–738

Ockwell, D., Watson, J., MacKerron, G., Pal, P., Yamin, F., Vasudevan, N. and Mohanty, P. (2007) *UK–India Collaborative Study on the Transfer of Low Carbon Technology*, University of Sussex, SPRU (Science and Technology Policy Research), Institute of Development Studies and the Energy Resources Institute, Brighton, UK

OECD (Organisation for Economic Co-operation and Development) (2010) *Measuring Innovation: A New Perspective*, OECD, Paris

Parthasarathy, S. et al., (2010) *A Public Good? Geoengineering and Intellectual Property*, University of Michigan, STPP Program, Ann Arbor, MI

Petrusson, U. et al., (2010) 'Global technology markets: The role of open intellectual property platforms', *Review of Market Integration*, vol 2, no 2–3, pp333–392

Rattray, L. (2009) *How Open Source Development Can Resolve the North–South Intellectual Property Conflict in UNFCCC Negotiations: A Bipartisan Technology Transfer Pathway*, School of Public Policy, Georgia Institute of Technology, Athens

Reichman, J. et al., (2008) *Intellectual Property and Alternatives: Strategies for Green Innovation*, Chatham House/DFID, London

Rimmer, M. (2009) 'The road to Copenhagen: Intellectual property and climate change', *Journal of Intellectual Property Law and Practice*, vol 4, no 11, pp784–788

Romano, C. and Burleson, E. (2011) 'The Cancun Climate Conference', *ASIL Insights*, vol 15, no 41, un-paginated

Sarnoff, J. D. (2011) *The Patent System and Climate Change*, Social Science Research Network (SSRN)

Shepherd, J. (2007) 'The future of technology transfer under Multilateral Environmental Agreements', *ELR News & Analysis*, vol 37, pp10547–10561

Sommer, T. et al., (2010) 'Climate change and intellectual property rights', *Nordic Environmental Law Journal*, vol 2, Special Issue

Srinivas, K. R. (2009) *Climate Change, Technology Transfer and Intellectual Property Rights*, RIS, New Delhi

Thomson, R. and Webster, E. (2010) 'The role of intellectual property rights in climate change: The case of agriculture', *WIPO Journal*, vol 2, pp133–141

Ueno, T. (2010) *Technology Transfer to China to Address Climate Change Mitigation*, Resources for the Future, Washington, DC

UNCTAD (United Nations Conference on Trade and Development) (2001) *Compendium of International Arrangements on Transfer of Technology: Selected Instruments*, UN/UNCTAD, Geneva

UNDESA/UNIDO (United Nations Department of Economic and Social Affairs/United Nations Industrial Development Organization) (2010) *Technology Development and Transfer for Climate Change: A Survey of Activities by United Nations System Organizations*, UNDESA/UNIDO, New York/Vienna

UNEP, EPO and ICTSD (United Nations Environment Programme, European Patent Organization and International Centre for Trade and Sustainable Development) (2010) *Patents and Clean Energy: Bridging the Gap between Evidence and Policy*, UNEP, EPO and ICTSD, http://ictsd.org/i/publications/85887/

UNESCO (United Nations Educational, Scientific and Cultural Organization) (2010) *World Science Report 2010*, UNESCO, Paris

van Overwalle, G. (2010) 'Turning patent swords into shares', *Science*, vol 330, no 17, December, pp1630–1631

Veugelers, R. (2010) 'EU Climate Change Policy: Mobilizing innovation?', Paper presented at the PIIE-Brueghel Conference, Brueghel

Watson, J. et al., (2011) *UK–China Collaborative Study on Low-Carbon Technology Transfer*, SPRU, Sussex, UK

WIPO (World Intellectual Property Organization) (2011) *Transfer of Technology*, WIPO, Geneva

Wolek, A. (2011) 'Biotech biofuels: How patents may save biofuels and create empires', *Chicago-Kent Law Review*, vol 86, no 1, pp235–257

World Bank (2008) *Global Economic Prospects: Technology Diffusion in the Developing World*, World Bank, Washington, DC

PART IV

Assessing Existing International Policy Mechanisms

7
Energy Pathways in Low Carbon Development

The Need to Go beyond Technology Transfer

Rob Byrne, Adrian Smith, Jim Watson and David Ockwell

Introduction

The relationships between energy and development are complex, compounded by increasingly differentiated situations amongst developing countries and within them. Moreover, the manner in which energy services are realized has consequences for our health, environment, wealth and social relations. Two important issues currently preoccupying the realm of international development are enhancing energy access while simultaneously addressing climate change, which are brought together in the term low-carbon development.

A recurring theme in studies and policies for energy and development is the role that innovation can play in improving sustainable energy access. International climate change negotiations place an emphasis on low-carbon technology transfer, which perpetuates a long history of expectations about technology providing solutions to energy and development challenges. While these expectations are not entirely unfounded, history indicates that solving the many problems associated with the provision of energy services involves a more complex set of interdependent processes than 'straightforward' transfer of technology. And yet international discussions are intensifying (once again) around innovation in the form of technology transfer – discussions that frame the issue, we argue, in terms of financing the flow of low-carbon technological hardware to developing countries. This 'hardware and finance' framing of low-carbon development has resulted in a limited number of general purpose policy instruments – such as the Clean Development

Mechanism (CDM) – that tend to neglect important details of how technology can be 'transferred' successfully and sustainably. We recognize that the financing of low-carbon technological hardware is an essential element in any approach to low-carbon development but – with reference to the assumptions and record of the CDM – question whether such policy instruments will create sustainable pro-poor development pathways.

Our main argument is that the CDM is based on a flawed conception of technology that, in effect, understands it as merely hardware. Based on this conception, the solution to the challenge of low-carbon development is simply to finance the deployment of low-carbon hardware in developing countries. An alternative conception of technology, rooted in the literature on innovation, emphasizes its systemic quality. From this perspective, technology is based on knowledge, skills and experience – for its design, manufacture, use and development. Moreover, the requisite knowledge, skills and experience – or capabilities – are distributed among firms and other organizations in a particular context. Based on this conception, the solution to low-carbon development becomes much more complex. If hardware is to work sustainably in a particular context, then the relevant capabilities must be cultivated. If hardware is to be adapted to a context, then capabilities may need to be developed or enhanced. The creation of new hardware requires innovative capabilities. All of these processes are time and resource consuming. Furthermore, the CDM is designed primarily to identify the cheapest and most profitable opportunities for reducing greenhouse gas (GHG) emissions in developing countries. Consequently, it encourages investments in those technologies and contexts that can offer lower technical and business risks. This reinforces static comparative advantages, marginalizing poorer countries and communities and further entrenching their poverty.

In order to address our critique of policy instruments such as the CDM, we conclude, there is a need to broaden the framing of low-carbon development by placing a systemic view of technology at its heart. However, we contend that this is not necessarily enough to enable developing countries to achieve pro-poor sustainability goals, although it may help them to achieve more self-directed development. The pursuit of sustainability, we argue, will require broadening the frame to include the relationships between social and political dimensions and technology. A broad framing of the challenge will help us to understand the transformational possibilities afforded by new technologies, and to develop policy responses that can exploit these transformational opportunities.

The remainder of the chapter begins with a discussion of the key conceptual ideas that inform our analysis of the CDM. We then give a brief account of the evolution of the mechanism. The bulk of the chapter then follows, giving an outline of the finance challenge in low-carbon development and providing a closer examination of the CDM. The chapter concludes with a discussion of the CDM in terms of our conceptual perspective, and reflects on what fundamental reforms and research needs flow from our analysis.

Key concepts for the analysis

In this section, we give a brief discussion of our conceptual perspective, which we have organized in two broad categorizations. The first is a 'pathways' approach to sustainability and the second recounts some key ideas from innovation studies.

Development pathways

Understanding that societal services or functions are realized dynamically out of the interplay of various co-evolving complex systems (social, technological, environmental), we can conceive of any particular unfolding of these dynamics as a development pathway[1] (Leach et al., 2007). Looking back, we can analyse how that pathway materialized and how it was maintained. Looking forward, we can analyse the trajectory of the pathway, how it is likely to unfold and who will benefit or not. But we also need to understand that each of the complex systems themselves, and their combination, can be framed in different ways. Each framing informs – and is informed by – a narrative that interprets the world in a particular way, reflecting and reinforcing the perspective of the narrator. As understood here, a narrative is used to 'suggest and justify particular kinds of action, strategy and intervention' (Leach et al., 2010, p3) and so attempts to enrol actors and their resources into particular ways of achieving development goals. If this enrolment is successful, then a particular direction of development is privileged, the result of which is an unfolding pathway co-evolving contingently and uncertainly in the interplay between these privileging forces and the various complex systeme noted above.

Implicit in this description is the notion that multiple framings, narratives and pathways are possible. Different groups of actors will interpret the world in different ways, arising from their own experiences, situations, understandings, values and interests. Favouring certain framings over others, they will seek to promote narratives that would help to create their preferred development pathways. As Byrne et al., (2011, p9) argue:

> The more persuasive a [narrative] (and the more influential its advocates), the more in tune it is with powerful perspectives on reality; and the greater the interests it mobilises, then the more likely it is to enrol support for its strategies, and become institutionalised into the … system, thereby becoming part of the reproduction of the system.

However, as Leach et al., (2010, p3) argue: 'other narratives … may not become manifested in actual pathways of intervention and change, remaining marginalised'. Such narratives are likely those of the already marginalized and powerless. We can begin to see, then, how framing and narrative devices serve to maintain uneven distributions of power and resources in self-reinforcing development pathways. And this alerts us to the idea that we need to attend not only to the direction a pathway takes, but also to the diversity of pathways available (and possible, desirable, etc.), as well as the distribution of benefits and disbenefits that each pathway affords (Stirling, 2009).

But this is not to argue that dominant narratives and pathways are immune to influences from the margins. As evidenced in the literature on socio-technical transitions, dominant socio-technical practices come under pressure from external dynamics, and experience internal tensions between the many dimensions (social, cultural, political, technical) that constitute those practices (see, for example, Geels, 2002; Raven, 2005; Smith, 2007). Climate change, for example, is creating increasing pressure on the dominant fossil fuel-based development pathway. And the climate change narrative has enrolled increasing numbers of actors and their resources; spawned the United Nations Framework Convention on Climate Change (UNFCCC) and instruments of climate governance such as the Kyoto Protocol; promoted certain strategies such as investment in renewable energy technologies; and argued for interventions such as carbon pricing. Of course, the fossil fuel-based development pathway remains dominant; but it is clearly under mounting pressure and we could argue that its dominance is beginning to erode.

In trying to analyse how dominant practices come to be eroded, we can draw from the socio-technical transitions literature once more. Here we see that there are various ways in which marginal, experimental or sometimes radical socio-technical practices can come to influence mainstream practices and even thoroughly transform them over time (Geels and Schot, 2007). Technology can play a central role in such transformations by affording opportunities for entirely new practices that create demands for widespread institutional change (Deuten, 2003). But if we are to make use of these transformational possibilities to realize normative goals, such as pro-poor low-carbon development, then we need to be careful how we understand technology itself (Watson et al., 2011). Our argument here is that an inadequate conception of technology will likely produce – at best – inadequate technology policy, such as with many 'technology transfer' efforts. Worse, such policy could be ineffective or even counterproductive (Byrne et al., 2011). For instance, inadequately conceived low-carbon technology transfer to developing countries could see the failure of those technologies, resulting in pressure to turn to carbon-intensive technologies instead, so that development pathways unfold in high-carbon directions. For insights on the nature of technology, and its role in helping to realize pro-poor self-determined development pathways, we can examine the innovation studies literature.

Technology and innovation systems

An important insight in the literature is that technology is not simply hardware. Embedded in the hardware is a reflection of the knowledge required to create it; and knowledge and skills – sometimes referred to as the software – are needed to adopt, use and adapt it (Bell and Pavitt, 1993; Ockwell et al., 2010). Extending this idea, some authors demonstrate that hardware is also embedded with assumptions such as those that could be described as social or cultural (Pacey, 1983; Agarwal,

1986; Wynne, 1995). An essential characteristic of this 'software' is tacit knowledge – a fundamental aspect of knowledge and skills that is difficult or impossible to articulate but can be cultivated through practice (Polanyi, 1966). Combining these ideas, we begin to form the notion of socio-technology, echoing the language of socio-technical transitions thinking discussed above. Flowing from these ideas, and demonstrated in the literature, we see that technologies are created, adopted and adapted within a systemic environment. This idea has long been studied in regard to innovation systems, with particular attention to the linkages between firms and other actors, and the institutional setting of policies, laws, regulations and norms (see, for example, Katz, 1987; Kim et al., 1989; Bell, 1990, 1997, 2009; Freeman, 1992; Bell and Pavitt, 1993; Hobday, 1995a, 1995b; Radošević, 1999; Ockwell et al., 2008; Watson et al., 2011). But, more recently, the broader dimensions have received attention in the socio-technical transitions literature (see, for example, Rip and Kemp, 1998; Geels, 2002; Berkhout et al., 2004; Raven, 2005; Geels and Schot, 2007; Smith, 2007; Smith et al., 2010; Byrne, 2011).

One way to understand the significance of some of these ideas is depicted in Figure 7.1, especially in regard to innovation systems and the ways in which the knowledge and skills required for self-directed development can be accumulated. Based on Bell (1990), the diagram shows three types of possible technology flow (A, B and C) during transfer projects into a local innovation system. Flow 'A' includes hardware, as well as the engineering and managerial services that are required for implementing such transfer projects. Flows of type 'B' consist of information about production equipment – operating procedures, routines, etc. – and training in how to operate and maintain such hardware. Bell (1990, p77) describes these flows as 'paper-embodied technology' and 'people-embodied knowledge and expertise'. Both flows 'A' and 'B' add to or improve the production capacity of a firm or economy, but do little or nothing for developing the skills needed for generating new technology. Flows of type 'C', however, are those that help to create the capability to generate new technology, what Bell calls 'technological capacity', or what we could also call innovation capacity or capability, as Bell (2009) does in an updated discussion. And the existing technological capabilities in the local context can be described as absorptive capacity, defined originally by Cohen and Levinthal (1990, p128) as the ability of a firm to 'recognize the value of new information, assimilate it, and apply it to commercial ends'. However, it has also been used to demonstrate the impact of individual firms' absorptive capacity on the ability of clusters of firms to adopt and adapt new technologies (Giuliani and Bell, 2005), and – within the low-carbon context and of particular relevance to us here – to explain the ability of countries to achieve technological learning through the Clean Development Mechanism (Doranova, 2009).

The diagram in Figure 7.1 does not show explicitly the importance of the institutional environment, although the innovation literature does so, especially with regard to formal national and international policies. These can help to enhance existing industrial activity – to raise the level of capabilities to increase competitiveness, for example – but are also important for fostering new industrial activity that

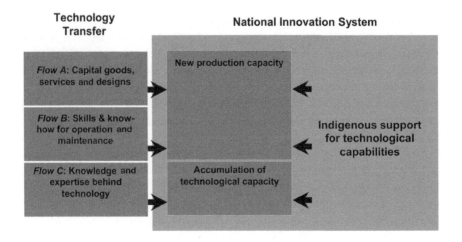

FIGURE 7.1 Technology transfer and indigenous innovation
Source: Adapted from Watson et al., 2011, p16, based on Bell, 1990

would otherwise not be pursued (see, for example, Cimoli et al., 2009). In the case of low-carbon technologies, this latter point is particularly relevant (Ockwell et al., 2010). Many of the current low-carbon alternatives are not yet competitive with carbon-intensive options; therefore market demand for many low-carbon technologies tends to be weak or marginal. But it is likely that we will need a range of low-carbon technologies, and the need is becoming increasingly urgent. In principle, appropriate policies could foster the improvement of low-carbon technologies, and the local capabilities and innovation systems that can sustain and develop them. The result could be – resonant with our comment on direction, diversity and distribution given above – a multiplicity of co-existing pathways, each appropriate to its context, promoting more equitable human development (Stirling, 2009).

As we will see later, the CDM is not helping to promote a multiplicity of pathways, and is doing little for diversity or equitable distribution. However, first, we will briefly describe the evolution of the CDM.

Evolution of the CDM

The CDM was created late in the process of the Kyoto Protocol negotiations following suggestions from Brazil for a Green Development Fund (an idea supported by the G77 and China but rejected by the industrialized countries) and the US for a market mechanism of emissions trading (Matsuo, 2003). Finally agreed within the protocol was a compromise between the two proposals that was expected to simultaneously promote sustainable development while assisting industrialized countries to meet their Kyoto commitments (Lecocq and Ambrosi, 2007, pp134–135). By investing in 'clean' projects in developing countries, from which Certified Emission Reductions (CERs) are then issued, industrializd-country firms can offset some of their own emissions where this is cheaper to do than invest in cleaner technologies in their own production processes (Hepburn, 2009). The CERs can then be traded in an Emission Trading Scheme (ETS), such as the EU ETS.

Although the CDM was hailed by both industrialized and developing countries as a 'win–win' agreement (Matsuo, 2003), it continues to attract criticism over a number of aspects. We will visit some of these criticisms later in this chapter. Here, we observe that the form the CDM has taken reflects longstanding and often opposing positions – in regard to technology and aid – of industrialized and developing countries (Byrne et al., 2011, pp61–66). Moreover, it reflects the still-dominant (neo-liberal) assumption that markets will deliver social benefits more efficiently than other forms of economic organization, and contributes an important element of what Newell and Paterson (2009, p81) call climate capitalism. Related to this economic assumption is, we can argue, an inadequate understanding of technology; one that sees technology as hardware and fails to understand the systemic nature of technology creation, adoption and development. A consequence of this narrow understanding is that 'clean development' is seen as a problem of financing the deployment of low-carbon or emissions-reducing hardware – a hardware-and-finance framing of the challenge of low-carbon development. The result is a pathway that privileges particular technologies and contexts over others, as we attempt to demonstrate below. Furthermore, although there are now many suggestions and debates about reform of the CDM, the framing itself of low-carbon development needs to be broadened if instruments of the climate governance regime are to contribute to achieving sustainability goals.

A hardware and finance framing of low-carbon development

We have argued that the dominant pathway for energy and development being pursued at the international level is the result of framing the challenge as one of financing the deployment of technological hardware. Technology certainly plays an important role in the provision of the energy services that underpin development processes. Historically, energy technologies have tended to be designed and manufactured in the industrialized economies. As a result, the developing countries have been preoccupied with gaining access to these technologies through development assistance. Since the early 1960s, several institutions have been created – multilateral and bilateral – to help developing countries in terms of science and technology, in general, and, since the 1980s, in energy technology, in particular. In the next section, we elaborate upon the CDM, which is one of the most successful of these institutions to date in deploying carbon finance. First, and accepting for now the hardware and financing frame, we examine the size of the problem and the record of trying to address the problem in terms of all sustainable energy finance mobilized.

The hardware-finance challenge and sustainable energy investments

Various estimates for the amount of finance necessary to meet the climate challenge in developing countries have been suggested. Stern (2006, p249), for example, gives

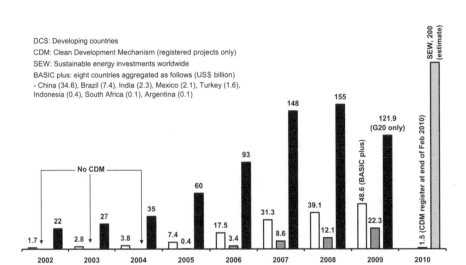

FIGURE 7.2 Estimated investments in low-carbon energy technologies (US$ billion), 2002 to 2010, comparing 'developing countries', the Clean Development Mechanism (CDM) and the world
Source: UNEP and NEF, 2009; UNEP Risø, 2010; and PCT, 2010

a range of US$350 billion to $400 billion per annum.[2] While investments in sustainable energies worldwide have been increasing rapidly during recent years, there is still a long way to go before they reach the requisite levels in developing countries. Figure 7.2 shows estimated world investments in sustainable energies (low-carbon technologies and energy efficiency) between 2002 and 2009, with a projection for worldwide investments in 2010 (PCT, 2010). These are disaggregated into an approximate developing-country group and registered CDM projects in order to compare the levels of finance with worldwide investments. The 'developing-country' group actually includes some of the richer countries[3] and therefore overstates the levels of finance to developing countries, and is likely to include the CDM (although it is not clear whether this would be registered projects only or include those further upstream in the CDM pipeline). However, the point here is to examine indicative figures rather than to perform statistical analysis, so the accuracy of the numbers is not critical to our needs.

Some disaggregation for 2008 of the developing country group is possible with the data given in UNEP and NEF (2009). This is shown in Figure 7.3 and clearly displays a heavy bias towards just three countries: China, Brazil and India.

This bias is reflected in the CDM; indeed, it is even more skewed. Figure 7.4 shows the accumulated investments made through the CDM since 2005, based on registered projects only. China has received over 70 per cent of the finance available so far, while India has been the next best beneficiary, with 13.5 per cent of the investments. Brazil has received 1.4 per cent, leaving the rest of the participating countries with 13.6 per cent among them.

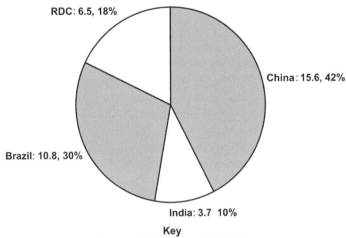

FIGURE 7.3 Sustainable energy investments in developing countries (US$ billion), 2008
Source: UNEP and NEF, 2009

Based on the figures given in UNEP and NEF (2009) for the developing-country group, the rate of growth of investments in low-carbon technologies in developing countries will need to be about 18 per cent per year in order to reach

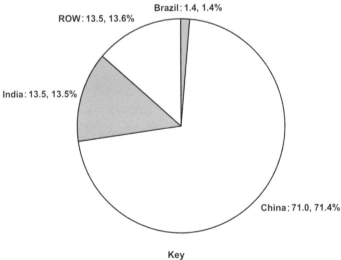

FIGURE 7.4 CDM-registered projects and accumulated investment value (US$ billion), as of end of May 2011
Source: UNEP Risø, 2011

estimated levels over US $300 billion by 2020. Questions about whether such growth in finance is achievable over an extended period and how investments are going to be administered (e.g. if they include a significant amount of multilateral aid and/or work through a CDM-like framework) are the subject of intense debate in international forums. There has been some movement in recent years, but there is still a long way to go.

This gives some indication of the finance challenge and the relative size of the CDM within it. We now use the rest of this section to examine the record of the CDM in more detail. The mechanism allows developers to finance projects in developing countries that are low-carbon and 'additional' (meaning that they would not have been financed without the CDM). The CO_2e emissions[4] saved are then converted into CERs that can be traded on the international market, such as in the EU ETS. Kyoto Protocol Annex I countries can use the CERs to help them meet their Kyoto targets while, in principle, non-Annex I countries receive investments that support their sustainable development. Although the CDM does not explicitly mandate 'technology transfer', it has become an important mechanism for investments in technological hardware in (some) developing countries.

The CDM pathway

Our argument here is that, in effect, the hardware and finance framing underpinning the CDM privileges certain pathways over others. Those privileged pathways are located in places with sufficient absorptive capacity to host low-carbon energy projects, and the projects themselves are limited to technologies that are already sufficiently developed for the profitable generation of CERs. These pathways do not do much for the development of low-carbon innovative capabilities in other settings, nor for the improvement of a wider portfolio of appropriate technologies. Those exceptional CDM projects that do manage to seed local innovative benefits underscore the weaknesses inherent in the more general pattern.

CDM-related pathways suffer from two interrelated difficulties. First, similarly to the hardware and finance framing, more generally, the CDM exploits static comparative advantages rather than building dynamic capabilities that can transform local contexts for development. Second, there remains a simplistic understanding of technology that has two effects: one, the focus turns to finance and hardware flows; and, two, learning and capabilities are under-supported in the mechanism. The two main arguments combine when we see that the participation of a few countries and use of just a few technologies means there is a lack of diversity from which to generate rich learning opportunities. Without such learning, it is likely that the co-benefits claimed of the CDM will not materialize.

As a market mechanism, the CDM creates incentives for firms to invest in low-carbon projects that are least-cost and/or will produce the highest returns through the sale of emissions credits. As such, 'mature' technologies, large-scale projects and low-risk investment environments tend to be the most attractive. Looking at the state of investments to date, we have already seen the bias towards

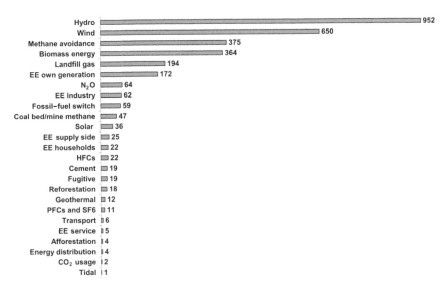

FIGURE 7.5 Number of registered CDM projects as of end of May 2011, disaggregated by project type (3145 total registered projects)
Source: UNEP Risø, 2011

a few countries – mainly China; but there is also a bias towards a small range of technologies. Figure 7.5 shows the number of registered projects by type, demonstrating clearly that the majority of projects are implemented using a small number of different technologies.

When examined in percentage terms, we can see that over 80 per cent of the registered CDM projects are implemented using just five types of technology, only one of which could be considered a *new* renewable energy technology – wind – although mature relative to other new renewables. An alternative explanation for this preponderance of just a few technologies resulting from their maturity could be that the methodologies needed for other kinds of projects have not been available until more recently. If this were the case, then it might be argued that the other technologies represent only a small share of the total but could grow in importance. However, Figures 7.6 and 7.7 show the numbers of projects at validation since 2004 (Figure 7.6 shows the top nine project types and Figure 7.7 the rest). No obvious pattern suggesting any methodology bottleneck is apparent in these figures. Therefore, although we might contend that some of the explanation for the clear bias in chosen technology types could be due to unavailable methodologies, the main reason appears more likely to be one of technology-maturity.

In general, this preference for just a few project types is reflected in the total capacity that might be installed through the CDM. Figure 7.8 shows the potential installed capacity of all projects (for which power is a meaningful measure) disaggregated by type. These are not only registered projects, but include those at validation and those awaiting registration. Finally, there is a dominance of large projects

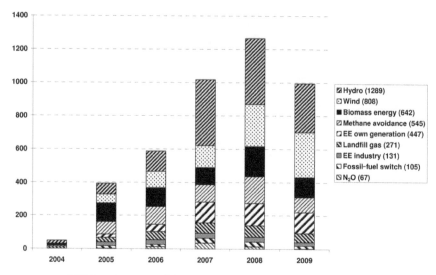

FIGURE 7.6 CDM projects at validation over time, disaggregated by project type (top nine types)
Source: UNEP Risø, 2011

in the CDM portfolio. Figure 7.9 shows average project size by type, and average size of all projects.

It should be clear from this brief examination of the CDM portfolio that, while we need to be careful not to over-interpret project distributions, there are strong biases that result directly from the desire among the industrialized nations for

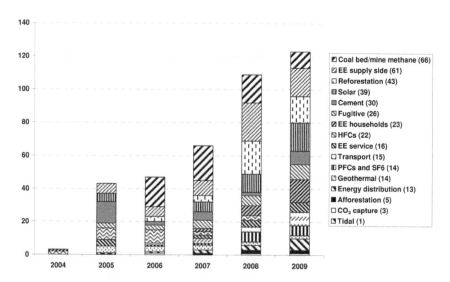

FIGURE 7.7 CDM projects at validation over time, disaggregated by project type (other 16 types)
Source: UNEP Risø, 2011

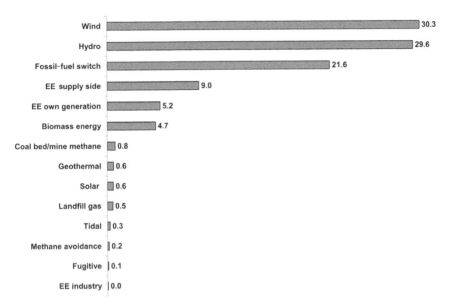

FIGURE 7.8 Total CDM project capacity by type (GW)

Note: Total capacity refers to all CDM projects for which power capacity is meaningful.
These include projects at validation, those awaiting registration and those already registered.
Ten project types are not classifiable using power as the measure: N_2O, HFCs, cement, forests,
EE households, PFCs and SF6, EE service, transport, energy distribution, and CO_2 capture.
Source: UNEP Risø, 2011

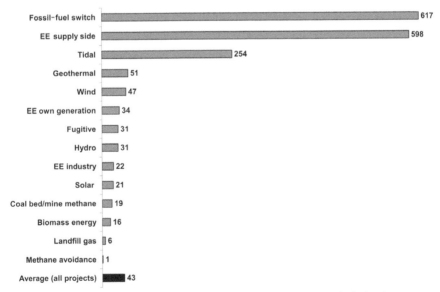

FIGURE 7.9 Average CDM project size (MW): All projects are included, where
power is used as a measure of size (those at validation, awaiting registration and
those already registered)
Source: UNEP Risø, 2011

economically efficient carbon reductions – the main determinant for the form the mechanism has taken. The significance of these biases is that diversity is being constrained – in terms of contexts where low-carbon technologies are being deployed, and in terms of the kinds of technologies being developed. In the short term, there are benefits of both a public and private nature. Global public goods benefits derive from (cheaper) climate change mitigation, but the private gains are likely to be skewed in favour of industrialized-country firms. In other words, the CDM reinforces static comparative advantages. It is not transforming local contexts in a way that makes a broadening geography of locations attractive for low-carbon investment.

Consequently, the least developed countries risk being marginalized. At worse, they may even be left with little option but to establish carbon-intensive development pathways. In the long term, there may be little absorptive capacity for low-carbon technologies in these countries. As the need to mitigate climate change becomes increasingly urgent, low-carbon technologies may be imposed upon the least-developed countries, undermining the hard-won development gains and good development practice of recent years (e.g. participatory development practice, ownership and governance). Moreover, with low absorptive capacity, there is a high likelihood that low-carbon technologies will fail in the least-developed countries, undermining climate change mitigation. This is where the first critique is interrelated with our second critique of the CDM, an argument to which we now turn.

A second critique of the CDM concerns the extent to which it fosters learning and the accumulation of technological capabilities, from which development can be enhanced. The critique stems from the observation that the current form of the CDM is fundamentally influenced by the narrow understanding of technology as hardware, although it accepts the need for *some* supporting 'software' (mainly, operation and maintenance skills). However, that supporting software, it is assumed, can be transferred with relative ease along with the hardware. This narrow view of technology partially explains the dominance of finance in international negotiations, although, as Young (2002) discusses in relation to the birth of the Global Environment Facility (GEF),[5] it is also about maximizing aid flows. Of course, finance is important – without it there would be no acquisition of hardware. But just buying equipment is not necessarily enough to realize 'technology transfer' in any deep sense of the term, and is far from the notion of low carbon innovation, let alone any possibilities for socio-technical transformation. The literature on technological capabilities shows that the degree to which absorptive capacity exists in the 'recipient' country (or sector) is an important determinant of the success of 'transfer'. If absorptive capacity is low, then it is difficult to adopt (and adapt, develop, design) new technological hardware, as well as to create the skills, knowledge, organizational changes, and institutional arrangements and linkages necessary to facilitate its sustainability. Even if absorptive capacity is high, it takes effort to adopt a *new* technology – that is, processes of learning must take place and these need to be resourced. In a context of low absorptive capacity, learning is much more difficult and requires additional support (e.g., longer-term training, subsidies to lower risk, complementary policies, and so on). None of this is included in the present form of the CDM, and few studies of the mechanism appear even to recognize the problem. Indeed, many studies of the CDM seem to be exclusively concerned with how much 'technology' is being 'transferred' from industrialized (or foreign) to developing countries, paying little critical attention to what this really means. An examination of the methodologies of a handful of studies reveals this preoccupation with technology *transactions* from which inferences are made about 'technology transfer' (e.g. de Coninck et al., 2007; Seres et al., 2007; Dechezleprêtre et al., 2008; Schneider et al., 2008; Seres and Haites, 2008). These studies analyse project design documents (PDDs) – written before a project is implemented and meant to persuade others that a project is worth supporting – for mention of technology transfer. Clearly, it is difficult to draw any robust conclusions about what kind of technology transfer is actually happening without investigating the results of projects rather than *ex ante* PDDs.

Two recent studies, however, use methodologies that are more sensitive to actual development outcomes. Doranova (2009) extends well beyond a review of PDDs in the CDM to include a survey of project-implementing companies, and makes use of science and technology indicators of host countries to incorporate absorptive capacity within the analysis. The research analyses projects registered by the end of January 2007 (497 projects in 41 countries). The survey, which attempts to capture progress on technological capacity-building, is focused on companies in four countries,[6] giving a final sample size of 104 firms. The study finds that 52 per cent of projects use locally sourced technology and expertise, 19 per cent use foreign sources and 22 per cent use a combination; and foreign participation tends to be in larger projects (perhaps because of higher returns in terms of carbon credits). It also finds that local capabilities are important for absorbing technological hardware when it is sourced from elsewhere. Once technological hardware has been acquired, there are processes of learning-by-doing that depend for their success upon this absorptive capacity. Learning, however, is highest in operational capabilities, is lower for process improvement capabilities and is lower still for the more advanced design and development capabilities (Doranova, 2009, pp40, 142). One other finding concerns the role of national policy in relation to technological learning. Across all three types of technological capability (operational, process improvement and innovative), the analysis finds that national policy has some positive statistical significance in determining technological capacity-building (Doranova, 2009, p130).

Disch (2010) concentrates on the so-called 'development dividend' claimed for the CDM, using PDDs supplemented with primary data from the websites of Designated National Authorities (DNAs) administering the CDM process. These data are then analysed using 15 indicators that are intended to describe the environmental, economic and social dimensions of sustainable development. As a consequence, and considering the large number of criteria, the study is focused on just six countries[7] (although it investigates 122 PDDs). Perhaps the most important conclusion it draws is that the main co-benefit of the CDM is job creation. In general, other benefits are usually weak or non-existent. The exceptional country in this regard is Peru, which has an unusual project assessment procedure that is highly proactive – that is, the Peruvian DNA engages local stakeholders at the design stage of a project and makes on-site visits during the assessment phase, rather than confine itself to a desk-based study only. The result appears to be a much higher quality CDM project portfolio than in other countries, with more co-benefits than those found elsewhere.

The upshot of all this is that we know very little about the development outcomes from technology transfer in the CDM, although there is an indication that some does occur and a little is relatively deep. What is clear is that, for many, 'technology transfer' still means the acquisition of hardware and how much it costs. In the case of the CDM, the priority is on economically efficient emissions reductions, with an assumption – or simply a hope – that transfer of technology will take place. As a result, we have many studies that measure variables such as number and type of projects, country of origin of technological hardware, and CERs issued. Very few studies assess the more qualitative aspects of the CDM, despite claims that such co-benefits will materialize. As with a lot of the international debates, the CDM takes a very narrow framing of the transfer of technology and, as a result, has no system that encourages or fosters technological capability-building and self-reinforcing low-carbon development processes. Consequently, any low-carbon development that does occur is more the result of good fortune than of framing or strategy.

Conclusions

The argument we have attempted to substantiate in this chapter is that the CDM is promoting a particular low-carbon development pathway that privileges a small number of contexts and technologies. As a result, it is unlikely that the mechanism will contribute meaningfully to development goals such as improving energy access amongst the world's poorest people and industrialization in the poorer countries, or to achieving widespread sustainability in the developing world.

One important reason for this pessimistic outlook, we have argued, is that the design of the CDM reflects a flawed conception that technology is simply hardware. In contrast, the conception adopted in this chapter is one in which technology has a systemic quality. Technological hardware reflects the knowledge and skills necessary for its design and manufacture; and knowledge and skills are necessary for the successful adoption of technology. Furthermore, the ease with which technology can be absorbed into any particular context depends upon the existing technological capabilities in that context: knowledge, skills and experience available in local firms, organizational linkages and the institutional environment – or, in other words, its innovation system. These technological capabilities can be improved and enhanced over time, increasing the possibilities for adapting and creating technologies that can support self-directed development. But this takes strategic focus – such as directed or encouraged by national and international policy environments – and the targeting of adequate resources. At present, the CDM directs investments to the cheapest and/or most profitable emissions reduction projects; it is only a weak (and partial) *development* mechanism.

If the CDM, or similar instrument of the climate regime, is to be reformed in order to address its many criticisms, then there is a need to broaden the framing of low-carbon development beyond its current form of financing the deployment of low-carbon hardware. A framing that includes a systemic conception of technology at its heart would be a useful beginning. It would help us to understand in what ways 'technology transfer' can facilitate the accumulation of low-carbon technological capabilities that could enable self-directed development in poor countries. If that conception were to understand how broader social and political dimensions relate to technology, then our frame would enable us to examine how wide-scale and long-term transformation could be encouraged. Such a transformational approach might enable poor countries to achieve not only self-directed development, but also be better suited to the achievement of sustainability goals.

But it should be clear from our discussion in this chapter that there is a need for research into the impacts of the CDM in terms of co-benefits – research that does not rely exclusively on the *ex-ante* information available in project design documents. It would also be useful to compare other development efforts to the actual record of the CDM, and especially those in the poorer developing countries. Candidates here could include the work of the GEF, particularly those projects implemented through the United Nations Development Programme (UNDP) that incorporate forms of capacity-building alongside the financing of hardware. But there have been many other efforts at technology dissemination from which to learn, not least the efforts of some developing countries independently of donor-driven agendas. The need for this kind of research becomes more urgent when we consider the intention of donors such as infoDev and the UK Department for International Development to trial Climate Innovation Centres, and the negotiations concerning the UNFCCC's technology mechanism that seek to implement a Climate Technology Centre and Network. And, finally, this kind of research could help us to develop more robust theoretical approaches to the role of technology in development – approaches that can themselves inform the creation and reform of the instruments and mechanisms of the climate regime in order to realize pro-poor low-carbon development pathways to sustainability in developing countries.

Notes

1   The pathways approach referred to here is being developed within the STEPS Centre (Social, Technological and Environmental Pathways to Sustainability) at the University of Sussex in Brighton, UK. For more on the pathways approach and the work of the STEPS Centre, see www.steps-centre.org/.

2   Stern's (2006) estimate assumes the cost of stabilising emissions between 500ppm and 550ppm CO_2e (450ppm CO_2) to be 1 per cent of global gross domestic product (GDP), which came to the range of US \$350 billion to \$400 billion at the time.

3   The figures are from the UNEP and NEF (2009) report, in which the world is divided into the following regions: Europe, North America, South America, Asia and Oceania, and the Middle East and Africa. The 'developing countries' shown in the charts are comprised of these regions excluding Europe and North America.

4   CO_2e emissions (carbon dioxide equivalent emissions) include all greenhouse gases under one measure by referring to the equivalent carbon dioxide impact each gas has upon the climate.
5   Young (2002) describes how, during the 1992 Rio Earth Summit, there were contentious negotiations between the industrialized and developing countries over the mechanism for financing technology transfer through the GEF and governance of the fund. The industrialized countries were able to get things much their preferred way and, in any case, as Young (2002, p67) observes, 'the main priorities of most Southern governments at Rio were reduced to simply maximising aid flows and technology transfer as far as possible'.
6   The countries selected are Brazil, China, India and Mexico, accounting for over 70 per cent of projects implemented and generating an initial sample size of 361 companies. The response rate of 28.8 per cent results in a final sample size of 104 (Doranova, 2009, p23).
7   These are China, India, Brazil, Peru, Malaysia and South Africa (Disch, 2010, p53).

References

Agarwal, B. (1986) *Cold Hearths and Barren Slopes: The Woodfuel Crisis in the Third World*, Zed Books, London

Bell, M. (1990) *Continuing Industrialisation, Climate Change and International Technology Transfer*, Report prepared in Collaboration with the Resource Policy Group, Oslo and Science Policy Research Unit, University of Sussex, Brighton, UK, December

Bell, M. (1997) 'Technology transfer to transition countries: Are there lessons from the experience of the post-war industrializing countries?', in D. Dyker (ed) *The Technology of Transition: Science and Technology Policies for Transition Countries*, Central European University Press, Budapest

Bell, M. (2009) 'Innovation capabilities and directions of development', *STEPS Working Paper* 33, STEPS Centre, Brighton, UK

Bell, M. and Pavitt, K. (1993) 'Technological accumulation and industrial growth: Contrasts between developed and developing countries', *Industrial and Corporate Change*, vol 2 no 2, pp157–210

Berkhout, F., Smith, A. and Stirling, A. (2004) 'Socio-technological regimes and transition contexts', in B. Elzen, F. Geels and K. Green (eds) *System Innovation and the Transition to Sustainability: Theory, Evidence and Policy*, Edward Elgar, Cheltenham, UK

Byrne, R. (2011) *Learning Drivers: Rural Electrification Regime Building in Kenya and Tanzania*, Ph.D. thesis, University of Sussex, Brighton, UK

Byrne, R., Smith, A., Watson, J. and Ockwell, D. (2011) 'Energy pathways in low-carbon development: From technology transfer to socio-technical transformation', *STEPS Working Paper* 46, STEPS Centre, Brighton, UK

Cimoli, M., Dosi, G. and Stiglitz, J. (eds) (2009) *Industrial Policy and Development: The Political Economy of Capabilities Accumulation*, Oxford University Press, New York, NY

Cohen, W. and Levinthal, D. (1990) 'Absorptive capacity: A new perspective on learning and innovation', *Administrative Science Quarterly*, vol 35, pp128–152

de Coninck, H., Haake, F. and van der Linden, N. (2007) 'Technology transfer in the Clean Development Mechanism', ECN-E-07-009, Energy Research Centre of The Netherlands, The Netherlands

Dechezleprêtre, A., Glachant, M. and Ménièrea, Y. (2008) 'The Clean Development Mechanism and the international diffusion of technologies: An empirical study', *Energy Policy*, vol 36, pp1273–1283

Deuten, J. (2003) *Cosmopolitanising Technologies: A Study of Four Emerging Technological Regimes*, Ph.D. thesis, University of Twente, Enschede, The Netherlands

Disch, D. (2010) 'A comparative analysis of the "development dividend" of Clean Development Mechanism projects in six host countries', *Climate and Development*, vol 2, pp50–64

Doranova, A. (2009) *Technology Transfer and Learning under the Kyoto Regime: Exploring the Technological Impact of CDM Projects in Developing Countries*, Ph.D. thesis, UNU-MERIT, Maastricht University, Maastricht, The Netherlands

Freeman, C. (1992) *The Economics of Hope*, Pinter Publishers, London and New York

Geels, F. (2002) 'Technological transitions as evolutionary reconfiguration processes: A multi-level perspective and a case-study', *Research Policy*, vol 31, pp1257–1274

Geels, F. and Schot, J. (2007) 'Typology of sociotechnical transition pathways', *Research Policy*, vol 36, no 3, pp399–417

Giuliani, E. and Bell, M. (2005) 'The micro-determinants of meso-level learning and innovation: Evidence from a Chilean wine cluster', *Research Policy*, vol 34, pp47–68

Hepburn, C. (2009) 'International carbon finance and the Clean Development Mechanism', *Smith School Working Paper Series*, Climates of Change: Sustainability Challenges for Enterprise, Smith School of Enterprise and the Environment, University of Oxford, Oxford, September

Hobday, M. (1995a) 'East Asian latecomer firms: Learning the technology of electronics', *World Development*, vol 23, no 7, pp1171–1193

Hobday, M. (1995b) *Innovation in East Asia: The Challenge to Japan*, Edward Elgar, Aldershot, UK

Katz, J. (1987) *Technology Generation in Latin American Manufacturing Industries*, Macmillan, London

Kim, Y., Kim, L. and Lee, J. (1989) 'Innovation strategy of local pharmaceutical firms in Korea: A multivariate analysis', *Technology Analysis and Strategic Management*, vol 1, no 1, pp360–375

Leach, M., Scoones, I. and Stirling, A. (2007) 'Pathways to sustainability: An overview of the STEPS Centre approach', *STEPS Approach Paper*, STEPS Centre, Brighton, UK

Leach, M., Scoones, I. and Stirling, A. (2010) 'Governing epidemics in an age of complexity: Narratives, politics and pathways to sustainability', *Global Environmental Change*, vol 20, pp369–377

Lecocq, F. and Ambrosi, P. (2007) 'The Clean Development Mechanism: History, status, and prospects', *Review of Environmental Economics and Policy*, vol 1, no 1, pp134–151

Matsuo, N. (2003) 'CDM in the Kyoto Negotiations: How CDM has worked as a bridge between developed and developing worlds?', *Mitigation and Adaptation Strategies for Global Change*, vol 8, pp191–200

Newell, P. and Paterson, M. (2009) 'The politics of the carbon economy', in M. Boykoff (ed) *The Politics Of Climate Change: A Survey*, Routledge, London and New York

Ockwell, D., Watson, J., MacKerron, G., Pal, P. and Yamin, F. (2008) 'Key policy considerations for facilitating low carbon technology transfer to developing countries', *Energy Policy*, vol 36, pp4104–4115

Ockwell, D., Haum, R., Mallett, A. and Watson, J. (2010) 'Intellectual property rights and low carbon technology transfer: Conflicting discourses of diffusion and development', *Global Environmental Change*, vol 20, no 4, pp729–738

Pacey, A. (1983) *The Culture of Technology*, Basil Blackwell, Oxford

PCT (The Pew Charitable Trusts) (2010) *Who's Winning the Clean Energy Race? Growth, Competition and Opportunity in the World's Largest Economies*, The Pew Charitable Trusts, Washington, DC, and Philadelphia, PA

Polanyi, M. (1966) *The Tacit Dimension*, Routledge and Kegan Paul, London

Radošević, S. (1999) *International Technology Transfer and Catch-Up in Economic Development*, Edward Elgar, Cheltenham, UK

Raven, R. (2005) *Strategic Niche Management for Biomass: A Comparative Study on the Experimental Introduction of Bioenergy Technologies in The Netherlands and Denmark*, Ph.D. thesis, Technische Universiteit Eindhoven, Eindhoven, The Netherlands

Rip, A. and Kemp, R. (1998) 'Technological change', in S. Rayner and E. Malone (eds) *Human Choices and Climate Change vol. 2: Resources and Technology*, Battelle, Columbus, OH

Schneider, M., Holzer, A. and Hoffmann, V. (2008) 'Understanding the CDM's contribution to technology transfer', *Energy Policy*, vol 36, pp2930–2938

Seres, S. and Haites, E. (2008) *Analysis of Technology Transfer in CDM Projects*, Report Prepared for the UNFCCC Registration and Issuance Unit, CDM/SDM, December

Seres, S., Haites, E. and Murphy, K. (2007) *Analysis of Technology Transfer in CDM Projects*, Final Report Prepared for the UNFCCC Registration and Issuance Unit, CDM/SDM, December

Smith, A. (2007) 'Translating sustainabilities between green niches and socio-technical regimes', *Technology Analysis and Strategic Management*, vol 19, no 4, pp427–450

Smith, A., Voß, J. and Grin, J. (2010) 'Innovation studies and sustainability transitions: The allure of the multi-level perspective and its challenges', *Research Policy*, vol 39, pp435–448

Stern, N. (2006) *The Economics of Climate Change*, Draft report posted online, HM Treasury and Cabinet Office, UK, www.hm-treasury.gov.uk/sternreview_index.htm, accessed 5 November 2009

Stirling, A. (2009) 'Direction, distribution and diversity! Pluralising progress in innovation, sustainability and development', *STEPS Working Paper* 32, STEPS Centre, Brighton, UK

UNEP and NEF (United Nations Environment Programme and New Energy Finance) (2009) *Global Trends in Sustainable Energy Investment 2009*, UNEP and NEF

UNEP Risø (2010) CDM pipeline website, www.cdmpipeline.org/, accessed March 2010

UNEP Risø (2011) CDM pipeline website, www.cdmpipeline.org/, accessed June 2011

Watson, J., Byrne, R., Mallett, A., Stua, M., Ockwell, D., Xiliang, Z., Da, Z., Tianhou, Z., Xiaofeng, Z. and Xunmin, O. (2011) *UK–China Collaborative Study on Low Carbon Technology Transfer*, Final report for UK Department of Energy and Climate Change, SPRU, University of Sussex, Brighton, and Laboratory for Low Carbon Energy, Tsinghua University, Beijing, April

Wynne, B. (1995) 'Technology assessment and reflexive social learning: Observations from the risk field', in A. Rip, T. Misa and J. Schot (eds) *Managing Technology in Society: The approach of Constructive Technology Assessment*, Pinter, London

Young, Z. (2002) *A New Green Order? The World Bank and the Politics of the Global Environment Facility*, Pluto Press, London

8

LOW-CARBON TECHNOLOGY TRANSFER UNDER THE CLIMATE CHANGE CONVENTION

Evolution of Multilateral Technology Support

Anne-Marie Verbeken

Background

A surge of international activity related to low-carbon technology development and transfer ensued the Bali Climate Conference,[1] held in 2007. For the first time since the creation of the United Nations Framework Convention on Climate Change (UNFCCC), technology transfer became a key issue in the negotiations in Bali, and led to the agreement on the establishment of a Technology Mechanism[2] at the Cancun Climate Conference in 2010. Following the Bali Conference, the Global Environment Facility (GEF), as the financial mechanism of the UNFCCC, was asked to elaborate a programme to 'scale up the level of investment in the transfer of environmentally sound technologies', which resulted in a US$50 million programme,[3] funding amongst other activities a number of pilot technology transfer projects in support of mitigation and adaptation (GEF, 2008a). Outside the UNFCCC, international and regional organizations, including the World Bank and the Asian Development Bank, launched new funds and stepped up their climate change technology programmes and lending.

While technology undoubtedly has a fundamental role in reducing CO_2 emissions, until the Bali Conference, it was a largely neglected and least implemented aspect of the UNFCCC (CIEL and South Centre, 2008). Embedded in several articles of the UNFCCC, the issue of technology transfer ended up being relegated to the Subsidiary Body for Scientific and Technological Advice (SBSTA) and its now-terminated Expert Group on Technology Transfer (the EGTT),[4] which mostly served as a fact-finding mechanism (CIEL and South Centre, 2008). The lack of concrete proposals and demand from developing countries for low-carbon technologies also contributed to the lack of action beyond a technology needs assessment programme and information-sharing activities.

With the adoption of the GEF's Poznan Strategic Programme and the creation of the new Technology Mechanism, for the first time activities with an explicit technology development and transfer objective[5] are being programmed and financed[6] under the UNFCCC. It is unclear at this point in time what the role of the GEF will be in implementing the Technology Mechanism, which is set to become operational in 2012, and which organization will host the Climate Technology Centre and Network (CTCN) component.[7] However, the GEF has profiled itself as a key player in multilateral technology transfer mechanisms and as a repository of extensive experience with the transfer of technologies. Since the Bali Conference, the GEF has published material describing its evolving strategies and policies with respect to the transfer of climate change mitigation and adaptation technologies, lessons learned and project examples of its experience (GEF, 2008b, 2010a). The GEF has also progressed some of the key elements of the new Technology Mechanism – for example, through the funding of an Asian Development Bank and United Nations Environment Programme (UNEP)-implemented project, which will establish a 'Pilot Asia-Pacific Climate Technology Network and Finance Center'.[8]

Despite the new focus on implementing the technology transfer objectives in the UNFCCC and GEF, and the GEF's own characterization as having played a catalytic role in supporting the transfer of environmentally sound technologies (ESTs), there is a surprising lack of substantive and reflective narrative premised on empirical knowledge and analysis of the issue by the GEF and the implementing agencies.[9] Nor have the GEF and its implementing agencies attempted to assess the impact of different approaches of past technology support upon technology transfer, or determined what works and what doesn't beyond a broad statement that technology is primarily transferred through markets, requires a long-term engagement and a comprehensive approach incorporating capacity-building at all levels (GEF, 2008b, 2010a). These broad conclusions equally apply to the goal and concept of market transformation and do not enlighten us on what may be required to design a technology transfer programme. The only material that provides an objective and in-depth assessment of impact and lessons learned from GEF projects are individual project and programme evaluations, and reviews by the Scientific and Technical Advisory Panel (STAP) of the GEF. More could be done to mine these evaluations and reviews and learn from the unique source that more than 17 years of GEF project experience offers, particularly since GEF projects undergo extensive monitoring and independent evaluation, with the view to inform future programmes and projects.

The aim of this chapter is to critically review the evolution of the technology transfer narrative of the financial mechanism of the Climate Change Convention, to highlight some experiences with technology support, and to raise some of the issues that need to be addressed. Four case studies of technologies that have been supported by GEF projects are presented from a technology transfer perspective, applying a conceptual framework that distinguishes different types of technology transfer support. They illustrate some of the challenges and also opportunities for future climate change mitigation technology transfer projects funded by multilateral mechanisms. This chapter does not aim to draw broad conclusions on technology transfer support, as it is only presenting a narrow window on this complex issue.

Climate change mitigation technology support by multilateral mechanisms

Intergovernmental organizations (IGOs) have supported a large number of activities assisting non-Annex I countries in the adoption and diffusion of climate change mitigation technologies, and, to a lesser degree, adaptation technologies, since the establishment of the UNFCCC.[10] In the wide range of projects and programmes implemented by IGOs, the primary objective has rarely been the transfer of technologies as such. It was not until the adoption of the *Bali Action Plan* in December 2007 that programmes with an explicit focus on technology transfer were created. However, technology-related support was a component of many IGO projects, a significant number of which received grant funding from the GEF.

The GEF Trust Fund[11] is the largest single source of grant funding for climate change projects in developing countries, amounting to US$2.7 billion in March 2010, which leveraged an estimated US$17.2 billion in co-financing (GEF, 2010a). Approximately US$250 million is allocated annually by the GEF to climate change activities (GEF, 2010a). Another important source of climate change mitigation technology finance is the Clean Development Mechanism (CDM). Activities funded from sources other than the GEF and the CDM, prior to the recent creation of the Clean Technology Fund by the World Bank and of dedicated climate change funding windows in other development banks, comprised a broad spectrum of often smaller-scale projects and a significant number of activities based on partnerships and networks, with varying degrees of technology content. Programmes and funds created following the Bali Conference include the World Bank's Climate Investment Funds (CIFs) and the GEF's Poznan Strategic Programme. The Climate Technology Programme is another recently launched initiative, involving a World Bank partnership with the United Nations Industrial Development Organization (UNIDO), which is piloting 'climate innovation centres' to accelerate the development and deployment of low-carbon technologies.

The portfolio of GEF-funded mitigation projects that has built up since the mid-1990s has undergone an evolution as the approach to supporting renewable energy and energy efficiency technologies changed over time. During the mid-1990s, many projects focused on the demonstration of renewable energy and energy efficiency technologies. In particular, off-grid rural photovoltaic systems (PV) received extensive support (GEF, 2004). Due to their low impact upon reducing GHG emissions, and their failure to lead to sustained demand for PV systems, support for this type of project was greatly reduced and replaced by a more successful community-based approach to rural renewable energy systems, under the GEF-funded United Nations Development Programme (UNDP) Small Grants Programme (SGP) and by market transformation projects, such as the Photovoltaic Market Transformation Initiative (PVMTI), a much publicized project covering India, Kenya and Morocco, which promotes the commercialization of PV technology by demonstrating business models that can be financed on a commercial basis (IFC, 1998). Rural PV projects are really about energy access, while the GEF focus is on CO_2 reductions (GEF, 2004). Instead, 'renewable energy for productive purposes and other renewable energy sources such as wind and biomass were given more attention' (GEF, 2004, p77). Another development was that more energy efficiency projects incorporated financing mechanisms and energy service company (ESCO) development (GEF, 2004).

By the late 1990s, the GEF's approach had shifted towards barrier removal projects for renewable energy and energy efficiency, and more business and market-oriented interventions (GEF, 2004). The overarching goal of climate change mitigation projects in the GEF became market development and transformation, achieved through barrier removal (GEF, 2005). A variety of measures were used, ranging from the introduction of standards and certification schemes, awareness-raising, capacity-building, the creation of ESCOs and the promotion of business models, to providing tailor-made financial products for investments in energy efficiency and renewable energy (GEF, 2004). Another development was the increased effort of financial IGOs in mobilizing and enhancing the capacity of local financial markets to support energy efficiency measures and renewable energy technologies (GEF, 2005).

The market transformation approach was much more successful in the energy efficiency portfolio than for renewable energy technologies, where cost and other barriers remained too high. As a result of allocating resources in areas where impact was feasible and at relatively low cost, the share of energy efficiency projects went up and the number of renewable energy projects declined. By June 2009, the share of GEF-4 energy efficiency projects was 77 per cent of the total mitigation funding, compared to 13 per cent for renewable energy (GEF, 2010b, p78). Despite market transformation becoming a central theme for the GEF, the climate change portfolio remained quite diverse until GEF-4, when the original operational programmes were retired and replaced by new strategic priorities that channelled resources into fewer areas.[12] The GEF initially also pursued the adoption of emerging, near-commercial low-GHG technologies, including fuel cell technologies and concentrated solar power (CSP); but as these projects proved to be difficult and were considered a failure, the programme was virtually terminated[13] at the start of GEF-4 in 2006, and it was not until GEF-5 in 2010 that the support of 'innovative low-carbon technologies' was reintroduced as a fully fledged strategic objective.

During GEF-4, the shift away from technology-specific support continued with the adoption of predominantly technology-neutral strategic objectives, and an increased focus on market transformation as an overriding concept (GEF, 2006a). The industrial energy efficiency strategic priority was the only programme in GEF-4 with an explicit technology transfer objective, with its focus on 'sector-specific technology transfer' (GEF, 2006a). The transport and renewable energy priorities emphasized a mostly technology-neutral approach, supporting planning and better public transport management, and policy reform and grid regulation, respectively.

In general, the concept of technology transfer played a minor role in the GEF's evolving climate change strategies until the Bali Conference. Technology transfer received cursory references in the GEF's initial Operational Programmes (OPs) in the context of the low-carbon near-commercial technologies OP, which allowed for pursuing technology transfer in cases where substantial cost reductions could be achieved by the use of local manufacturing (GEF, 1995). The only programming document in which technology transfer is addressed is a 2001 Climate Change Programme Study undertaken by the GEF's monitoring and evaluation (M&E) team. One of the emerging lessons in 2001 was that 'technological know-how transfer is more difficult than projects anticipate given problems with technology acquisition and application to domestic conditions' (GEF, 2001a, p3). The lesson comes from two China projects for efficient boilers and efficient refrigerators, which provided direct support to manufacturers to acquire technological know-how. In both cases 'this acquisition is proving more difficult than originally expected, suggesting that attempts in other GEF projects to transfer technological know-how directly to domestic manufacturers may prove difficult' (GEF, 2001a, p6).

Technology transfer is not mentioned at all in the context of climate change mitigation in the second and third Overall Performance Studies of the GEF,[14] completed in 2002 and 2006, respectively. A 2004 Climate Programme Study conducted by the GEF's former M&E team contains a few references to technology transfer in project examples, but the issue is not given further consideration in the team's analysis and recommendations. However, it reappears in the fourth Overall Performance Study (OPS4) prepared after Bali, where technology transfer is placed in the context of market transformation:

> It is of interest to note that growing attention is placed in both energy efficiency and renewable energy on the process of market transformation, in particular, barrier removal and technology transfer. As part of this approach, there is an emphasis on the need to strengthen institutions, build capacity, and create the right enabling environments required for successful long-term market transformation processes, as well as the introduction of cost-effective technologies.
>
> (GEF, 2010b, p86)

This suggests that, since Bali, the GEF sees technology transfer as part of the process of market transformation, alongside barrier removal.

The brief review of the place that technology transfer took in the narrative of the GEF's evolving climate change mitigation strategic objectives and programmes provides clear evidence that it was not a primary objective of GEF programmes, and where it was an explicit focus, as under the industrial energy efficiency strategic objective in GEF-4, the concept is not further elaborated upon. Nevertheless, by asserting that technology is primarily transferred through markets, and that for those markets to operate efficiently, barriers need to be systematically removed, implies that market transformation will lead to technology transfer by private actors.

Whether the GEF's market transformation approach catalysed technology transfer is hard to ascertain as it was not an outcome that was assessed in projects, and positive outcomes may, in any case, be difficult to attribute with certainty to specific interventions, as opposed to changes in the broader context of a project. It is therefore nearly impossible to evaluate the GEF's role in technology transfer except in those projects and programmes where the adoption of a specific technology was an explicit objective.

Indeed, in its 2008 official report to the UNFCCC with the title *Elaboration of a Strategic Program to Scale up the Level of Investment in the Transfer of Environmentally Sound Technologies*,[15] the GEF recognizes the lack of reporting on its activities on technology transfer, and the lack of systematic efforts to draw on project experience to formulate lessons learned and to carry out an in-depth analysis (GEF, 2008a, p9). The GEF also points to the limited understanding of the process of technology transfer in different national contexts and markets, and of the various roles of different actors, including its own role. The GEF further admits that its function as a technology transfer mechanism can be improved and strengthened, and identifies gaps in its support to date – namely, '(1) the weak link between GEF project development and Technology Needs Assessments (TNAs)[16] and national communications; (2) a lack of adequate reporting and knowledge management on technology transfer activities; (3) an uneven engagement with the private sector; and (4) the limited synergy with the carbon market' (GEF, 2008a, p8). In the follow-up to the recognition of the gap in reporting and knowledge management, two years later, the GEF stated in its report to the Cancun Conference[17] that it has 'recently launched a project on dissemination of GEF experiences and successfully demonstrated ESTs'[18] (GEF, 2010c, p21). The aim of the project is to generate a number of case studies related to technologies demonstrated through GEF projects that would analyse GEF experience and articulate lessons learned (GEF, 2010c).

In conclusion, the GEF has so far not articulated the difference between catalysing market transformation and catalysing technology transfer, and has not yet set out its possible role in technology transfer processes. In the following section, the different dimensions of the concept of technology transfer are further examined and different types of technology transfer support are codified.

Defining technology transfer and codifying technology transfer support

There are many definitions of technology transfer, but the GEF has explicitly adopted the Intergovernmental Panel on Climate Change (IPCC) definition of technology transfer. The IPCC defines technology transfer as (IPCC, 2000):

> ... a *broad set of processes covering the flows of know-how, experience and equipment* for mitigating and adapting to climate change amongst different stakeholders such as governments, private sector entities, financial institutions, non-governmental organizations (NGOs) and research/education institutions ... the broad and inclusive term 'transfer' *encompasses diffusion of technologies and technology cooperation* across and within countries. It covers technology transfer processes between developed countries, developing countries and countries with economies in transition. It *comprises the process of learning to understand, utilize and replicate the technology, including the capacity to choose and adapt to local conditions* and integrates it with indigenous technologies.

The definition gives a broad interpretation to technology transfer that also covers diffusion of technology within and across countries, and encompasses a wide range of actors in the transfer and diffusion process. It does not restrict technology transfer to the firm-level processes of technology and know-how acquisition between a technology recipient and supplier, but extends it to the diffusion process within sectors and the wider economy. The learning process of using and replicating technology and the capacity to adapt technology to local conditions is also considered part of technology transfer.

By conflating technology diffusion and the transfer of know-how and technology and learning processes at the firm level, any type of indirect support, such as the promotion of building codes, could be considered technology transfer support. It therefore becomes difficult to distinguish market transformation support from technology transfer support, and all GEF support can thus be interpreted as potentially catalysing technology transfer. This is, indeed, what the GEF asserts: that since its inception, it has facilitated technology transfer and has supported technology transfer activities in almost 100 developing countries (GEF, 2010a).

Although the GEF does not appear to differentiate between the Poznan Strategic Programme projects and other projects in terms of their approach, a cursory review reveals that the former have an explicit focus on technology and appear to support local technological capacity and manufacturing, in contrast to the majority of other mitigation projects. The role of technological capabilities, in particular, in enabling technology transfer and diffusion is underplayed by the GEF. Technological capacity development appears to be lumped together with other forms of capacity-building. Yet, technological capabilities are considered as being central to technology transfer. Indeed, Rosenberg (1982, p271) observed that 'the most distinctive single factor determining the success of technology transfer is the early emergence of an indigenous technological capacity'. The sole actor in the GEF to have repeatedly pointed out the importance of technological capabilities is the STAP.[19]

In view of the range of processes and actors that technology transfer (as defined by the IPCC) comprises, it is useful to organize the different levels and aspects of technology transfer support in a conceptual framework. Within the complex set of processes, three broad types of approaches to technology transfer support can be identified (Haum, 2010, p69):[20]

- *Direct technology transfer support:* at the level of technology and know-how acquisition from the technology supplier (including the know-how to operate, maintain and adapt technology). It entails the acquisition of technology through the financing thereof and/or increases the quality of the know-how flow to the technology recipient, supplier or manufacturer (Bell, 1990).Within this category, varying degrees of technology transfer can be distinguished: from the assimilation of skills to operate and maintain technology through to adapting and replicating technology (Bell, 1990). The technology transfer may be the result of a commercial sale, licensing, joint ventures, foreign direct investment, subcontracting or technology cooperation (Bell, 1990). The support, for example, may comprise the financing of the technology investment, or be related to the acquisition of know-how, or support licensing.

- *Indirect support of technology transfer through strengthening of technological capabilities:* strengthening of the absorptive capacities of technology recipients (but may also target suppliers) or strengthening of the opportunities for technology learning through training, technology cooperation and partnerships; strengthening the links between firms and technology intermediaries; strengthening the national innovation system; and the establishment and strengthening of research and development (R&D) and technology centres (Ockwell, 2007). The outcome is increased technological capabilities of technology recipients.

- *Market-pull and technology policy support for technology transfer:* targeted support in the broader institutional environment to promote technology adoption and diffusion and support investment. Includes creating incentives to invest in and adopt renewable energy and energy efficiency technologies through laws and regulation (the adoption of building codes, for example, or energy efficiency standards). It usually requires the development of investment delivery mechanisms and capacity-building of different institutional actors. It may also include awareness-raising and technology information. This category is unrelated to an increase in the technological capabilities of technology recipients. However, it leads to better understanding of technology opportunities amongst government institutions, financial institutions, technology users, distributors and entrepreneurs, and it creates an enabling environment for technology transfer.

The first two types of technology transfer support are focused on technology acquisition, the flow of technology know-how, and increasing technological capabilities. The third category of support comprises a broad range of measures, policies, tools, capacity-building and investment mechanisms that have to be adapted to the institutional environment of each country. It is in this third category of support that most experience has been gained through the implementation of GEF projects and programmes. However, since projects and programmes were not designed to monitor and evaluate technology transfer results, the impact of various measures and support is uncertain. Generating knowledge about the effect of different types of project support on technology transfer would require a focused effort, involving the gathering of additional relevant information and suitable analysis.

Although the importance of technological capabilities is stressed here, having the right policies, incentives and market conditions in place is a prerequisite for investment and further diffusion of a technology. Technology transfer does not occur in a vacuum, but needs a supportive complementary organizational and policy context. It is embedded in sector- and economy-wide processes of technological change and institutional development, and it is mediated by markets and government policies (Martinot et al., 1997; Ockwell, 2007; Bell, pers comm). Unless direct technology transfer support is reinforced with complementary policies, investment and further adoption is unlikely.

However, at its core, technology transfer is a firm-level process where technological capabilities play a key role. Key questions are therefore: are market pull measures sufficient; do they lead automatically to technology transfer; can technology transfer be left to the market and private actors in the case of low-carbon, renewable and energy efficient technologies, or does it require support at the level of technology and know-how acquisition and the strengthening of technological capabilities? The nature of the pilot projects in the Poznan Strategic Programme suggests that technological capabilities are important, at least for certain categories of technologies, and that support for technology and know-how acquisition may be necessary and valuable. The questions as to where capacity-building support has generally been targeted and which approaches were most effective in catalysing technology transfer, in particular, deserve greater examination. In the next section four technology case studies are presented in which GEF and other IGO technology support for different categories of technology is reviewed from a technology transfer perspective, applying a conceptual framework.

Technology case studies

Four categories of technologies are considered in the case studies: emerging technologies (namely, CSP and fuel cells for transport); renewable energy (namely, wind power and geothermal energy, each considered as a separate category); and industrial energy efficiency. The review draws on available information sources, particularly evaluations and reviews by the evaluation units of the implementing agencies (IAs) of the GEF, the GEF and its Evaluation Office, and STAP, where available, and other relevant information material prepared by the implementing agencies. The selection of projects and programmes was determined by the availability of reviews and evaluations and therefore comprises mostly projects that are in an advanced stage of implementation or are completed. The review does not seek to replicate evaluations and assessments of projects and programmes already undertaken, but instead synthesizes available findings, observations and recommendations relevant to technology transfer. The review further applies the conceptual framework presented in the previous section. Special attention is paid to observations made and lessons learned with regard to technological capabilities, opportunities for technology learning, and the quality of technology and know-how transfer.

Case study 1: Emerging technologies

Few GEF projects have supported emerging low-carbon technologies. Both the GEF concentrated solar power (CSP) and fuel cell bus (FCB) programmes provide interesting case studies and illustrate the challenge of promoting high-cost energy technologies for which no market exists at the time of project inception. The CSP programme started in 1996 with four proposed projects, in India, Mexico, Morocco and Egypt (GEF, 2001b). The aim of the multi-country programme was to accelerate the process of cost reduction of CSP technology, and to demonstrate its technical performance in a range of climate and market conditions (GEF, 2001b). Conceived as demonstration projects, the GEF would subsidize the cost of developing the solar fields in four locations. No progress was made in implementing the projects until recently, with the commissioning in 2009 of the Kuraymat Integrated Solar Combined Cycle Power Plant (ISCC) in Egypt. The Kuraymat BOOT (build–own–operate–transfer) project is financed by the Egyptian New and Renewable Energy Authority (NREA), the GEF, which provided US$50 million, and the Japan Bank for International Cooperation (JBIC) (World Bank, 2011, p92). The project in India was cancelled in 2004, while a plant is under construction in Morocco and the Mexico project is moving forward (GEF, 2010a).

It took 13 years for the first plant to be commissioned. The project was initially criticized for its lack of involvement of the CSP industry, insufficient attention to the hardware, an initial inappropriate insistence on an Independent Power Producer (IPP) project, an underestimation of the cost, and a lack of buy-in by the countries of the primary objective of developing a technology (STAP, 2003, 2004; IEG, 2010). They were technology-push projects in a vacuum, lacking a supportive policy context, and lacking in-country industry stakeholder champions and matching technological competence (STAP, 2004). A renaissance of CSP and declining technology costs were instrumental in moving the stalled projects forward (IEG, 2010).

The Kuraymat plant is a hybrid power plant with 150MW capacity using solar thermal energy and natural gas to generate electricity (World Bank, 2011). The solar field consists of parabolic trough *collectors* and is part of a hybrid power plant (World Bank, 2011). The solar technology for the project has been provided by Flagsol GmbH, the company that designed the solar field, delivered the control for the solar field, and was responsible for supplying important key components, primarily the parabolic mirrors and absorber pipes (Solar Millennium, 2011). The solar field was built and is put into operation in cooperation with the Egyptian company Orascom Construction Industries (Solar Millennium, 2011). The total investment was approximately 250 million Euros, of which 30 per cent was for the solar field (Solar Millennium, 2011). After two years of operation of the solar part of the plant by Flagsol and Orascom, the plant will be handed over to its owner, NREA (Solar Millennium, 2011). About 60 per cent of the value for the solar field at Kuraymat is generated locally (World Bank, 2011).

In terms of technology transfer, the participation of local industry is critical, and a promising approach could combine international cooperation to facilitate knowledge transfer, international support for R&D, the involvement of local companies where possible, and funds to compensate for the potential extra costs related to using local components (World Bank, 2011). The Kuraymat project also offers experience on the issue of integrated project contracting versus separate contracts for different components, which could, for example, provide useful guidance for carbon capture and sequestration projects (World Bank, 2011).

The CSP projects were conceived as demonstration projects with unclear technology transfer objectives. Although the GEF subsidized the Kuraymat solar field, and therefore the commercial purchase of the CSP technology, which is direct technology transfer support, no information is available about the flow of know-how related to the development of the solar field, its operation and maintenance, and the adaptation of this relatively simple technology to local conditions. A positive aspect is the 60 per cent of local content of the engineering, procurement and construction (World Bank, 2011).

Similarly, the GEF FCB programme supported the commercial demonstration of fuel cell buses and associated refuelling systems in the largest bus markets in five countries: Beijing, Cairo, Mexico City, New Delhi, São Paulo and Shanghai (GEF, 2006b). GEF intended to catalyse the commercialization of FCB technology for urban areas of developing countries, and 'to assist developing countries in gaining experience with FCBs early in its product cycle through partnerships with technology developers' (GEF, 2006b, p3). Here the focus was much more on increasing the technological competence of the project countries and on adapting the technology to local needs compared to the CSP projects. Yet, the project went to the demonstration stage in China only, a country with a strong domestic hydrogen and fuel cell R&D support programme, and world-class technology centres. In this case, the failure of advancing the technology in industrialized countries and of bringing down the cost, an evolution not anticipated in 2001, impacted heavily upon stakeholders' interests and upon the sustainability of the projects.

In its advice to the GEF on the fuel cell bus project, STAP felt that 'developing countries cannot be passive recipients of the technologies, and financial support from GEF can be legitimate if the firms benefiting from the "learning by doing" process in manufacturing the cells can be identified in the client countries of the GEF' (STAP, 2000, p4). STAP's view was that local ownership is central to the success of GEF projects in developing countries, and said that clarification would need to be provided on 'how the GEF would approach the complex patent and licensing procedures that local manufacture would require' (STAP, 2000, p4). No detailed information is currently publicly available on the nature of the technology partnership and flow of technology know-how in this direct technology transfer project.

Although both programmes were characterized by delays and failures, important lessons were learned with upstream high-risk technology projects. A STAP review concluded that emerging technology projects need in-country stakeholder champions who subscribe to the development of new technologies and are willing to build the necessary constituency (STAP, 2004). These stakeholders can be national policy-makers; but a strong national constituency can also be built if there is an in-country industry willing to push for the development of the technology (STAP, 2004). The role of the GEF could be to facilitate technology partnerships between interested participants in developed and developing countries in order to move technology forward more rapidly (STAP, 2004).

Both programmes underscore the importance of technological capabilities, country ownership and market developments to successfully demonstrate a near-commercial technology in a developing country so that it can be further adopted and adapted. They also highlight the inherent riskiness of advancing technologies at the global level, an action which may require the development of a clear technology roadmap (IEG, 2010). Capacity-building would be an important aspect of implementing technology roadmaps. An example of a technology capacity-building programme is the World Bank Carbon Capture and Sequestration (CCS) Capacity Building Trust Fund established in 2009 (IEG, 2010). The fund is aimed at creating opportunities for developing countries to explore CCS potential, realize the benefits of domestic technology development, and progress and facilitate appropriate policy initiatives (Global CCS Institute, 2009).

The case study focuses on large-scale high-tech energy technologies. A different approach should be adopted with supporting the development of small-scale technologies that reduce poverty and are easy to replicate, but in which there is little private-sector interest – for example, cookstoves and farming techniques such as biochar (IEG, 2010).

Case study 2: Wind power

IGO support for wind energy has been relatively limited compared to solar technologies, and earlier projects[21] have proven difficult to move beyond an initial barrier removal phase and into a pilot or demonstration phase, which has not always led to further investment (UNDP, 2008). Direct World Bank Group (WBG)[22] investments in wind power have been modest, and it is through indirect catalysis, rather than investment in individual power plants, that the WBG believes it can affect a large enough volume of investment (IEG, 2010). Both the UNDP and WBG have supported wind power barrier removal and demonstration projects, often with a GEF grant, while UNEP has focused on wind resource assessments in a number of countries. GEF is often the main contributor for the barrier removal part of UNDP–GEF projects, while pilot wind farms are generally mainly financed by the private sector (UNDP, 2008, p14). Technology transfer is left to the private sector and technical capacity-building is a component of some but not all projects. Where pilot wind turbines are turned over to state utilities as turnkey projects, training to improve the operation and maintenance of the wind turbines is crucial to ensure continued power generation and optimal capacity utilization (UNDP, 2008; IEG, 2010). However, a review of the 2003 to 2008 wind portfolio by the World Bank's Independent Evaluation Group (IEG) found that only a 'quarter of wind investment components included training and capacity building for installation or maintenance' (IEG, 2010, p19). A few UNDP projects, in countries with a high potential for wind, make provision for assessing the potential for local manufacturing of some components, and for capacity-building for local companies interested in entering the wind energy technology markets (UNDP, 2008, p49).

World Bank-supported policy advice and piloting have been helpful in China and Mexico in catalysing large-scale installations of wind facilities (IEG, 2010). However, policy advice may take years to bear fruit and success is uncertain (IEG, 2010). In Mexico, World Bank-supported dialogue 'together with demonstration wind projects, contributed to a renewable energy law that overcomes previous policy biases against renewable power' (IEG, 2010, p23). In 2009, total installed wind generation capacity in Mexico was 415MW and is projected to increase to 2.5GW in 2012 (IEA, 2009, p121). La Venta III is the first IPP wind energy project in Mexico. The contract awarded includes a complement to the electricity buy-back price of about US$0.015 per kilowatt hour (kWh) that will be granted by GEF through the World Bank (IEA, 2009, p121).

In addition to barrier removal and pilot projects, Mexico's wind technology infrastructure was supported by the UNDP–GEF with the installation of Certe-IIE, Mexico's first wind turbine test centre, and the first small wind energy power producer in Mexico (IEA, 2009). The UNDP–GEF provided further technology support through the implementation of a Regional Wind Technology Centre (WETC) set up by the Instituto de Investigaciones Electricas (IEA, 2009). The Poznan programme extends this technological capability support to Mexico's domestic wind turbine production through an Inter-American Development Bank (IADB) project to help develop a value chain for the domestic production of wind turbines adapted to local conditions (GEF, 2010a).

In China, the World Bank is applying a successful technology development approach first developed for rural solar home systems through the Renewable Energy Development Project (REDP), which emphasizes domestic research and development and manufacturing competence rather than licensing of foreign IPRs (IEG, 2010). REDP used a combination of technical assistance, incentives and subsidies to boost the capabilities of the small-scale SHS manufacturers (IEG, 2010). The REDP model is now being applied to the goal of promoting wind turbine improvements under the China Renewable Energy Scale-Up Project (IEG, 2010).

The technology transfer content and impact of the wind portfolio is hard to assess on the basis of available information and without independent assessments. However, recent projects in countries with large markets for wind power or with a strong commitment to wind energy development indicate a barrier removal and market development approach combined with support for targeted R&D and domestic industry capabilities in some countries. These projects appear to take a more integrated approach better linked to investment and local industry. The Mexico projects illustrate the need for having to adapt technology to local conditions because of the nature of the wind resources in Oaxaca and the necessary infrastructure required to support this. In general, wind projects tend to be complex, involving different phases, and may involve policy dialogue that can take years to yield results, market development support, co-financing of pilot plants or subsidizing the electricity cost differential. Technology transfer support predominantly occurs through market and policy support. In some projects there is also direct support through training for operation and maintenance of wind turbines, and indirect technology transfer support through financing of targeted R&D and technical assistance of the manufacturing industry.

The technology transfer issues will be very different depending upon the size of the market, the existence of relevant manufacturing capacity, the interest to develop a wind industry, experience with existing wind projects, and whether wind resources are developed by an IPP or a state utility. One issue that is important for all countries with potential resources is resource mapping and detailed assessments of sites with good wind potential (IEG, 2010).

Case study 3: Geothermal energy technologies

Although geothermal energy can be a least-cost clean energy source and provides base-load electricity, it remains largely unexploited in the developing world (Chandrasekharam and Bundschuh, 2002). Only about a dozen developing countries are currently generating electricity from geothermal resources. The GEF has supported a number of countries in exploiting their geothermal energy potential, either for heat or electricity generation. A major barrier for the development of geothermal resources is the cost of confirming the quality and exact location of exploitable resources (GEF, 2008b). Siting makes a huge difference in economic returns (IEG, 2010). The productivity of wells varies strongly and is a major factor determining the cost of the electricity generated. In order to address the resource confirmation barrier, the GEF has established several contingent funding mechanisms to reimburse the costs of drilling non-productive wells (GEF, 2008b). In addition, the GEF has provided funding and technical assistance for above-ground resource confirmation and for capacity-building and training.

A recently evaluated GEF project in Kenya, 'Joint Geophysical Imaging (JGI) Methodology for Geothermal Reservoir Assessment', implemented by UNEP, provides an interesting example of technology transfer by public-sector actors. The project was also singled out by the OPS3 team as 'an example of the GEF's ability to target limited resources in an area where there is specific technical, financial, or other risk that is preventing independent market-driven action, and where the GEF role could be limited, cost-effective and catalytic' (GEF, 2005, p110). The objective of the JGI project was to transfer and adapt joint geophysical imaging (JGI) methods to the unique conditions of the Rift Valley, with the view of improving the 'siting' of geothermal wells and thereby increasing the productivity of the wells (UNEP, 2002). JGI refers to the combination of geophysical and seismic methods used for the mapping of geothermal resources. KenGen, a parastatal with long experience in geothermal energy development, had not previously used this combination method for mapping, which is highly complex, but was very keen to adapt and test the method. A major component of the project was the purchase, adaptation to the field conditions of Kenya, and testing of seismic and other equipment. The 'soft' component involved the method and software development, data collection, interpretation and testing, and training. The research was carried out in partnership with Duke University. The method and equipment was used to locate six wells, with very good results. The predicted production of the wells was an average of 5.6MW per well, compared to 1.4MW and 2.8MW per well for KenGen's two existing geothermal fields (UNEP, 2009). The project was heralded as a success in the international press, and expectations were created for replication in other parts of the Rift Valley.

This was a small-budget but very ambitious project that tried to develop and apply a very complex method and techniques in a difficult environment, with the ultimate objective of bringing down the cost of electricity. Much was achieved, although the sustainability of the achievements is in doubt because of the slim human resource base upon which the continued use and future development of this method rests, insufficient involvement of KenGen in drawing up equipment specifications, and because of disagreement between Duke University and KenGen about the ownership of computer codes (UNEP, 2009). The evaluation points out that the capacity of KenGen to implement a project of this nature successfully was limited, and overall the complexity of the project, and the resources it required, were underestimated by all partners (UNEP, 2009).

The JGI project is linked to a larger geothermal development initiative supported by the GEF – the African Rift Valley Geothermal Development Facility (ARGeo) – which aims to support the development of geothermal resources in Kenya, Djibouti, Ethiopia, Eritrea, Uganda and Tanzania, and which started implementation in 2011. A key project component, apart from contingent guarantee funding, is a regional network of geothermal and geological agencies in the six countries, which will be linked to geothermal agencies in Iceland, Germany and the US, amongst others (UNEP, 2006). Key objectives are building capacity, promoting South–South and North–South knowledge-sharing and transfer, sharing information and equipment, and facilitating exchange of personnel for training purposes (UNEP, 2006).

The importance of technological and institutional capabilities for exploiting a difficult natural resource such as geothermal energy is illustrated by the failure of a 5MW pilot plant at Aluto Langano in Ethiopia. Built by Ormat as a turnkey plant in 2000, it was handed over to Ethiopia's state electricity generating company after minimal training of its hydro engineers in Iceland. After a year of operation, and many technical problems, the plant was shut down following substantial damage to the wells. The plant is still not back in operation, and the failure has caused setbacks in the development of Ethiopia's extensive geothermal resources.

The JGI project relied on a combination of direct and indirect technology transfer through a partnership with a public research organization to successfully develop and apply a complex technology and achieve its objective of improving the siting of wells. It also illustrates the need to carefully consider the design of partnerships to avoid disputes, and the importance of commitment to the technology partnership by all parties. Other projects illustrate the importance of technological capabilities in the operation and maintenance of geothermal fields.

Case study 4: Industrial energy efficiency

IGOs have supported a growing portfolio of industrial energy efficiency projects. By 2008, the GEF had funded more than 30 projects in the industrial sector to promote technology upgrading and the adoption and diffusion of energy-efficient technologies (GEF, 2008b). Some projects focus on the development of market mechanisms, such as energy service companies, the creation of dedicated financing instruments, and technical assistance to stimulate investment in new technologies (GEF, 2008b). Other projects identify one or more sub-sectors and specific technologies. The range of industries includes construction materials (brick, cement and glass), steel, coke-making, foundry, paper, ceramics, textiles, food and beverages, tea, rubber, and wood (GEF, 2008b). A number of projects also promote energy-efficient equipment, such as boilers, motors and pumps, as well as cogeneration in the industrial sector (GEF, 2008b). In some projects, the GEF is promoting South–South technology transfer, as in the transfer of energy-efficient brick kiln technology from China to Bangladesh. The technology was developed, adopted and disseminated in China, and is being transferred to Bangladesh (GEF, 2010a).

The World Bank Group's efforts in energy efficiency are increasingly focusing on the development of improved energy efficiency investment delivery mechanisms that can operate in the market and help accelerate investment in energy efficiency (World Bank, 2008). However, development and operation of energy efficiency investment delivery mechanisms is an institutional development issue, which requires a sustained effort, ensuring that the planned institutional solutions match the institutional environment in which they are expected to function (World Bank, 2008). The results achieved in developing energy efficiency lending schemes in Hungary or ESCOs in China, for example, have been the result of about eight years of persistent effort (World Bank, 2008, p142). The WBG has also found that in most cases, steady and strategic government support is a very important enabling factor for the type of institutional development required to truly improve delivery of energy efficiency financing (World Bank, 2008). Projects that target industrial sectors directly, generally implemented by the UNDP and UNIDO, increasingly seek policy engagement and utilize policy instruments to achieve their objectives and involve different forms of technical know-how transfer. They are among the most technology transfer-oriented projects in the GEF climate change mitigation portfolio.

Two projects, one in Malaysia, completed in 2008, and a project implemented by the UNDP in India, entitled Energy Efficiency in Steel Re-Rolling Mills, offer valuable lessons on the importance of targeting capacity-building efforts and of long-term support for technology transfer and innovation. The steel re-rolling mill (SSRM) sector in India is characterized by the use of outdated technologies and practices, and low energy efficiency (UNDP, 2007). Unlike the large plants in the steel sector, the SRRM sector has not received government support to improve energy efficiency; as a result, it is a major consumer of energy. Key components of the UNDP project include establishing technical and financial feasibility of EcoTech options and technology packages;[23] the use of benchmarks;

a demonstration component; capacity-building; the development of business networks; a Technology Information Resource and Facilitation Centre; and facilitating and establishing ESCO services (UNDP, 2007). In order to address the slow response of the SRRM manufacturers to the opportunities offered by the project, a mid-term evaluation recommended a number of changes, including the promotion of a step-wise package of technology instead of promoting state-of-the-art technologies first because the SRRM technologies are so outdated; and to extend support and training for domestic equipment providers and consultants so as to develop capacity for local manufacturing, instead of focusing on SRRM units only (UNDP, 2007, p35).

The evaluation further notes the importance of capacity-building efforts coupled with long-term support to technology innovation through the involvement of research, design and development institutions, both within India and outside the country (UNDP, 2007, p38). The project experience shows that, in the absence of energy efficiency regulation and when incentives for technology upgrading are lacking, because business is thriving and electricity prices are very low, a technology support approach may not be successful.

Another UNDP project, Malaysian Industrial Energy Efficiency Improvement, concludes that promotion of energy efficiency is a long-term policy issue and needs an achievement horizon of decades, depending on the government's willingness to put substantial resources behind such efforts on a continuing basis (UNDP, 2004, p12). It also cautions not to exaggerate the potential of certain instruments such as ESCOs to deliver results (UNDP, 2004). The project experience also points to the key role of energy efficiency regulation and the enforcement that is needed to sustain an energy efficiency programme.

The industrial energy efficiency projects show the importance of mutually reinforcing sustained efforts and measures that combine market development and policy support comprising institution-building, policy instruments and financial products, and direct and indirect technology transfer support in the form of know-how transfer, technological capacity-building of technology users and suppliers, and R&D support where appropriate.

Conclusion

Any assessment of the reviewed GEF projects and programmes in the three technology categories should take into account the high level of ambition of their goals of transforming markets, pushing near-commercial technologies down the learning curve, demonstrating and applying new technologies in difficult environments, and removing barriers to the investment and diffusion of mitigation technologies. Success is often only partial, takes time to materialize and depends heavily upon an enabling environment. There are also failures in some cases from which more can be learned than from successes. Transpiring from all the projects is the importance of a long-term sustained engagement and efforts on the part of the project agencies, recipient government and other parties. Only limited information is available on the technology transfer impact of the projects and, hence, on the success factors, which constrains the quality and depth of the lessons that can be learned from the projects. In some cases, the projects that combined the three types of technology transfer identified earlier were the most successful – for example, the development of wind power in Mexico.

Nevertheless, a number of lessons can be distilled from the project experience. Projects that promote the adoption of emerging near-commercial technologies need supportive domestic programmes, committed national project partners, and technology partnerships between developing and industrialized countries. The existence of prior R&D links, or knowledge networks and adequate domestic capacity, are important success factors, and national champions for the technology are essential. Without these, there is no basis to build from and demonstration projects are bound to fail. IGO support should be adapted to the level of national capacity and commitment in countries. Demonstration projects may have to be preceded by capacity-building, participation in knowledge networks, and targeted R&D support where there is a sufficient research capacity to build on, provided through partnerships to create the necessary enabling conditions for their implementation.

Supporting technology transfer and diffusion requires a broad approach to capacity-building and may involve extending membership of international networks to developing country public and private actors. In order to sustain project results, capacity-building efforts should be coupled with long-term support to technology innovation through the involvement of R&D institutions and relevant government agencies. The projects illustrate the difficulty of targeting capacity-building support effectively in the right places in the network of technology suppliers and users, as well as in the larger complementary institutional and business context of a technology. This is an aspect where better understanding and analysis are needed.

Project experience further underscores the importance of a good understanding and analysis of the market, policy and technology barriers that projects aim to address. A prior assessment of capacity gaps and needs to provide baseline information may be critical to avoid underperformance. Transferring technology requires a multifaceted approach that has to be tailored to each technology and implementation context. Projects that did not benefit from or help to create a supportive complementary organizational and policy context failed to achieve their technology objectives. Overall, project experience points to the importance of mutually reinforcing sustained efforts and measures that combine technology-pull policy support, comprising institution-building, policy instruments and financial products, and indirect technology transfer support targeted at technological capacity-building and innovation of technology users, manufacturers and technology suppliers.

Some of the weaknesses in the projects could have been overcome with a more explicit recognition at the outset of the various dimensions of technology transfer, at least for those projects with a technology transfer objective. A better articulation of different types of technology transfer support for mitigation and their rationale would help with programme and project design. This is particularly urgent as the new Technology Mechanism becomes operational and new technology initiatives are launched. There is a need to analyse and codify the knowledge generated by more than 17 years of GEF project experience. The technology transfer narrative is lagging behind the development of new programmes and projects in the GEF, UN agencies, development banks and bilateral agencies.

The fact that technology transfer processes are not yet well understood, the lack of technology expertise in multilateral agencies and uncertainty about the role that multilateral mechanisms could play in technology transfer appears to be hampering progress with designing coherent programmes. Supporting technology transfer for climate change mitigation is a difficult proposition. Technology is owned mainly by private actors, companies want to keep technologies proprietary, and engaging the private sector and catalysing investment has not been entirely successful in the past.[24] The need for better understanding of technology transfer needs and issues, and of the role that various mechanisms could play, is well recognized. It is therefore a very positive development that the GEF is taking the lead in advancing the understanding of technology transfer in support of climate change mitigation and in codifying knowledge through the learning component in the Poznan Strategic Programme and other planned knowledge products.

Notes

1 The 13th session of the Conference of the Parties (COP) of the United Nations Framework Convention on Climate Change (UNFCCC) took place in Bali in December 2007.
2 The Technology Mechanism is to facilitate the implementation of enhanced action on technology development and transfer to support action on mitigation and adaptation.
3 The Poznan Strategic Programme on Technology Transfer.
4 The mandate of the EGTT was terminated following the establishment of the Technology Mechanism, and replaced by the Technology Executive Committee (TEC).
5 The Technology Mechanism is to facilitate the implementation of actions on technology development and transfer, which include the enhancement of endogenous capacities and strengthening of national innovation systems (UNFCCC, 2010).
6 There is no agreement yet among the parties to the convention on how the Technology Mechanism will be financed.
7 The Technology Mechanism comprises the Technology Executive Committee (TEC) and the Climate Technology Centre and Network (CTCN). The latter is to provide in-country support to developing country parties.
8 The project proposal was approved by the GEF Council in May 2011.
9 GEF projects are implemented by the 'implementing agencies', originally the UNDP, UNEP and the World Bank, but later on also comprising other UN agencies and the regional development banks.
10 The United Nations Framework Convention on Climate Change (UNFCCC).
11 The GEF started as a World Bank pilot initiative in 1991 and was restructured in 1994 to become a partnership between the World Bank, the UNDP and UNEP, later expanded to include other UN agencies and the regional development banks. The GEF is the financial mechanism of the UNFCCC. It has a secretariat located in the World Bank in Washington, DC.
12 GEF-4 refers to the fourth phase of the replenishment of the GEF Trust Fund, which began in July 2006 and ended in July 2010.

13 The programme was given low-priority status but not completely terminated. Support was still provided for targeted research projects to keep abreast of new technology developments.

14 At the end of each four-year phase, the GEF undergoes a comprehensive independent evaluation of its performance.

15 The programme that became the Poznan Strategic Programme after the Poznan climate conference in December 2008.

16 TNAs are one of the five pillars of the Technology Transfer Framework of the UNFCCC, which provides support to developing countries to assess their technology needs. TNAs can be found on the UNFCCC website.

17 The Cancun Conference refers to the 16th session of the Conference of the Parties to the UNFCCC, held in December 2010 in Cancun.

18 ESTs are defined as environmentally sound technologies.

19 In 1998 STAP advised the GEF on the technology transfer of clean and low-GHG technologies in which it emphasizes the importance of technological capabilities. It does so again in its advice on fuel cell buses and on OP-7.

20 Adaptation of Haum's categorization of options for designing technology transfer policies.

21 More recent projects appear to have been more successful in catalysing investment, possibly because of better policy environments. A 2008 UNDP review of its portfolio of wind projects concludes that 20 projects out of 34 were cancelled, leaving only 14, of which one was completed at the time of publication. Two more were cancelled since then and others were added.

22 The two institutions referred to here are the International Bank for Reconstruction and Development (IBRD) and the International Finance Corporation (IFC).

23 Energy-efficient and/or environmentally friendly technology packages.

24 The 2010 Earth Fund Review by the Evaluation Office of the GEF presents a brief overview of the factors that have hampered a more successful engagement.

References

Bell, M. (1990) *Continuing Industrialisation, Climate Change and International Technology Transfer*, SPRU, University of Sussex, Brighton, UK

Chandrasekharam, D. and Bundschuh, J. (eds) (2002) *Geothermal Energy Resources for Developing Countries*, A. A. Balkema, Lisse, The Netherlands

CIEL and South Centre (2008) *Intellectual Property Quarterly Update*, Fourth Quarter, Geneva

GEF (Global Environment Facility) (1995) *Revised Draft Operational Strategy – October 1995*, GEF Council Meeting, Washington, DC, http://www.thegef.org/gef/sites/thegef.org/files/documents/GEF.C.6.3.pdf

GEF (2001a) *Climate Change Program Study Synthesis Report*, GEF/C.17/Inf.5., Washington, DC

GEF (2001b) *Thematic Review of GEF-Financed Solar Thermal Projects, Monitoring and Evaluation Working Paper 7*, Washington, DC

GEF (2004) *Office of Monitoring and Evaluation, 2004, Climate Change Program Study 2004*, Washington, DC, http://www.gefweb.org/COUNCIL/GEF_24/C.24.ME.Inf.2.pdf.

GEF (2005) *OPS3: Progressing Toward Environmental Results: Third Overall Performance Study of the GEF*, Office of Monitoring and Evaluation of the Global Environment Facility, Washington, DC

GEF (2006a) *Focal Area Strategies for GEF-4 – Working Drafts and Proposed Process*, http://www.thegef.org/gef/sites/thegef.org/files/documents/C.30.5%20Focal%20Area%20Strategies.pdf

GEF (2006b) *UNDP–GEF Fuel Cell Bus Program: Update*, GEF/C.28/Inf.12, Washington, DC

GEF (2008a) *Elaboration of a Strategic Program to Scale Up the Level of Investment in the Transfer of Environmentally Sound Technologies*, GEF/C.34/5.Rev.1, Washington, DC

GEF (2008b) *Transfer of Environmentally Sound Technologies: The GEF Experience*, Washington, DC

GEF (2010a) *Transfer of Environmentally Sound Technologies: Case Studies from GEF Climate Change Portfolio*, Washington, DC

GEF (2010b) *OPS4: Progress Towards Impact – Fourth Overall Performance Study of the GEF*, Washington, DC

GEF (2010c) *Implementation of the Poznan Strategic Program on Technology Transfer: Report of the GEF to the Sixteenth Session of the Conference of the Parties to the United Nations Framework Convention on Climate Change*, Washington, DC

Global CCS Institute (2009) Web site, http://www.globalccsinstitute.com/institute/media-centre/media-releases/global-ccs-institute-injects-funding-drive-project-deployment

Haum, R. (2010) *Transfer of Low-Carbon Technology under the United Nations Framework Convention on Climate Change: The Case of the Global Environment Facility and Its Market Transformation Approach in India*, PhD thesis, SPRU – Science and Technology Policy Research, University of Sussex, Brighton, UK

IEA (International Energy Agency) (2009) *Wind Energy: Annual Report*, IEA/OECD, Paris

IEG (Independent Evaluation Group) (2009) *Climate Change and the World Bank Group. Phase I: An Evaluation of World Bank Win–Win Energy Policy Reforms*, World Bank, Washington, DC

IEG (2010) *Climate Change and the World Bank Group. Phase II: The Challenge of Low-Carbon Development*, World Bank, Washington, DC

IFC (International Finance Corporation) (1998) 'India, Kenya, and Morocco: Photovoltaic Market Transformation Initiative (PVMTI)', in *Project Document*, IFC, Washington, DC

IPCC (Intergovernmental Panel on Climate Change) (ed) (2000) *Methodological and Technological Issues in Technology Transfer: A Special Report of the Intergovernmental Panel on Climate Change*, IPCC Special Reports on Climate Change, Cambridge University Press, Cambridge

Martinot, E., Sinton, J. E. and Haddad, B. M. (1997) 'International technology transfer for climate change mitigation and the cases of Russia and China', *Annual Review of Energy and the Environment*, vol 22, pp357–401

Ockwell, D. et al., (2007) UK–India Collaboration to Identify the Barriers to the Transfer of Low-Carbon Energy Technology, Final Report, Department for Environment, Food and Rural Affairs, London

Rosenberg, N. (1982) *Inside the Black Box: Technology and Economics*, Cambridge University Press, Cambridge

Solar Millennium (2011) Website, http://ssef3.apricum-group.com/wp-content/uploads/2011/02/2-Solar-Millennium-Fuchs-2011-04-04.pdf

Solar Thermal Magazine (2011) 19 July, http://www.solarthermalmagazine.com/2011/01/02/egypt's-first-solar-thermal-plant-goes-into-operation-in-kuraymat/

STAP (Scientific and Technical Advisory Panel) (2000) *Strategic Advice on Commercialization of Fuel Cell Buses: Potential Roles for the GEF*, UNEP, Nairobi

STAP (2003) *Compilation of Presentations at the STAP/GEF Brainstorming Session on OP7*, UNEP, Nairobi

STAP (2004) *Report of the Brainstorming Session on OP7*, UNEP, Nairobi

UNDP (United Nations Development Programme) (2004) *Achieving Industrial Energy Efficiency in Malaysia*, UNDP, Kuala Lumpur

UNDP (2007) *Energy Efficiency in Steel Re-Rolling Mills: Mid-Term Review*, UNDP, New York, NY

UNDP (2008) *Promotion of Wind Energy: Lessons Learnt from International Experience and UNDP–GEF Projects*, UNDP, New York, NY

UNEP (United Nations Environment Programme) (2002) *Joint Geophysical Imaging (JGI) for Geothermal Reservoir Assessment*, UNEP/GEF Project Document

UNEP (2006) *African Rift Geothermal Energy Development Facility (ARGeo)*, UNEP Project Document

UNEP (2009) *Terminal Evaluation of the UNEP/GEF Project 'Joint Geophysical Imaging (JGI) for Geothermal Reservoir Assessment'*, Evaluation and Oversight Unit, UNEP

UNFCCC (United Nations Framework Convention on Climate Change) (2010) *Report of the Conference of the Parties on its Sixteenth Session held in Cancun from 29 November to 10 December 2010, Addendum. Part Two: Actions Taken by the COP at its Sixteenth Session*, Decisions adopted by the COP, FCCC/CP/2010/7/Add.1, http://unfccc.int/resource/docs/2010/cop16/eng/07a01.pdf#page=18

World Bank (2008) *Financing Energy Efficiency: Lessons from Brazil, China, India and Beyond*, ESMAP Report, World Bank, Washington, DC

World Bank (2011) *North Africa and Middle East Assessment of the Local Manufacturing Potential for Concentrated Solar Power CSP Projects*, ESMAP Report, http://arabworld.worldbank.org/content/dam/awi/pdf/CSP_MENA__report_17_Jan2011.pdf

9

TECHNOLOGY TRANSFER AND THE CLEAN DEVELOPMENT MECHANISM (CDM)

Erik Haites, Grant A. Kirkman, Kevin Murphy and Stephen Seres[1]

Background

The Clean Development Mechanism (CDM), established by the Kyoto Protocol, enables projects that reduce greenhouse gas (GHG) emissions in developing countries to earn credits that can be used by developed country (Annex I) parties to help meet their national emissions limitation commitments. Although the CDM does not have an explicit technology transfer objective, it contributes to technology transfer by financing emission reduction projects that use technologies currently not available in the host countries.

Participants must prepare a project design document (PDD) for a proposed project. The proposed project must be approved by the 'designated national authority' of the host country. An independent 'designated operational entity' (DOE) must ensure that the proposed project meets all of the requirements of a CDM project. As part of this validation process, the DOE must solicit public comments on the proposed project. Once a proposed project is posted for public comment it is considered to be 'in the pipeline'. A proposed project approved by the CDM Executive Board is 'registered' and can earn credits for the emission reductions that it achieves.

Project participants are required to provide details of the technology to be used for their projects in the PDDs from which technology transfer information can be derived. The CDM Executive Board does not define 'technology transfer' in its glossary (UNFCCC, 2006), so the statements in the PDDs reflect the implicit definitions of the project participants. It is clear from the PDDs that project participants almost universally interpret technology transfer as meaning the use by the

CDM project of equipment and/or knowledge not previously available in the host country.

There is no universally accepted definition of technology transfer.[2] Technology transfer is generally understood to involve some transfer of knowledge.[3] Experts differ on how much knowledge must be transferred to qualify as technology transfer: enough to operate and maintain the equipment, to produce similar equipment domestically, to adapt the technology to local conditions, or to further develop the technology. In practice, more countries will use a technology than produce the associated equipment, and fewer still have the capacity to develop the technology. For example, virtually every country has the capacity to operate and maintain electricity-generating equipment, but any given type of generating equipment – coal, oil, natural gas, nuclear, hydro, wind, solar, geothermal, etc. – is manufactured by a relatively small number of countries and development of the technology occurs only in a few.

CDM project participants naturally focus on the technology transfer needed to implement their project – equipment and the knowledge needed to install, operate and maintain it – rather than the host country's ability to manufacture or develop the technology. Approximately one third of the projects that claim technology transfer expect it to involve only imported equipment.[4] A follow-up survey found that almost 90 per cent of the projects that claimed technology transfer in the PDD involved technology transfer and that transfer of both knowledge and equipment is much more common than indicated by the PDDs.[5] In practice, about 25 per cent of the projects that claim technology transfer involve only equipment imports.[6]

Several papers have analysed technology transfer by CDM projects for registered projects (de Coninck et al., 2007; Dechezleprêtre et al., 2008; Das, 2011) or projects in the pipeline (Haites et al., 2006; UNFCCC, 2007, 2008, 2010; Seres et al., 2009) based on statements in the PDDs. This chapter examines the technology transfer for 4984 projects in the CDM pipeline in 81 countries as of 30 June 2010.[7] The second section reviews the patterns of technology transfer for CDM projects, while the third discusses the origins of the technology, the links between technology supply and credit purchases, and the diversity of technology supply. The fourth section presents the results of a statistical analysis undertaken to identify the characteristics of CDM projects and host countries that influence the rate of technology transfer. Barriers to technology transfer are analysed in section five and conclusions are drawn in section six.

Patterns of technology transfer in the CDM

The results reported in this chapter are based on the most recent United Nations Framework Convention on Climate Change report (UNFCCC, 2010). Project data were compiled from lists maintained by the UNFCCC Secretariat and the United Nations Environment Programme (UNEP) Risø Centre, which assigns a project type to each project. Information on the technology transfer associated with projects was collected from the individual PDDs. The UNEP Risø Centre revised its project type definitions in 2009 to better reflect the diversity of projects being proposed. The results presented here reflect the new definitions. Since many projects were reclassified, results relating to project types should not be compared with those presented in earlier studies.

Information about technology transfer was collected from the individual PDDs. Sections where the project participants describe the technology and know-how to be used, whether it is to be transferred to the host country and demonstrate the additionality of the project, were reviewed thoroughly to determine whether the project involves technology transfer. In addition, each PDD was searched for a number of keywords relating to technology transfer.[8] In many cases, the PDD explicitly states that the project involves no transfer of technology.

A survey of the 3296 projects covered by the 2008 study (UNFCCC, 2008) was conducted to check the accuracy of the technology transfer codes based on the PDDs. Participants for 370 projects (about 11 per cent) responded. The survey responses were representative of the 3296 projects covered by the 2008 study. The survey responses indicate that the code assigned on the basis of the PDD was correct for 88 per cent of the projects that involve no technology transfer and 89 per cent of the projects that involved technology transfer.

Of the 4984 projects used in the current (2010 data) study, 1206 PDDs do not mention technology transfer. The survey of 2008 projects found that 58 per cent of projects that did not mention technology transfer in the PDD actually involved technology transfer, while 41 per cent did not. Earlier studies (Haites et al., 2006; UNFCCC, 2007, 2008; Seres et al., 2009) assumed that projects that do not mention technology transfer in their PDD involve no technology transfer. The survey results indicate that that assumption would understate the rate of technology transfer. The results presented here exclude the 1206 projects for which the PDD does not mention technology transfer.

Using only the 3778 (4984 minus 1206) projects which explicitly state that they will, or will not, involve technology transfer, 40 per cent of projects accounting for 59 per cent of estimated emission reductions involve technology transfer (see Table 9.1). The distribution of projects across project types is not uniform: approximately one third of the project types have 20 or fewer projects, while another third have over 100 projects each, with hydro, wind, biomass energy and methane avoidance dominating the totals. The average size varies widely by project type from less than 10ktCO$_2$e per year for CO$_2$ capture to 3696 ktCO$_2$e per year for hydrofluorocarbon (HFC) reduction projects. The overall average is 140ktCO$_2$e per year.

Technology transfer is very heterogeneous across project types. The percentage of projects that claim technology transfer ranges from 13 to 100 per cent for different project types. International transfer of technology is unlikely if the technology is already available in the host country. Thus technology transfer is low for mature technologies already widely available in developing countries, such as hydroelectric generation (13 per cent) and cement production (21 per cent). Projects that claim some technology transfer are, on average, larger than those that do not involve technology transfer. This is true for most project types except for energy

TABLE 9.1 *Technology transfer by project type*

Project type	Number of projects	Average project size (tCO_2e/year)	Technology transfer claims as percent of		Percentage of projects where technology transfer could not be determined
			Number of projects	Annual emission reductions	
Afforestation	10	26,839	33%	47%	40%
Biomass energy	643	67,974	34%	45%	33%
Cement	32	152,152	21%	19%	41%
CO_2 capture	3	9675	100%	100%	33%
Coal bed/mine methane	65	585,532	53%	68%	15%
EE households	32	40,852	38%	58%	50%
EE industry	126	29,481	65%	64%	46%
EE own generation	421	132,383	39%	63%	24%
EE service	20	12,052	83%	96%	70%
EE supply side	75	388,323	70%	74%	43%
Energy distribution	17	301,221	25%	15%	6%
Fossil-fuel switch	109	397,817	78%	89%	32%
Fugitive	35	477,864	47%	55%	46%
Geothermal	15	222,085	91%	97%	27%
HFCs	22	3,696,440	91%	97%	0%
Hydro	1372	109,965	13%	11%	17%
Landfill gas	297	154,841	82%	87%	25%
Methane avoidance	566	46,200	78%	81%	26%
N_2O	70	711,373	100%	100%	6%
PFCs and SF_6	17	291,838	75%	91%	53%
Reforestation	42	101,433	25%	23%	43%
Solar	47	22,402	60%	43%	26%
Tidal	1	315,440	100%	100%	0%
Transport	24	100,435	82%	93%	54%
Wind	923	91,732	34%	38%	19%
Grand total	4984	139,925	40%	59%	24%

distribution and solar projects, where the projects that involve technology transfer are much smaller than similar projects that do not claim technology transfer.

The host country has a significant impact upon the rate of technology transfer. As part of its approval process, the host country government may choose to impose technology transfer requirements. In addition, factors such as tariffs or other barriers to imports of relevant technologies, perceived and effective protection of intellectual property rights, and restrictions on foreign investment can affect the extent of technology transfer in CDM projects. Table 9.2 presents data on technology transfer rates for the ten countries with the most CDM projects. The rate of technology transfer varies widely from 13 per cent of projects in India to 83 per cent of projects in Mexico. Reasons for differences in the rate of technology transfer by host country are a focus of the statistical analysis presented in this chapter's fourth section.

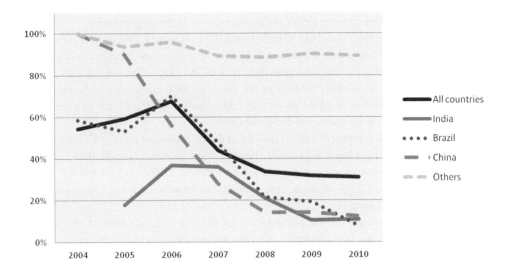

FIGURE 9.1 Trend in technology transfer as a percentage of projects

Trends in the rate of technology transfer for CDM projects are shown in Figure 9.1 for the three countries that host the most projects – China, India and Brazil – as well as for all other host countries and for all projects. The rate of technology transfer has declined over time.[9] The decline in the rate of technology transfer has been greater than the decline of the overall average in China, India and Brazil, while the

TABLE 9.2 *Technology transfer for projects in selected host countries*

| Host country | Number of projects | Estimated emission reductions (tCO$_2$e/year) | Average project size (tCO$_2$e/ year) | Technology transfer claims as percentage of | | Percentage of projects where technology transfer could not be determined |
				Number of projects	Annual emission reductions	
Brazil	338	34,269,995	101,391	25%	54%	32%
Chile	69	7,182,105	104,088	52%	72%	43%
China	1993	387,496,440	194,429	19%	47%	9%
India	1254	117,940,808	94,052	13%	23%	40%
Indonesia	100	11,207,814	112,078	59%	43%	38%
Malaysia	127	7,009,300	55,191	60%	65%	35%
Mexico	165	14,588,291	88,414	83%	84%	16%
Republic of Korea	72	19,607,821	272,331	50%	69%	42%
Thailand	115	6,369,257	55,385	80%	79%	16%
Viet Nam	106	6,772,217	63,889	73%	60%	24%
All others	645	84,944,348	131,697	59%	64%	31%
Grand total	4984	697,388,396	139,925	40%	59%	24%

rate of technology transfer for all other host countries has declined only modestly and remains much higher than the average rate of technology transfer.

Several factors contribute to these trends. First, as more CDM projects of a given type are implemented in a country, the rate of technology transfer declines. This indicates that a transfer of technology to a CDM project creates capacity in the country that allows later projects to rely more on local knowledge and equipment. The large number of CDM projects in China, India and Brazil leads to a declining rate of technology transfer to those countries.

Second, development and/or transfer of the technologies used by CDM projects appear(s) to have been happening through other channels as well, thus reducing the rate of technology transfer via CDM projects. Haščič and Johnstone (2009) find that the CDM explains only part of the transfer of wind technologies to developing countries.

Finally, changes to the composition of the pipeline, in terms of the mix of project types and host countries, affect the average rate of technology transfer. New host countries typically have a higher rate of technology transfer, so the growing number of host countries moderates the decline in the overall rate of technology transfer.

Patterns of technology transfer

Most innovation on climate mitigation technologies occurs in developed countries. Sixty per cent of patents for 13 climate mitigation technologies originate in the US, Japan or Germany (Johnstone et al., 2010; Dechezleprêtre et al., 2011; Popp, 2011). As shown in Figure 9.2, the top five technology suppliers for CDM projects are Germany, the US, Japan, Denmark and China.[10] About 85 per cent of the CDM projects that involve technology transfer get their technology from developed countries.

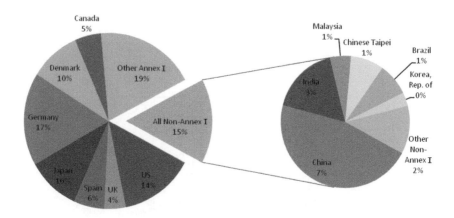

FIGURE 9.2 Leading sources of transferred technology as a percentage of projects

Most project types draw on technology from several countries, as shown in Table 9.3. Although Germany supplies technology for the largest number of projects, it is the main supplier only for energy efficiency households (EE households), wind, N_2O destruction and HFC (tied with Japan). The US is the largest technology supplier for nine project types, Japan for four project types, and Denmark for two project types. China is the main supplier of technology for hydro projects.

A large market share for a few technology suppliers might indicate that the technology is controlled by a few sources (an oligopoly), which could restrict the distribution of the technology and/or keep the price relatively high. Table 9.3

TABLE 9.3 *Diversity of technology supply by project type*

Project Type	Number of projects	Projects with no technology transfer	Number of projects that claim technology transfer	Number of known technology suppliers	Share of four largest suppliers*	Share of largest supplier*	Largest supplier
Afforestation	10	4	2	2	–	50%	
Biomass energy	643	284	147	28	52%	20%	Denmark
Cement	32	15	4	5	88%	25%	
CO₂ capture	3	–	2	1	–	100%	Denmark
Coal bed/mine methane	65	26	29	8	88%	37%	US
EE households	32	10	6	2	–	93%	Germany
EE industry	126	24	44	18	56%	29%	Japan
EE own generation	421	196	126	10	95%	54%	Japan
EE service	20	1	5	2	–	50%	
EE supply side	75	13	30	15	59%	29%	US
Energy distribution	17	12	4	1	–	100%	US
Fossil-fuel switch	109	16	58	19	74%	39%	US
Fugitive	35	10	9	5	94%	61%	US
Geothermal	15	1	10	11	79%	27%	US
HFCs	22	2	20	5	89%	28%	Japan/ Germany (Tie)
Hydro	1372	986	148	24	75%	48%	China
Landfill gas	297	40	182	26	45%	16%	US
Methane avoidance	566	91	328	26	56%	20%	US
N₂O	70	–	66	14	85%	40%	Germany
PFCs and SF₆	17	2	6	4	–	25%	
Reforestation	42	18	6	4	–	33%	
Solar	47	14	21	9	77%	27%	US
Tidal	1	–	1	1	–	100%	Austria
Transport	24	2	9	5	86%	29%	Sweden
Wind	923	495	253	9	95%	34%	Germany
Total	4984	2262	1516	44	51%	17%	Germany

Note: * As a share of total projects.

presents the number of supplier countries and the shares of the largest supplier country and four largest supplier countries as percentages of the annual emission reductions for projects that claim technology transfer. A high percentage for the largest or four largest supplier(s) might indicate that an oligopoly dominates supply of the technology.

Of the 14 project types with at least 10 projects that claim technology transfer, only for EE own generation does the share of the largest foreign supplier country exceed 50 per cent. For the 13 other project types, the share of the largest supplier country ranges from 16 to 48 per cent of the projects that involve technology transfer. The combined share of the four largest supplier countries across these 14 project types ranges from 45 to 95 per cent. The number of supplier countries is nine or more, except for HFCs and perfluorocarbons (PFCs), where it is five.

The market shares of technology suppliers are lower than indicated by these figures. Since the firm(s) supplying the technology often is not specified in the PDD, the figures are calculated on the basis of supplier countries. Some countries have a few firms that export a given technology, so the firm shares would be lower. In addition, for most of these project types there is at least one domestic supplier in many host countries. Thus, the project types where there is sufficient data suggest that project developers have a choice among a number of domestic and/or foreign suppliers with no dominant supplier able to restrict the distribution of the technology and/or keep the price high.

Data for 231 projects for which the technology suppliers are known and credits have been transferred to a buyer's account were analysed to examine whether technology supply and credit purchases are related. For all buyers, except Austria and Switzerland, more of the credits purchased come from projects for which the country is a technology supplier. For some countries, the links are very strong – 91 per cent of Denmark's credit purchases come from projects for which it is a technology supplier.[11] The corresponding percentages for Spain, Germany and Japan are 50, 40 and 37 per cent, respectively. Whether the Certified Emissions Reduction (CER) purchases are contractually linked with the technology supply is not known.

Statistical analysis

The statistical analysis seeks to identify the characteristics of CDM projects and host countries that influence the rate of technology transfer. Previous studies have conducted such analyses by estimating a single equation (Haites et al., 2006; UNFCCC, 2007; Dechezleprêtre et al., 2008; UNFCCC, 2008; Doranova et al., 2010; Flues, 2010) or two equations (UNFCCC, 2010). Results for both approaches are reported here. The equations reported here are estimated using a much larger number of projects than any previous study and are the first to explicitly include changes in the variables over time.

With a single equation, technology transfer for a CDM project is related to the project type, project characteristics (size, unilateral, small scale, etc.), year the project enters the pipeline, and various host country characteristics (population, per capita GDP, etc.). A project either involves technology transfer (TT = 1) or does not involve technology transfer (TT = 0). With a dependent variable that can only take the values 0 or 1, the appropriate form of regression is a binomial logit model.

With a two equation approach, the first equation relates technology transfer to project type, project characteristics, year the project enters the pipeline and host country, but not the country characteristics. This equation is also a binomial logit regression. This estimated equation is then used to predict the probability of technology transfer for each combination of project type, host country and year based on the characteristics of the CDM projects in the host country that entered the pipeline that year, yielding a value between 0 and 1. For example, the first equation predicts that wind projects in India that entered the pipeline in 2005 have a probability of 0.2836 (28.36 per cent) of involving technology transfer.[12] The second equation relates those predicted probabilities to country characteristics to identify those that influence the rate of technology transfer.[13]

Data

The variables and sources of the data used in the analysis are specified in Table 9.4. A number of variables, including the project type, host country and year, are 'dummy' variables; they take a value of 1 if the project has that characteristic, and a value of 0 otherwise. For example, the China variable has a value of 1 for each project located in China and 0 for any project located in another host country. For a given dummy variable, one category must be used as the reference category to avoid problems with the regression analysis. EE own generation is used as the reference project type because it has a rate of technology transfer approximately equal to the average for all projects. Brazil is used as the reference host country, again because it has a rate of technology transfer approximately equal to the average for all projects. The reference year is 2008 because it has the largest number of projects and is near the middle of the period.[14]

Linear relationships among the variables − collinearity − is problematic for the regression analysis. Population, gross domestic product (GDP), foreign direct investment (FDI), fossil fuel generation and renewable generation are strongly correlated. To address the collinearity we use the log of population, per capita GDP, express FDI as a percentage of GDP, and calculate the renewable share of electricity generation. Correlation coefficients indicate that collinearity is not a problem for the other variables.

For the global climate change mitigation technology patent data variables used, the technologies covered by the patent applications need to be matched to the CDM project types. In the case of biomass energy, CO_2 capture, coal bed/mine methane, geothermal, HFCs, hydro, N_2O, PFCs and SF_6, solar, tidal and wind, the patent technology categories matched the project types closely. For cement, EE households, EE own generation, EE supply side, energy distribution, landfill gas and methane avoidance, the patent categories cover technologies used by some of the

TABLE 9.4 *Variables used in the regression analysis*

Variable name	Description
Project characteristics	
Size	Estimated annual emission reduction in tCO_2/year.
Unilateral	Dummy variable. 1 if the project has been approved only by the host country at the time it is posted for public comment; 0 otherwise.
Small scale	Dummy variable. 1 if the project uses a small-scale methodology. The size of projects eligible to use small-scale methodologies has changed over time.
Number	Number of projects of the same type in the host country. First project = 1, second project of the same type in the same country = 2, etc.
Project type	Dummy variable for each project type except EE own generation.
Year	Dummy variable for the year the PDD is posted for public comment. Variables for 2004, 2005, 2006, 2007, 2009 and 2010.
Host country characteristics	
Host country	Dummy variable for each host country except Brazil.
Population[a]	Natural logarithm of the population.
Per capita GDP[a]	Per capita GDP in 2005 US dollars using purchasing power parity exchange rates.
FDI[a]	Inward foreign direct investment as a percentage of GDP.
Capital formation[a]	Gross fixed capital formation as a percentage of GDP.
Imports[a]	Imports as a percentage of GDP.
Exports[a]	Exports as a percentage of GDP.
Tariff[b]	Tariff rate in 2007/2008.
ODA[c]	Official development assistance per capita.
Renewable[d]	Percentage of electricity generated from renewable sources in 2004.
Business[e]	Host country's rank on the ease of doing business index.
Democracy[f]	Host country's score on the democracy index (0 = least, 10 = most democratic).
Knowledge[g]	Host country's discounted stock of patent applications – current year = previous year's stock less 10% plus new applications. Covers all patents.
Technology[g]	Discounted stock of patent applications relating to the technology used by the CDM project.
Transfer[g]	Number of applications by foreign patent holders relating to the technology used by the CDM project.

Source: (a) World Bank (2011) World Development Indicators, http://data.worldbank.org/;
(b) WTO's World Tariff Profiles 2009, http://www.wto.org/english/res_e/publications_e/
world_tariff_profiles09_e.htm; (c) OECD 2011 statistics, http://www.oecd.org/document/0,3746,
en_2649_201185_46462759_1_1_1_1,00.html; (d) US Energy Information Administration
(2011) Data, http://www.eia.doe.gov; (e) World Bank (2010) Doing Business 2010,
http://www.doingbusiness.org/economyrankings/; (f) Hadenius, A. and Teorell, J. (2005) 'Assessing
alternative indices of democracy', Committee on Concepts and Methods Working Paper Series 6;
(g) Patent application data provided by Nick Johnstone and Ivan Haščič, OECD (for a description of
the data and the calculation of the discounted stock of patent applications, see Johnstone et al., (2010)

projects. Patent data for technologies used by afforestation, EE industry, EE service, fossil fuel switch, fugitive and transport projects are not available, so those project types are excluded from some regression equations.

A larger, richer country with a more open economy and more technology transfer via other channels is expected to have greater technological capacity and therefore less likely to need technology transfer for CDM projects. Population and GDP per capita are measures of the host country's size and income. Imports, exports, ease of doing business and the democracy index are all measures of an open economy. In contrast, higher tariffs reflect a less open economy. More FDI is an indicator both of an open economy and technology transfer via other channels. Official development assistance (ODA) projects can include, implicitly or explicitly, technology transfer. Patent applications by foreigners are a measure of technology transfer. The discounted stock of patent applications and, since many CDM projects involve renewables, the share of electricity generated by renewables are measures of the country's technological capacity. All of these variables, except the tariffs, are expected to have negative coefficients. A higher rate of growth, as measured by capital formation, is expected to create a demand for more technology and to have a positive coefficient.

Results: One equation approach

The estimated equations are presented in Table 9.5. With a single equation, collinearity is a problem if the country characteristics (population, GDP per capita, etc.) and host country dummy variables are included, so the dummy variables are excluded. Exclusion of the afforestation, EE industry, EE service, fossil fuel switch, fugitive and transport reduces the number of projects by 326. The statistical software drops any variable for which prediction is perfect. This happens if there is only one project in a category – project type, host country or year – or all projects in a category claim (or do not claim) technology transfer. As a result, the 69 CO_2 capture, N_2O and tidal projects are dropped, as well as 171 projects in 53 host countries and 4 projects in 2003. Finally, 34 observations are dropped due to missing values for some independent variables, leaving 3174 observations for analysis.

The statistical performance for the single equation is quite good, with a pseudo r^2 of 0.52 and correct prediction of over 88 per cent of the observations.[15] The results indicate that:

- Larger projects are more likely to involve technology transfer.
- Small-scale projects are less likely to involve technology transfer.
- Technology transfer falls as the number of projects of the same type in a host country increases.
- Although the effect is not statistically significant (at the 5 per cent level), unilateral projects are less likely to involve technology transfer.

TABLE 9.5 *Coefficients of the estimated regression equations (coefficients that are significant at the 0.05 level or more are highlighted in the table)*

Variable	Single equation			Two equation approach			
	Coefficient[a]	Marginal effect[b]	Lag	Equation 1 coefficient[a]	Marginal effect[b]	Equation 2 coefficient	Lag
Size	1.16E-06	1.59E-07		1.00E-06	1.67E-07		
Unilateral[c]	-0.150	-0.021		-0.097	-0.017		
Small scale[c]	-0.859	-0.114		-0.836	-0.136		
Number	-0.005	-7.40E-04		-0.006	-0.001		
Year 2004[c]	0.423	0.066		0.644	0.126		
Year 2005[c]	1.021	0.183		0.688	0.134		
Year 2006[c]	1.340	0.248		1.140	0.235		
Year 2007[c]	0.519	0.079		0.545	0.100		
Year 2009[c]	-0.100	-0.013		0.022	0.004		
Year 2010[c]	-0.834	-0.092		-0.003	-0.001		
Argentina[c]				2.206	0.497		
Armenia[c]				1.948	0.441		
Chile[c]				3.352	0.675		
China[c]				-0.551	-0.092		
Colombia[c]				2.420	0.539		
Ecuador[c]				3.748	0.709		
El Salvador[c]				1.202	0.260		
Guatemala[c]				2.911	0.620		
India[c]				-0.830	-0.120		
Indonesia[c]				3.958	0.729		
Iran[c]				1.699	0.383		
Kenya[c]				3.543	0.692		
Korea (South)[c]				2.619	0.575		
Lao PDR[c]				2.734	0.594		
Malaysia[c]				3.333	0.676		
Mexico[c]				3.897	0.732		
Nigeria[c]				1.376	0.303		
Pakistan[c]				2.748	0.596		
Peru[c]				3.224	0.660		
Philippines[c]				1.034	0.217		
South Africa[c]				2.309	0.518		
Sri Lanka[c]				3.850	0.716		
Thailand[c]				4.016	0.736		
Uganda[c]				0.963	0.201		
United Arab Emirates[c]				2.350	0.526		
Uruguay[c]				3.839	0.716		
Viet Nam[c]				5.744	0.796		
Afforestation[c]				-3.879	-0.207		
Biomass energy[c]	-1.167	-0.118		-1.295	-0.159		
Cement[c]	-1.308	-0.114		-1.673	-0.165		
Coal bed/mine methane[c]	0.100	-0.014		-0.125	-0.020		
EE households[c]	0.023	0.003		0.062	0.010		
EE industry[c]				1.380	0.302		
EE service[c]				1.154	0.248		
EE supply side[c]	0.100	0.014		-0.022	-0.004		
Fossil-fuel switch[c]				0.626	0.122		
Fugitive[c]				-3.184	-0.203		
Geothermal[c]	0.638	0.106		0.143	0.025		
HFCs[c]	1.429	0.285		1.228	0.266		
Hydro[c]	-2.507	-0.279		-2.426	-0.317		
Landfill gas[c]	0.875	0.152		0.662	0.129		
Methane avoidance[c]	-0.012	-0.002		-0.197	-0.031		

TABLE 9.5 *Coefficients of the estimated regression equations (cont'd)*

Variable	Single equation Coefficient[a]	Marginal effect[b]	Lag	Two equation approach Equation 1 coefficient[a]	Marginal effect[b]	Equation 2 coefficient	Lag
Reforestation[c]				-3.756	-0.209		
Solar[c]	-0.909	-0.091		-0.780	-0.102		
Transport[c]				1.850	0.419		
Wind[c]	0.808	0.128		0.357	0.063		
Population	-2.008	-0.275	3			-0.188	2
Per capita GDP	-1.49E-04	-2.04E-05	1			-1.01E-05	1
FDI	-0.040	-0.006	2			-0.009	1
Capital formation	0.115	0.016	3			-0.003	1
Imports	0.030	0.004	2			-0.005	3
Exports	-0.098	-0.014	1			-0.004	3
Tariff	-0.167	-0.023				-0.020	
ODA	-0.216	-0.030	1			-0.018	2
Renewable	-0.007	-0.001				-0.003	
Business	-0.032	-0.004				-0.002	
Democracy	-1.205	-0.165				-0.138	
Knowledge	-2.99E-05	-4.09E-06				-2.61E-06	
Technology	5.57E-04	7.62E-05	1			-1.19E-04	1
Transfer	-0.006	-7.90E-04	2			8.49E-04	1
Constant	28.712			0.355		3.822	
Observations	3,174			3,530		364	
Pearson's Chi[2d]	2,047			2,367		30.6[e]	
Probability > Chi[2d]	0.00			0.00		0.00[e]	
Pseudo R^{2f}	0.52			0.51		0.53	
Correctly classified[g]	88.15%			86.63%			

Notes: [a] Binomial logit regression. The coefficients describe the effects of the independent variables on the predicted logarithmic odds of technology transfers. [b] The marginal effect is the effect of a one unit change in the variable on the predicted probability of technology transfer. This is the same as the coefficient estimated using the ordinary least squares method. Thus, the marginal effects for the single equation can be compared with the coefficient for equation 2. For example, the marginal effect of exports is −.014. This implies that if the host country's exports as a percentage of GDP rise 1 percentage point, the model predicts a drop in probability of technology transfer of 1.4 percentage points, other things equal. The marginal effect of project size is 1.59E-07, which implies that if a project's estimated annual emission reductions increase by 1 tCO$_2$/year, the model predicts an increase in probability of technology transfer of 1.59E-07, other things equal. In the case of a dummy variable, the marginal effect is the effect of the presence of the characteristic on the predicted probability of technology transfer. For example, the single equation predicts that the probability of technology transfer is 27.9% lower for a hydro project, other things equal. [c] Dummy variable. [d] The value of the Pearson Chi^2 is used to test the null hypothesis that the coefficients of all of the variables are equal to zero. The probability of a Chi^2 value greater than the value calculated for each of the equations is less than 0.0000, indicating that at least some of the variables are statistically significant. That is confirmed by tests for the individual coefficients. The shaded values in the table indicate coefficients statistically significant at the 5% confidence level. [e] In the case of ordinary least squares estimation, the F test is used to test the null hypothesis that the coefficients of all of the variables are equal to zero. The value of F (14, 349) indicates that at least some of the variables are statistically significant, which is confirmed by tests for the individual coefficients. [f] The pseudo R^2 is an indicator of the explanatory power of the equation. A value of 0 indicates no explanatory power, while a perfect explanation would have an R^2 = 1. [g] The percentage of observations correctly classified is another indicator of the explanatory power for the binomial logit equations. If the equation predicts a probability of technology transfer greater than 0.5 for a project, given its characteristics, it is correctly classified if technology transfer was claimed and incorrectly classified if no technology transfer was claimed. Similarly, if the predicted probability is less than 0.5, it is correctly classified if no technology transfer was claimed and incorrectly classified if technology transfer was claimed. Since the dependent variable for Equation 2 can have any value between 0 and 1 inclusive, the correctly classified calculation cannot be applied to that equation.

The year variables indicate that the rate of technology transfer has changed significantly over the relatively short life of the CDM (2004 to 2010). Technology transfer was more common during the early years of the CDM and has become less frequent since 2008. Of the 13 project types included in the equation, biomass energy, cement and hydro projects are less likely than average to involve technology transfer, while landfill gas and wind projects are more likely than average to involve technology transfer.

The single equation model has statistically significant coefficients with the expected negative signs for population, exports as a percentage of GDP, ODA per capita, ease of doing business rank, democracy index score, and stock of knowledge (as measured by discounted stock of patent applications). A higher rate of capital formation, as expected, is associated with a higher rate of technology transfer in CDM projects.

The knowledge stock specific to the CDM project as measured by the discounted stock of patent applications for the relevant technology(ies) and the number of applications by foreign patent holders for technologies relevant to the CDM project – a measure of technology transfer via all channels for those technologies – are found to be jointly significant in statistical tests. But the statistical significance and signs of the individual coefficients are not consistent. In the single equation model the knowledge stock specific to the CDM project has the wrong (positive) sign and is not statistically significant, but it is statistically significant with the expected negative sign in the two equation approach. The number of applications by foreign patent holders for technologies relevant to the CDM project is statistically significant in both the one and two equation approaches and has the expected negative sign in the single equation model, but a positive coefficient in the two equation approach. The inconsistent results may reflect variability in the precision of the match between the technologies covered by patent applications and the CDM project types – for example, a close match for wind but less precise match for landfill gas.

Where we have time series data for the country characteristics, we test different lags for their impact upon the rate of technology transfer in CDM projects. The single equation results suggest that population growth in a host country reduces the rate of technology transfer three years later. Most of the host country characteristics have lags of one or two years, which suggests that their effect on the rate of technology transfer is relatively quick. The estimated lags should be considered approximations. A lag that is a year longer or shorter often yields similar statistical results. Hence, the estimated lag for a given characteristic often differs between the one and two equation approaches.

Results: Two equation approach

A two equation approach allows the analysis of more project types and inclusion of host country dummy variables in the first equation. Due to perfect prediction, the statistical software drops 53 of the 80 host countries (most with only one or two projects) as well as the CO_2 capture, N_2O and tidal project types, leaving 3530 observations. The 27 countries that remain host 88 per cent of all projects. These projects represent 535 combinations of project type, host country and year the project was posted for public comment. When those for afforestation, EE industry, EE service, fossil fuel switch, fugitive and transport are deleted, we are left with 364 observations for estimation of the second equation.[16]

The statistical results for the two equation approach are an improvement over those reported in UNFCCC (2010) with a pseudo r^2 of 0.51 and correct prediction of over 86 per cent of the observations for the first equation and an adjusted r^2 of 0.53 for the second equation.[17] The improved results are due, with roughly equal contributions, to the use of time series data for many of the independent variables, the use of lags for those variables, and the inclusion of knowledge stock specific to the CDM project and number of applications by foreign patent holders for technologies relevant to the CDM project.

The results reported above for the project characteristic and year variables are confirmed by the estimated coefficients for equation 1 shown in Table 9.5. The estimated coefficients also show that the host country has a significant influence on the rate of technology transfer; 20 of the 27 host country variables are statistically significant. With a strictly random distribution, about 5 per cent of the host countries – 1 or 2 – would be statistically significant, so 20 is a particularly strong result. Consistent with Figure 9.1, almost all host countries, except China and India, have a rate of technology transfer higher than the average for all projects.

Afforestation, biomass energy, cement, fugitive, hydro, and reforestation projects are less likely than average to involve technology transfer, while EE industry, landfill gas transportation and wind projects are more likely than average to involve technology transfer. These results are consistent with those of the single equation approach.

The estimated coefficients for equation 2 in Table 9.5 relate host country characteristics to the rate of technology transfer for CDM projects. The results are similar to those reported for the one equation approach. A host country with a larger population, higher tariffs, more ODA per capita, a higher percentage of renewable generation, a higher rank for ease of doing business, a higher score on the democracy index, and a larger stock of knowledge (as measured by discounted stock of patent applications) is likely to have a lower rate of technology transfer for CDM projects. The export and capital formation variables are not statistically significant, while the tariff and renewable variables are significant.

Comparison with other results

Our results are generally consistent with those of previous studies (Dechezleprêtre et al., 2008; Doranova et al., 2010; Flues, 2010).[18] The coefficient for the log of population is negative and statistically significant. Dechezleprêtre et al., (2008) find a negative coefficient that is not statistically significant, while Doranova et al., (2010) find a significant positive coefficient. Flues (2010) finds population to be a significant positive influence on the number of CDM projects that a country is likely to host.

Our coefficient for per capita GDP is negative and not significant; Deche-zleprêtre et al., find that it has an insignificant negative coefficient, while Doranova et al., find an insignificant positive coefficient. Our coefficient for FDI inflows as a percentage of GDP is negative and not significant. Dechezleprêtre et al., also find an insignificant negative coefficient, while Doranova et al., exclude this variable due to collinearity concerns.

Dechezleprêtre et al., (2008) use total trade – imports plus exports as a percent-age of GDP – and find that it has a significant positive coefficient. Doranova et al., (2010) find that this variable has an insignificant positive coefficient. We separate imports and exports and find a significant negative coefficient in the one equation approach and insignificant negative coefficients in the two equation approach.

The coefficient for the renewables share of electricity generated is negative and significant in the two equation approach. Doranova et al., (2010) find that it has an insignificant positive coefficient. Dechezleprêtre et al., (2008) use the ArCo index[19] as a measure of a country's technological capacity and find a significant positive coefficient, but note that the sign and its statistical significance differs by sector. Our measure of a country's technological capacity – the discounted stock of patent applications – has a significant negative sign indicating that the greater a country's capacity, the lower the rate of technology transfer for CDM projects.

Barriers to technology transfer

Under the UNFCCC, developing countries can request funding to conduct a Technology Needs Assessment (TNA). TNAs follow a country-driven approach, bringing together stakeholders to identify technologies needed by the country to reduce greenhouse gas emissions and barriers to the implementation of those technologies.[20] The UNFCCC Secretariat groups the barriers identified into nine categories – economic/market; human capacity; information and awareness; insti-tutional; regulatory; policy related; technical; other; and infrastructure – in declining order of frequency. All of the barriers identified by the TNAs are general barriers, not barriers specific to CDM projects.

For 43 countries it is possible to relate the barriers identified to specific technol-ogies and, hence, to CDM project types. This allows barriers to technology trans-fer to be included in the analysis of factors that influence the rate of technology transfer in the CDM for the first time. These 43 countries have a higher rate of technology transfer for CDM projects than that for other CDM host countries. So the barriers in these 43 countries may not be as daunting as for other CDM host countries and the results may not apply to other CDM host countries.

Barrier variables are not included in the results presented in Table 9.5 due to multicollinearity problems. Inclusion of the knowledge stock specific to the CDM project and the number of applications by foreign patent holders for technologies relevant to the CDM project substantially reduces the number of host countries and project types. The overlap with the barrier data then becomes quite small: five host countries with 1852 observations (90 per cent in China) for the single equation approach and six host countries and 120 observations for equation 2 in the two equation approach. In both approaches, multicollinearity is a problem.

In UNFCCC (2010) the barriers are included in equation 2 of the two equation approach. Barrier data are available for only a few of the 27 host countries covered by the statistical analysis, so the number of observations is reduced from 494 to 177.[21] Of the nine categories of barriers, three – economic/market, information and intellectual property rights – are found to be statistically significant. Including the barriers changes the sign and level of statistical significance of several of the other country characteristic variables.

Countries and project types that face an information[22] barrier have a lower rate of technology transfer for CDM projects, while countries and project types that face economic/market or intellectual property right (IPR)[23] barriers have a higher rate of technology transfer for CDM projects. These results can be interpreted as indicating that the CDM helps to overcome some barriers to technology transfer (Schneider et al., 2008). CDM projects, for example, earn revenue from the sale of credits, thus helping to overcome some economic/market barriers. Concerns about a host country's IPR regime that inhibit technology transfer could be addressed for specific CDM projects. For example, licences or other agreements with the CDM project participants covering the imported knowledge and/or equipment could provide better protection than the country's IPR regime and therefore lead to more technology transfer through CDM projects.

Conclusions

Technology transfer is not an explicit objective of the CDM, but the CDM contributes to technology transfer by financing emission reduction projects using technologies currently not available in the host countries. It is clear from the PDDs that project participants overwhelmingly interpret technology transfer as meaning the use of equipment or knowledge not previously available in the host country.

Of the projects that explicitly state they will, or will not, involve technology transfer, 40 per cent accounting for 59 per cent of estimated emission reductions involve technology transfer. Previous studies underestimate the rate of technology transfer because they assume no mention of technology transfer in the PDD means the project does not involve technology transfer, which is incorrect.

Technology transfer is more common for larger projects. Technology transfer is very heterogeneous across project types. The host country has a significant impact upon the rate of technology transfer. Almost all countries, except China, India and Brazil, have a higher than average rate of technology transfer.

The rate of technology transfer for all projects has declined over time. The decline has been steeper than the overall average in China, India and Brazil, the countries that host most of the CDM projects. The rate of technology transfer for all other host countries is much higher than the overall average and has declined only modestly.

The technology transferred mostly (58 per cent) originates from Germany, the US, Japan, Denmark and China. Each of these countries is the leading technology supplier for one (China) to nine (US) project types. Most Annex I countries tend to buy credits from projects for which they are a technology supplier, although the nature of this relationship is not known.

The project types for which we have sufficient data suggest that project developers have a choice among a number of domestic and/or foreign suppliers with no dominant supplier able to restrict the distribution of the technology and/or keep the price high.

The statistical analysis indicates that the host country has a significant impact upon the rate of technology transfer for CDM projects. Technology transfer via the CDM is less likely if the host country already has a larger technological base. This is the case for a host country with a larger population, more ODA per capita, a higher rank for the ease of doing business, a higher score on the democracy index, and a larger stock of knowledge (as measured by discounted stock of patent applications). Other characteristics may also play a role.

The statistical analysis is the first to explicitly include time lags, knowledge stock specific to the CDM project type and host country, and a measure of overall technology transfer specific to the project type and host country (applications by foreign patent holders). All of these additions contribute to the improved statistical results.

Notes

1 We would like to express our appreciation to Ana Pueyo for insightful comments on an earlier draft of this chapter.
2 See Popp (2011, p136).
3 UNCTAD (1985, Chapter 1, para 1.2) excludes the mere sale or lease of goods from technology transfer.
4 Equipment imports could involve some training that is not mentioned as a transfer of knowledge in the PDD.
5 See UNFCCC (2010, p36).
6 See UNFCCC (2010, Table A-9, p37). If equipment imports involve some training that is not mentioned as a knowledge transfer, the percentage would be lower.
7 Although some proposed projects will be rejected or withdrawn, a statistical test indicates that projects yet to be registered do not differ from registered projects (UNFCCC, 2010). Using all projects in the pipeline provides a larger population for the analysis.
8 Keywords included technology, transfer, import, foreign, abroad, overseas, domestic, indigenous, etc.
9 Projects for which technology transfer is not known are excluded. The decline is much larger when measured in terms of estimated annual emission reductions than in terms of number of projects (UNFCCC, 2010).
10 Many PDDs indicate that there will be a transfer of technology, but do not specify the source of the technology. This is, at least partly, due to projects for which the technology has not yet been sourced because the project has not yet been implemented. The source of the technology is unknown for about 20 per cent of the projects that involve technology transfer. If more than one country supplied technology to a project, each country is credited with a fraction of the project.
11 The UK and Switzerland are listed as buyers of 33 and 30 per cent, respectively, of the credits. This reflects the purchases by funds and other financial intermediaries located

in these countries. For these countries, the percentage of projects for which it is a technology supplier from which it buys credits is a better indicator. The UK purchases credits from about 52 per cent of the projects to which it supplies technology, while Switzerland buys credits from less than 1 per cent of the projects for which it is a technology supplier.

12 If there is only a single project of a given type that enters the pipeline during a year in a specific country, the predicted probability would be approximately equal to the claimed technology transfer ($TT = 1$ or $TT = 0$) for that project. With multiple projects of the same type in a host country entering the pipeline in a given year, the probability will be an average for those projects and so have a value between 0 and 1.

13 Since the predicted probabilities can have any value between 0 and 1 inclusive, the second equation is estimated using the ordinary least squares method.

14 The year (2004, 2005, etc.) and increasing numerical values ($2004 = 1$, $2005 = 2$, etc.) were tested in lieu of the dummy variables, but they did not work as well because the time trend is not linear. Those definitions would yield only a single coefficient for the time trend. The dummy variables yield six coefficients, some of which are positive while others are negative.

15 The pseudo r^2 and percentage of observations correctly classified are indicators of the explanatory power of the equation. If the equation predicts a probability of technology transfer greater than 0.5 for a project, given its characteristics, it is correctly classified if technology transfer was claimed and incorrectly classified if no technology transfer was claimed. Similarly, if the predicted probability is less than 0.5, it is correctly classified if no technology transfer was claimed and incorrectly classified if technology transfer was claimed.

16 These are the project types with no matching technologies with patent data and, hence, no data for the patent variables.

17 Since the predicted probabilities, which are the dependent variable for the second equation, can take any value between 0 and 1, it is not possible to calculate the percentage of observations classified correctly.

18 Flues (2010) examines the factors that determine the probability that a country will host a CDM project.

19 See Archibugi and Coco (2004).

20 TNAs also cover adaptation technologies, but the analysis here is limited to mitigation technologies.

21 See UNFCCC (2010, Tables B-14 and B-15).

22 Examples of information barriers in TNAs include, in rank order, shortage of information on energy efficiency and ecological safety of technology equipment; shortage of information about governmental structures; difficulties in obtaining information on organizations and companies that deal with energy efficient and modern climate change mitigation technologies; lack of information among investors on the potential technology market; and lack of information about financing.

23 An IPR barrier means that potential technology suppliers are concerned about transferring technology to the country due to the perceived weakness of the IPR regime.

References

Archibugi, D. and Coco, A. (2004) 'A new indicator of technological capabilities for developed and developing countries (ArCo)', *World Development*, vol 32, no 4, pp629–654, http://www.danielearchibugi.org/papers/index.htm

Das, K. (2011) *Technology Transfer under the Clean Development Mechanism: An Empirical Study of 1000 CDM Projects*, Working Paper 014, The Governance of Clean Development Working Paper Series, School of International Development, University of East Anglia, Anglia, UK

de Coninck, H..C., Haake, F. and van der Linden, N. (2007) 'Technology transfer in the Clean Development Mechanism', *Climate Policy*, vol 7, no 5, pp444–456

Dechezleprêtre, A., Glachant, M. and Ménière, Y. (2008) 'The Clean Development Mechanism and the international diffusion of technologies: An empirical study', *Energy Policy*, vol 36, pp1273–1283

Dechezleprêtre, A., Glachant, M., Haščič, I., Johnstone, N. and Ménière, Y. (2011) 'Invention and transfer of climate change mitigation technologies on a global scale: A study drawing on patent data', *Review of Environmental Economics and Policy*, vol 5, no 1, pp109–130

Doranova, A., Costa, I. and Duysters, G. (2010) 'Knowledge base determinants of technology sourcing in Clean Development Mechanism projects', *Energy Policy*, vol 38, pp5550–5559

Flues, F. (2010) *Who Hosts the Clean Development Mechanism? Determinants of CDM Project Distribution*, Working Paper 53, Centre for Comparative and International Studies (ETH Zurich and University of Zurich), Zurich

Haites, E., Duan, M. and Seres, S. (2006) 'Technology transfer by CDM projects', *Climate Policy*, vol 6, no 3, pp327–344

Haščič, I. and Johnstone, N. (2009) *The Clean Development Mechanism and International Technology Transfer: Empirical Evidence on Wind Power Using Patent Data*, OECD, Paris

Johnstone, N., Haščič, I. and Watson, F. (2010) *Climate Policy and Technological Innovation and Transfer: An Overview of Trends and Recent Empirical Results*, OECD, Paris

Popp, D. (2011) 'International technology transfer, climate change, and the Clean Development Mechanism', *Review of Environmental Economics and Policy*, vol 5, no 1, pp131–152

Schneider, M., Holzer, A. and Hoffmann, V. H. (2008) 'Understanding the CDM's contribution to technology transfer', *Energy Policy*, vol 36, pp2920–2928

Seres, S., Haites, E. and Murphy, K. (2009) 'Analysis of technology transfer in CDM projects: An update', *Energy Policy*, vol 37, pp4919–4926

UNCTAD (United Nations Conference on Trade and Development) (1985) *Draft International Code of Conduct on the Transfer of Technology, as at the Close of the Sixth Session of Conference on 5 June 1985*, Document no TD/CODE TOT/47, 20 June, United Nations, Geneva

UNFCCC (United Nations Framework Convention on Climate Change) (2006) *Guidelines for Completing the Project Design Document (CDM-PDD), and the Proposed New Baseline and Monitoring Methodologies (CDM-NM)*, Version 05, http://cdm.unfccc. int/Reference/Documents/copy_of_Guidel_Pdd/English/Guidelines_CDMPDD_ NM.pdf

UNFCCC (2007) *Analysis of Technology Transfer in CDM Projects* (lead author S. Seres), http://cdm.unfccc.int/Reference/Reports/TTreport/index.html

UNFCCC (2008) *Analysis of Technology Transfer in CDM Projects* (lead author S. Seres), http://cdm.unfccc.int/Reference/Reports/TTreport/index.html

UNFCCC (2010) *Analysis of the Contribution of the Clean Development Mechanism to Technology* (S. Seres with E. Haites and K. Murphy), UNFCCC, Bonn, http://cdm. unfccc.int/Reference/Reports/TTreport/TTrep10.pdf

10

PROJECT-BASED MARKET TRANSFORMATION IN DEVELOPING COUNTRIES AND INTERNATIONAL TECHNOLOGY TRANSFER

The Case of the Global Environment Facility and Solar Photovoltaics

Rüdiger Haum

Introduction

Low-carbon technologies hold the potential to mitigate climate change but also to contribute to national economic development through their contribution to improving technological and, hence, productive capacities, as well as benefits for improving energy security and energy access. Considering the continuing increase in carbon dioxide emissions worldwide and continuing widespread poverty in developing countries, the international transfer of low-carbon technologies seems more necessary than ever before (German Advisory Council on Global Change, 2011).

While the majority of low-carbon technologies have been developed within industrialized countries, recent research suggests that their international transfer through private actors, especially to developing countries, has been disproportionally slow (Dechezleprêtre et al., 2011). Political intervention in the form of bilateral official development assistance (ODA) targeted at the energy sector and in the form of multilateral dedicated efforts under the United Nations Framework Convention on Climate Change (UNFCCC) exists, most notably in the form of Joint Implementation (JI), the Clean Development Mechanism (CDM) and the Global Environment Facility (GEF).

However, research on the contribution of public funding to the international transfer of low-carbon technology is only recent (Peterson, 2008).[1] While initial efforts have been made to research the contribution of the CDM, almost no research exists on the subject of the GEF and international transfer of low-carbon technologies (Haites et al., 2006, Haum, 2011). The GEF is a multilateral finance institution aiming to solve global environmental problems through supporting projects in developing countries. A large share of these projects aims to mitigate climate change by supporting international technology transfer. In 2009, the GEF described itself 'as the largest public sector funding source supporting the transfer of environmentally sound technologies to developing countries' (GEF, 2009b, p1).

This chapter will add to understandings of the potential contribution of the GEF technology transfer approach to economic development. In order to do so, it reconstructs the GEF approach to designing technology transfer support projects and investigates one GEF project supporting solar photovoltaic technology (SPV) in India. The chapter will proceed as follows. It first explains how international technology transfer may contribute to economic development by expanding the technological capabilities of manufacturing firms. It then gives a short introduction to innovation in solar photovoltaic technology, as the GEF project aims to support this particular low-carbon technology. Third, the GEF is introduced and its market transformation approach is discussed, which is the main approach after which GEF climate technology projects are designed. The chapter then undertakes a case study of the Photovoltaic Market Transformation Initiative (PVMTI) in India, which was designed according to the market transformation approach. Finally, conclusions are drawn on market-based approaches and international technology transfer within multilateral finance institutions.

This chapter is based on document analysis and eight expert interviews with representatives of the GEF Secretariat, the GEF implementing agencies (the United Nations Environment Programme (UNEP), the United Nations Development Programme (UNDP), the World Bank and the International Finance Corporation (IFC)) and the UNFCCC Secretariat to understand the GEF market transformation approach. In order to understand the PVMTI project in India and its results in general, 11 interviews were conducted with relevant stakeholders of the PVMTI project. These include representatives of the agencies planning and financing the project (the GEF and IFC), the firms that were implementing the project (IMPAX Capital Management and IT Power India), the firms in India receiving PMVTI funding, as well as Indian photovoltaic (PV) experts.

In order to understand any effects of PVMTI India on international technology transfer activities, 16 representatives of the Indian photovoltaic industry were interviewed. Interviews aimed to establish existing technological capabilities of the industry (in the form of rough proxies), current international technology transfer activities and its drivers, as well as any role that PVMTI might have played. Research took place from July to October in 2008, ten years after the start of PVMTI and one year before its official closure.

Although the GEF's primary mission is to achieve environmental goals, understanding the GEF's contribution to economic development is justified for two reasons. First, developing countries are very much interested in the economic effects of the international transfer of low-carbon technology and have expressed this position in negotiations under the UNFCCC. In fact, some developing countries made access to technology the condition to participate in international climate policy as the UNFCCC aims to solve a problem for which they have no historical responsibility (Ockwell et al., 2010). Second, the political discourse abounds with references to economic effects of actions that are primarily aiming to reduce carbon emissions. Getting more clarity for the GEF is also desirable in order to inform political debates.

International technology transfer to developing countries and low-carbon technology

This section briefly conceptualizes international technology transfer as an industrial learning process leading to the expansion of technological capabilities of firms in developing countries. It is against this background that the GEF activities on low-carbon technology transfer will be discussed.

Technological capabilities

In its most basic sense, international technology transfer relates to the geographical relocation of knowledge between nation states (Bozeman, 2000). Research on the subject has underlined that technology may be understood as knowledge and international technology transfer as a dynamic learning process between a technology supplier and a technology recipient (Dahlmann and Westphal, 1981; Reddy and Zhao, 1990; Ivarson and Alvstam, 2005). Central to the transfer process is the notion of technological capabilities (Aggarwal, 2001).

Within a transfer process the supplier sources knowledge that is either embodied in machinery, codified (e.g. in manuals or blueprints) or tacit within a person or a group. The recipient learns how to make use of the supplied knowledge and expands its technological capabilities (Rosenberg and Fritschak, 1985). Technological capabilities within a firm are the skills that firms need to 'utilize efficiently the hardware (equipment) and software (information) of technology' (Morrison et al., 2008). They can be technical, managerial or organizational.

Technological capabilities may be divided into operational and innovative capabilities. Operational capabilities relate to the capability to use and operate a technology (Lall, 1992; Bell, 2009). Innovative capabilities relate to the capability to 'create new configurations of production technology and to implement changes and improvements to technologies already in use' (Bell, 2009, p2). Other classifications differentiate even further between capabilities. For the purpose of this research, I will distinguish between operational capabilities, intermediate capabilities (ability to perform incremental innovations) and innovative capabilities (ability to undertake equipment design at the technological frontier).[2] Intermediate capabilities are important because the successful operation of technology received within the transfer process often requires adaptation to local circumstances and needs in order to operate effectively. Innovative capabilities are important because in order to contribute to the long-term competitiveness of the recipient, they must learn to go further from the technology received and be able to develop innovative production technology, leading to more innovative products than its technology suppliers. Otherwise, they will be dependent upon external technology supply from others in cases where technology development advances elsewhere (Amsden, 2001; Figueiredo, 2002).[3]

The international transfer of low-carbon technology differs in two respects from conventional technology. First, as low-carbon technology (like most environmental technology) suffers from market failures and little demand exists in many countries, its international transfer as well as its widespread application must be supported through political intervention (Rennings, 2000; Smith, 2009). Second, as the primary goal of international transfer of low-carbon technology is the reduction of carbon dioxide, intermediate capabilities for local adaptation need to be built for successful transfer since these support effective operation and emission reduction. To achieve emission reductions, low-carbon technology might be transferred as an end-use product (e.g. advanced coal combustion technology) that might be installed into existing or newly-built power plants. Innovative capabilities to develop and manufacture new low-carbon technologies gain more importance if international transfer of low-carbon technology contributes to industrial development via the production of low-carbon technology (Haum, 2011).

Evolutionary economists stress that the process of transferring knowledge between firms is costly, requires specific learning efforts and is not necessarily successful. The main reasons mentioned in evolutionary economics are that knowledge is partly tacit and firms differ in their skills and competencies. Tacit knowledge refers to knowledge that individuals or groups of individuals cannot fully express as it derives from accumulated experience and includes rules of thumb, intuition, experience, etc. It may only be acquired through time-consuming face-to-face contacts or may be impossible to transfer at all, or the recipient has to acquire it through imitating the learning process of the technology supplier (Dosi, 1988; Cantwell, 1993; Rose et al., 2009). Firms differ with regard to their skills and competencies because their management make choices over time to produce certain goods and/ or provide certain services, which requires coherency and concentration on innovation related to the competencies and skills for that particular product or service. In the interpretation of the evolutionary economist, deciding and learning to do one thing will, to a certain extent, predispose the direction of future innovation as 'the learning and complementary strengths developed in the former effort provide a base of the next round'. Acquiring certain skills and competencies in one area, however, excludes the development of skills in other areas and firms will develop what Nelson calls 'core capabilities' that differ among firms to varying degrees (Nelson, 1991, p68). This means, in practice, that the level of differences in technological capabilities between technology supplier and recipient matters. The greater the difference, the greater the learning effort on the side of the technology recipient (Radosevic, 1999).

This has led to what some have referred to as the technology paradox. Technological capabilities on the recipient side are a condition for successful transfers. From the perspective of the supplier, if a recipient has sufficient technological capabilities, these will reduce transaction costs; they offer financial returns on transfer insofar as they allow the transfer in the first place (Teece, 1981). There is empirical evidence that some firms wanted to transfer 'more' technology in developing countries but effectively did not as recipients were lacking to use it (Scott-Kemmis and Bell, 1988). Companies in industrialized countries therefore tend 'to supply technology to potential competitors in other industrialized countries than to firms in the developing countries', which is somewhat paradoxical as technology transfer is needed mostly where technological capabilities are low (Baark, 1991, p911).

Conceptualizing international technology transfer as an industrial learning process leading to the formation of technological capabilities means that the relocation and application of a certain technology does not necessarily constitute a case of technology transfer. International technology transfer from this perspective has effectively taken place in cases where the recipient acquired new skills and expanded technological capabilities (Archibugi and Iammarino, 1999; Freeman and Soete, 2007; Ockwell et al., 2009). From the perspective of long-term competitiveness, firms must hold the operational as well as innovative capabilities with regard to the technological frontier.

Access to technology

The primary condition for international technology transfer leading to an expansion of the technological capabilities of the recipient is the availability of technology (Wei, 1995). Firms supply technology usually according to their business goals. These goals vary and include an increase in market share, to lower production cost, to diversifying product outreach, to gaining strategic advantages over competitors, and to gaining knowledge about local markets or special assets such as technological knowledge (Goulet, 1989; Narula and Dunning, 2000; Bruun and Bennett, 2002). While firms have a number of incentives to transfer parts of their technology internationally, at the same time they also have an interest in controlling their technological assets. The interest in control over technology stems from the importance of technological advantage for the economic performance of the business firm (Porter, 1985; Kogut and Zander, 1993). For Schumpeter (1961), innovation incorporating technical change is the most important source for quasi-monopolistic profits. Suppliers therefore might not only critically select recipients and countries, but also engage in 'international technology transfer management' by trying to retain some parts of their specific knowledge despite all inherent interest in the success of the transfer process (Liebeskind, 1996; Cannice et al., 2003). From a supplier perspective, the problem of losing a technological advantage through international technology transfer is framed in the terms of techno-economic security or technology leakage. Many firms transferring technology are aware of this problem and react with different strategies: transferring outdated technology, retaining core technology, increasing R&D efforts, changing ownership structures of a subsidiary, accepting leakage, and attempting to keep personnel (Bruun and Bennett, 2002; Cannice et al., 2003). In summary, reviewed theory suggests that firms have complex and diverse goals to internationally transfer technology that do not necessarily coincide with the goals of technology recipients.

Markets

The notion of markets can be evoked in two different meanings in relation to international technology transfer. In the first meaning, the notion of markets is used to categorize the different forms in which international technology transfer takes place. Market-mediated (also market-based) international technology transfer comprises those transfer activities that include the negotiation of a payment between supplier and recipient, such as licensing of a certain technology or purchase of machinery. Non-market-mediated transfer activities comprise, for example, technical assistance or imitation (reverse engineering) (Kim, 1991; Hoekman et al., 2005). The second meaning of markets in the context of international technology transfer relates to the motivation of technology owners to transfer its technology abroad. According to Narula and Dunning (2000), firms transfer technology mainly for four reasons: to seek natural resources; to gain entry into new markets; to restructure existing production; and to seek local strategic assets (e.g. highly trained engineers). German wind turbine technology manufactures, for example, chose to transfer parts of their technology mainly to gain access to the Chinese markets for wind power technology (Haum, 2004). Likewise, access to new markets may be a motivation for companies to engage in the acquisition of technology from abroad (technology import).

Summary

The theory reviewed here suggests that the effects of the GEF project on international technology transfer should be discussed with regard to the technological capabilities of Indian solar photovoltaic manufacturing firms, as well as their access to technology. This includes understanding their present technological capabilities and what role international technology transfer played in their formation, how far they are able to innovate, and what kind of access they have to technology.

Innovation in PV technology and technological capability for PV manufacturers

Photovoltaic systems convert sunlight into electricity. They usually consist of modules that contain a certain number of cells, an inverter, a support structure and a battery for stand-alone systems. The value chain in SPV comprises silicon ingot production, wafer slicing, PV cell production, PV module production, system assembly, sales, and after-sales services. This section considers innovation in PV cells and module manufacturing.

The conversion of sunlight takes place in the photovoltaic cell. PV cells can be classified in first-, second- or third-generation cells. First-generation technologies are based on crystalline silicon. Second generation includes thin film technologies, while third generation comprises concentrator photovoltaics, organics and other technologies at the developing stage (Jäger-Waldau, 2011). Around 80 per cent of PV cells are still silicon crystalline wafer based, a technology first used during the 1970s and now considered mature (Jäger-Waldau, 2011). The technological capabilities to manufacture comprise a complex set of activities, including the maintenance of a clean production environment (as airborne contaminants may enter the cell and reduce performance), the operation of complex production equipment, the control of complex production systems, and quality control.

Cell efficiencies for commercially produced cells have increased over the past decades and now range between 12 and 22 per cent for silicon cells and between 7 and 12 per cent for thin film cells (Greenpeace and EPIA, 2011). Indicators reflecting the technological capabilities of a PV cell manufacturer are cell technology produced and efficiency of the cell. An increase in intermediate capabilities would therefore be reflected in producing silicon cells of 15 instead of only 14 per cent efficiency. The acquisition of innovative capabilities would be reflected in learning to produce silicon cells of the highest efficiencies or thin film rather than silicone cells.

The technological capabilities that a firm must possess to manufacture PV modules comprise testing and sorting the cells into similar lots to determine the amount of current the cell may produce. The next step is the assembly, which means linking the cells to produce a cell string. Strings of cells are then vacuum laminated between protective glass to form the module. Finally, the module is tested for performance (Bruce, 2007).

The module manufacturing process has not changed significantly over time despite changes in input materials (PV cells) and production technology. It was therefore assumed that module manufactures vary, if at all, in their intermediate capabilities and no indicator for innovative capabilities was developed (Bruce, 2007).

The Global Environment Facility

The Global Environment Facility (GEF) was founded in 1991 and currently serves as the financial mechanism to four international environmental conventions (the Convention on Biological Diversity, the United Nations Framework Convention on Climate Change, the United Nations Convention to Combat Desertification, and the Stockholm Convention on Persistent Organic Pollutants). It is the only multilateral finance institution exclusively aimed at solving global environmental problems (biological diversity, climate change, international waters and ozone layer depletion, land degradation and persistent organic pollutants). The central institution of the GEF is the GEF Secretariat, which formulates projects and oversees their implementation. The GEF Secretariat is governed by the GEF Assembly (consisting of representatives of its currently 182 member countries), which reviews the GEF's work every three to four years. The GEF Assembly as well as the secretariats of the four conventions give guidance to the GEF Council, which oversees the GEF Secretariat. GEF projects are implemented by GEF implementing agencies. The number of GEF implementing agencies increased from initially three (the World Bank, UNEP and UNDP) to currently ten.

In its role as a financial mechanism to the UNFCCC, the GEF aims, with regard to mitigation, to 'support developing countries and economies in transition towards a low-carbon development path' (GEF, 2011, p17). In order to achieve such goals, the GEF operates 'the largest and most comprehensive global portfolio of investments in energy efficiency, renewable energy and other climate-friendly projects' (Eberhard et al., 2004, p1). According to the last publicly available GEF internal evaluation, the climate programme consisted of 659 projects from 1991 to 2009, for which it has allocated US$2.74 billion. Other publications mention 208 renewable energy technology projects and 131 energy efficiency projects for the period from 1991 to 2008 (GEF, 2009a, 2009b). However, these figures do not include any GEF climate change projects unrelated to either of the categories (e.g. transport or land use). Until 2003, GEF climate change projects were organized according to four Operational Programmes (OPs) – OP5: Removing Barriers to Energy Conservation and Energy Efficiency; OP6: Promoting the Adoption of Renewable Energy by Removing Barriers and Reducing Implementation Costs; OP7: Reducing the Long-Term Costs of Low Greenhouse Gas-Emitting Technologies; OP11: Promoting Sustainable Transport (GEF, 2003). The Operational Programmes were replaced by so-called GEF Focal Area Strategies from 2004 onwards.

The achievements of the GEF in the field of climate change are difficult to consider for a number of reasons. First, little independent evaluation exists and the published official GEF reviews are cautious in their wording (GEF, 2005, 2010). In addition, evaluators tend to describe unexpected project outcomes in terms of lessons learned rather than project failure (Martinot, 2002). Third, clear performance indicators and their consistent application across GEF projects are missing (Eberhard et al., 2004). Fourth, a large number of GEF climate change projects are not yet fully implemented and only 29 projects were closed by 2004 (Eberhard et al., 2004, p19). Particular projects also might not be as successful as planned and independent evaluation of particular GEF climate change projects is limited (Mulugetta et al., 2000; Kapadia, 2004; Heggelund et al., 2005). Officially, the GEF estimates the avoided direct and indirect emissions to be 224 million metric tonnes of carbon dioxide at an incremental cost of US$194 million (Eberhard et al., 2004). More recent figures are not publicly available. No comprehensive academic review of the GEF technology transfer activities exists.

The GEF and solar PV

The GEF has financed projects supporting off-grid solar photovoltaic technology since 1991, aiming primarily at applications for rural households (Martinot et al., 2000). The GEF selected PV technology as one of the 'GEF set of technologies', which comprised a range of renewable energy and energy efficiency technologies, which would not diffuse to developing countries without intervention (Gan, 1993).[4]

From a climate change perspective, rural electrification through SPV opens the possibility of displacing or avoiding carbon dioxide emissions as rural households typically use kerosene lanterns, batteries charged through diesel generators or paraffin lamps to satisfy their lighting needs. If SPV are used as an alternative to a grid connection, they also avoid carbon dioxide emissions as almost all electricity supplied through the grid is based on fossil fuel-based power stations in India (Drennen et al., 1996; Kaufman et al., 2002; Duke and Kammen, 2003; Taele et al., 2007).

The GEF and technology transfer: From technology supply to market transformation

International technology transfer has been part of the GEF's activities since its foundation. However, the GEF understanding on how international technology transfer is best supported through GEF projects has changed over time. An official GEF definition of technology transfer does not (according to interview sources) exist. The GEF understanding of technology transfer and how it evolved was established through document analysis and interviews with GEF staff.

The GEF operational strategy, which was approved in 1995 and lays the overall foundation for the design of GEF programmes, mentions two different types of programmes for 'promoting technologies'. The first type aims 'to expand, facilitate, and aggregate the markets for the needed technologies and improve their management and utilization, resulting in accelerated adoption and diffusion' (GEF, 1995, p36). The second type of programmes 'will pursue technology transfer, local procurement, and the development of appropriate industrial infrastructure' (GEF, 1995, p37). The operational strategy, in other words, stipulates the creation of demand for technology, but also the direct supply of technology. The former was supported through GEF projects in OP5 and OP6, and the latter in OP7. Around 2004, the GEF approach to international technology was narrowed down to the support of market creation and the so-called market transformation approach was promoted.[5]

In the context of energy policy, market transformation denotes an umbrella term describing a range of different policy instruments. These aim at enhancing the diffusion of clean (or cleaner) energy technologies by increasing demand. A significant feature of market transformation projects is that they aim to transform the market in such a way that the demand increases even after the project intervention, which has only limited duration (Schlegel et al., 1997; Blumstein et al., 2000). Relevant instruments include standards, labels, end-user subsidies, voluntary agreements, procurement incentives, etc. (Krause, 1996; Mahlia, 2004).

According to interviewees, the notion of market transformation as employed by the GEF has its roots in the above-mentioned policy context, but is used in a much broader sense. It captures the idea that the overall goal of the GEF was to develop markets. As one interviewee put it: 'We are trying to set the framework for markets, to get markets for technology in place that move efficiently.' Interviewees underlined that there was no deep conceptual development from the GEF concerning the term.[6] It was more often used 'as an operational framework' that guided project design through the past experience of the GEF.

Interviewees stated that the GEF currently speaks of the five pillars of market transformation. These five pillars are policy environment; the availability of financing; business models and management skills; information and awareness; and technological factors. According to GEF staff, the appropriateness of each of these pillars is a 'necessary' condition for market development. Otherwise it may impose barriers[7] to market development. These five pillars 'evolved from the five categories of barriers that emerged from ten years of portfolio experience' (interview source). A further central feature is the idea of catalysing investments. Within the GEF logic, the market transformation projects serve as role models for private investors to undertake similar activities, after the GEF project has demonstrated its feasibility (GEF, 2006a, 2006b).

The practical implication of adopting the market transformation approach is that the GEF is not directly supplying technology, but rather removing barriers to its diffusion by supporting the institutions considered necessary for the establishment of a market. Barriers vary with regard to technology and the context of project implementation. With regard to SPV applications, the GEF considers from a market-based perspective lack of end-consumer finance, lack of business models, lack of a trained workforce with managerial as well as technical skills, lack of awareness amongst end-consumers, and lack of a conducive policy environment (import taxes for systems, subsidies for other fuel sources, etc.) as particular barriers to their diffusion (Martinot, 2002, p314).

With regard to technology transfer, it is assumed that technology will be supplied by the private sector as a reaction to growing demand. Interviewees stated that there was a clear split between the public sector taking care of the framework conditions and the private-sector doing the investment and transfer. As one interviewee put it: 'Our goal is stimulating market demand. The supply question should take care of itself. It should follow the demand.' All interviewees agreed that technology transfer within GEF projects is effectively left to the private sector. As GEF projects aim at paving the way for technology transfer and for supporting the national diffusion of technology, the GEF argues that it is justified in claiming that all GEF projects 'involve technology transfer'.

Interviews revealed four main 'drivers' for the increased application of the market transformation approach (and the accompanying neglect of direct technology supply) within the GEF: lacking financial resources for direct supply of technology; missing skills to pursue other forms of technology transfer; negative project results when attempting to supply low-carbon technology directly; and donor influence. All interviewees agreed that the main driver for pursuing the market transformation approach is the lack of financial resources. Considering the large number of countries that the GEF was serving and the number of operational programmes

adopted, the support of a market environment to give incentives for private investments is considered more 'do-able' than providing technology directly. One interviewee stated that adopting the market transformation approach was 'a concession
to the reality of the climate change markets' as the GEF had substantial financial
resources but 'not enough to solve the climate change problem'.

Members of the GEF Secretariat and the World Bank also pointed out that early
projects under OP7, including direct technology transfer, 'were not particularly
successful'. From their perspective, the GEF and their implementing agencies were
'not good at such projects'. Getting a technology ready for the market, as outlined
in OP7, including technology transfer, could not be achieved by the World Bank
as they had no 'competence in technology management and commercialising
technologies'. Another interviewee expressed reservations against the capabilities of the GEF implementing agencies to support technology supply directly by
stating that the World Bank 'is not a hardware procurement agency'. Interviewees
disagreed on the importance of donor influence, stating that it ranged from minor
to substantial. More critical perspectives from the implementing agencies as well
as the UNFCCC Secretariat suggested that the chosen approach was also a result
of particular interests in the majority of the donors funding the GEF.

The GEF market transformation approach in practice: The PVMTI project in India

This section introduces the PVMTI project, its goals and its context in India.

Project goals

The Photovoltaic Market Transformation Initiative (PVMTI) project is one of the
earliest projects financed through the GEF and follows the market transformation
approach. The PVMTI took place from 1998 until 2009 and was implemented in
Kenya, Morocco and India. The results of PVMTI India are presented here. The
Global Environmental Facility bore the US$30 million cost of the project and the
International Finance Corporation (IFC) served as the implementing agency.

The main goal of the PVMTI has been to 'stimulate PV business activity in
selected countries and to demonstrate that quasi-commercial financing can accelerate its sustainable commercialization and financial viability in the developing world'
(IFC, 1998, p5). A further goal has been the replication of such business activity by
other investors without receiving PVMTI funding (IFC, 1998).

In order to achieve both goals, the PVMTI lowered the cost and risk for private
businesses to develop necessary market infrastructure as a basis for increased sales
through subsidized finance in the form of equity, soft loans and bank default
guarantees. Offered finance should be used as working capital for the development
of sales and distribution systems, or to subsidize downstream end-user finance. It
was further hoped that the sub-projects would be commercially viable by the end
of the project period in 2009 and in a position to repay part of the IFC/GEF

capital. Companies having received PVMTI finance are here referred to as PVMTI sub-projects or investee companies.

The GEF and IFC chose this approach as a reaction to five perceived main constraints on the diffusion of SPV applications in developing countries: the absence of successful business models; the lack of finance for business; the lack of relevant know-how; the lack of service support; and the absence of private commercial actors of a size which would have an interest in business activities for commercial rural PV (Gunning, 2003; Aboufirass, 2006). Finance was, however, considered to be the main bottleneck, as lack of commercial finance hinders any business activity and lack of end-user finance restricts purchases of PV systems (Derrick, 1998).

The GEF allocated about US$15 million funding (representing 50 per cent of the overall PVMTI budget), including technical assistance and cost for project execution, to PVMTI India. The IFC expected that 11 sub-projects (4 of them with a low probability) would receive funding in response to the PVMTI request for proposals. The number of possible investments was actually expected to exceed PVMTI funds (IFC, 1998). The IFC further expected that PVMTI-funded projects (plus those companies imitating PVMTI-funded business models) would install a total of 10MW PV generation capacity from 1998 to 2003 in addition to all other Indian PV installations (which were projected to amount to 8MW installed capacity) (IFC, 1998). The targeted users are private end-users able to afford electricity but unlikely to receive a grid connection and commercial users in need of non-grid energy sources.

The Indian context

In 2001, about 745 million Indians (72.2 per cent of the Indian population) lived in rural areas (Bhattacharyya and Srivastava, 2009). It is estimated that 70 to 80 million households still depend upon kerosene for lighting purposes, of which 92 per cent are located in rural areas (Srivastava and Rehman, 2006).

Rural households without connection to the electricity grid are considered to constitute the largest demand for off-grid SPV applications. As they lack, on average, the economic resources to purchase such systems, this particular demand can only be addressed through some form of public support. There is also private demand in urban areas as Solar Home Systems (SHS) are increasingly used as backup systems due to the unreliability of the electricity grid and regular power failures. The market is considered to be growing but not yet quantified (TERI, 2006).

The PVMTI project in India with its mission to support the diffusion of off-grid SPV application partly overlaps with Indian government efforts to electrify rural areas by means of off-grid SPV. The Indian government maintains a large rural electrification programme, which generally consists of the expansion of the main electricity grids and a programme supporting the development and diffusion of renewable energy technologies, including SPV. In 1975, the Department of Science and Technology initiated the Solar Photovoltaic Programme. The programme aimed at developing commercially viable PV applications, the creation of a strong manufacturing base, and the diffusion of PV applications to remote and rural areas of India. It covers different SPV applications such as solar streetlights, SHS, solar water pumping, solar lanterns and SPV power plants.

The Solar Photovoltaic Programme initially worked on a pure project basis through which renewable energy agencies on the state level designate villages to be electrified with SHS, solar lanterns, solar streetlights, solar water pumps and small PV-powered generators. Villagers are then offered to buy such systems and receive a subsidy depending upon location. Over time, the Indian government also opened stores selling SHS and other small-scale SPV applications directly to consumers without any subsidies and allowed private companies to sell directly to end-consumers with the state subsidy (interview sources). Since the most successful applications sold under PVMTI are SHS, the following discussion focuses on SHS. Until March 2008, roughly 403,000 SHS were installed under the government programme (MFNARE, 2008).

The Indian PV industry started to develop during the 1970s when the Indian government mandated state electronics companies with the manufacture of PV cells and modules. The industry grew significantly during the early 1990s when the Indian government liberalized the Indian economy and foreign firms opened PV manufacturing facilities in order to produce for the Indian market, which was considered (with the markets of many other developing countries) to be substantial. The industry grew further (Indian and foreign investment) from 2000 onwards in order to manufacture for growing demand abroad, especially in developed countries (Haum, 2011).

The India PV industry consists of government, joint-sector and private-sector companies. Private firms include firms of complete Indian ownership, joint ventures with multinationals and fully foreigner-owned subsidiaries (Srinivasan, 2007). In 2007, 19 companies operated in India producing wafers, cells, modules and systems. The number of PV cell and module manufactures seems to have increased to 15 by 2009; however, this figure is an Indian government estimate and not confirmed by primary research (ISA, 2010). There is no Indian manufacturer of solar cell production equipment. The Indian PV companies vary in their level of integration. Some companies just cover one step of the PV value chain while others cover up to three. Sales and service companies, however, do not seem to manufacture at all, although some assemble systems.

In 2009, 7.2GW photovoltaic-generation capacity was installed globally, of which 0.24GW (or roughly 1.8 per cent) were produced in India. The industry is strongly export driven. In 2005, the cumulative PV production (domestic and export) was 248.8MW. Aggregate installed capacity in India was roughly 89MW in various PV applications. A total of 160MW went into export (TERI, 2006). The annual export share has grown from 15 per cent in 1999 and 28 per cent in 2000, to more than 65 per cent in 2005 (TERI, 2001, 2006). In 2009, the cumulative share was 66 per cent (ISA, 2010). Annual figures are not publicly available. Indian PV manufactures supply the same PV systems to the Indian government, Indian private demand, as well as to PVMTI-founded companies. According to interviews, the technology has not changed due to PVMTI.

PVMTI results and effects on international technology transfer

PMVTI India funded four companies that aimed at increasing sales through the expansion of sales infrastructure. In total, the four companies sold 70,000 SHS, 8000 PV lanterns and 1000 streetlights from 2003 to 2007.[8] This result is well below initial expectations but represents an expansion of Indian domestic demand of SHS by almost 30 per cent, as an estimated 224,615 SHS were installed through the solar photovoltaic programme during the same period (Haum, 2011).[9]

In this context it is important to understand how far the sales of PVMTI-funded companies might have resulted in any form from international technology transfer. As PVMTI-funded companies (distributors) were not allowed to invest in manufacturing, any form of learning within the Indian PV manufacturing industry (producers) would have taken place among those companies manufacturing SPV, located in India and supplying to PVMTI-funded sales companies. The brief theoretical framework introduced at the beginning of this chapter introduced knowledge and international technology transfer as a learning process of knowledge supplied from abroad. It also introduced the concept of technological capabilities and distinguished between operational and innovative capabilities.

The following findings of the study of the Indian PV industry in relation to PVMTI are therefore presented under four sub-headings. Firstly, the operative, intermediate and innovative capabilities of Indian PV and module manufactures are established by the time PVMTI India was close to an end (2008). As stated before, they are defined by the ability to either produce silicon cells at various efficiency levels or thin film cells. As the PV module manufacturing process is very standardized, there is little room for significant innovation (they were interviewed for intermediate capabilities reflected in incremental improvements such as increased resource efficiency or lower module failure rates). Second, the role of international transfer activities in the formation of capabilities is discussed according to the interviews conducted. Third, the industry's perception of access to PV technology through international technology transfer is summarized. Fourth, the role that PVMTI India played in international technology transfer is discussed.

Current technological capabilities

All *cell* manufactures stated that they had the operational capabilities to manufacture silicon PV cells with an efficiency of between 12 and 13 per cent. Three manufacturers produce cells with 16 per cent efficiency. All six cell manufactures displayed intermediate capabilities by stating that they could repair, adapt and improve the production technology they acquired to a minor extent. Two had actually built their whole production technology by themselves. Incremental improvements through intermediate capabilities included, for example, the slight decrease of the thickness of the silicon layer or the increase of cell efficiency from 13 to 14 per cent.

Most cell manufacturers also stated that they were lacking innovative capabilities as their capabilities were insufficient to undertake what they referred to as breakthrough innovation, meaning the production of either thin film cells or silicon cells with an efficiency of 18 or 20 per cent, or more. Producing cells of 20 per cent efficiency (or more) or thin film cells would require a completely new set of production technology which they would have to acquire from abroad.

All six *module* manufactures stated that they could adapt and incrementally improve their production technology and therefore displayed intermediate capabilities. Some had built at least parts of their production technology themselves. Breakthrough technology that would significantly alter the cost or the quality of modules did not exist in PV module production by that time. Two companies stated that they planned to introduce automatic production lines. While automated production would be a considerable process innovation, it would not alter the product substantially.

The role of international technology transfer in the Indian PV industry

The Indian PV industry has historically (and at the time this research was undertaken) been engaged in a range of international technology transfer activities as knowledge recipients to integrate knowledge within existing production systems. The state companies, which were the first to produce PV cells and modules in India from the 1970s onwards, had built production technology partly themselves with the help of Indian research institutes, and partly imported it from abroad. Some had also sent engineers abroad for technical training (interview sources). Representatives of state companies tended to downplay the importance of international technology transfer.

Companies entering cell production during the early 1990s were importing their production technology and stated that international technology transfer was important by that time. Two accessed the technology via joint ventures and one bought it directly from the US. The remaining cell and module manufactures had purchased their production technology abroad. Most cell manufactures also had research links to R&D institutions in Europe and the US to improve and modernize their production technology during the PVMTI project's duration. These activities helped the cell manufacturers to maintain and expand their intermediate capabilities, as almost all acknowledged that they had never aspired to be at the technological frontier.

One cell manufacturer was on the way to acquiring innovative capabilities. It stated that it had acquired companies in the US and Europe to access knowledge on thin film technology and to undertake research on nanotechnology cells.[10] It had also bought a research cell plant in Europe to lower the cost in silicon production. If successful, the company would consider internalizing the production to India. All of these activities were undertaken to narrow the perceived gap in technological capacities between the company's Indian staff and the employees of the companies within technologically more advanced countries. The same manufacturer also stated that they had hired staff from abroad to increase the technological competences of the firm. By the time of this research, these capabilities were not put into use as no production site for thin film cells, for example, was operational.

The *module* manufactures imported at least parts of their production equipment. Only one sourced the equipment necessary to set up PV module production completely from the Indian market. In addition, most module manufacturers stated that foreign PV cell supplier links would be used to upgrade their technological and process knowledge.

Access restrictions to PV technology from abroad

Most cell manufacturers stated that they felt no access restriction to the technology they needed for their current business models. Four expressed certainty that they could acquire what was, at the time of interview, considered breakthrough technology in the future. The fact that it would have to be acquired from abroad as there was no Indian cell production manufacturer seemed of no particular concern to them. The four cell manufacturers considered upgrading their capabilities in the future; currently, despite lacking the capabilities to produce high-efficiency or thin film cells, their companies were still successful.[11] They acknowledged, however, that the technology 'needed to be a bit more in the market'. The main restriction appeared to be sufficient finance. One of the four companies stated (despite not experiencing access constraints) that companies in Europe, the US and Japan were in a much better position regarding access to R&D institutions generally.

One cell manufacturer expressed constraints to access and stated that there might be a problem in obtaining the latest technology from the technologically more advanced companies in the West. Technologically advanced companies, with their ability to produce cells with 22 per cent efficiency, would not sell this technology at this time. In addition, if the company wanted to introduce wafer slicing, it could not acquire the latest technology (so-called ribbon slicing) as companies would not sell it. The same company also stated that their R&D links to European research institutions would, in practice, not be as good as the links between European companies and the same research institutes. European companies would be favoured for access to latest knowledge.

The only cell manufacturer that had accessed breakthrough technology stated that it had made a huge investment as their competitors in Europe, the US and Japan were in a much better position to access state-of-the-art research knowledge. Making these huge investments was considered the only way to overcome access restrictions and to 'catch up'.

The contribution of the PVMTI

All interviewees (cell and module manufactures alike) stated that no PV production company in India had made any investment in production technology due to the demand created through the PVMTI. Their first and foremost explanation was that the demand created through PVMTI sales was too small to stimulate further investment in production technology or any other form of learning to innovate, which might have included international technology transfer to India. Although not all

PV manufacturers interviewed were familiar with PVMTI, almost all remaining manufacturers agreed that PVMTI had no impact upon the industry's development.

The large majority of investments made by cell and module manufacturers during the period that PVTMI India was active, which included, in many cases, the acquisition of production technology from abroad and upgrading of the production process with support from abroad, occurred because of the booming export markets in Germany, Japan, Spain, Italy, etc. Representatives from the government, the PVMTI stakeholders and local experts all shared this view. Interviewees stated that lacking energy supply in India also played a role in their company strategy, but that it was less important than exports. Companies emphasized that they exported to developing countries, especially Africa; but this aspect was also minor in relation to the demand in industrialized countries, especially during 2004 and 2007.

Discussion of PVMTI India

The case study undertaken here suggests that PVMTI India has not served as an incentive for the Indian PV manufacturing industry (cells and modules) to upgrade their technological capabilities by means of international technology transfer. Stopping at this sole conclusion would, however, not do justice to PVMTI and its underlying approach.

One can conclude that technology could have been transferred to India if the demand created through PVMTI was larger. There are several arguments supporting this assumption. First, no Indian manufacturer of PV production technology exists and the cell manufacturer who stated that they had built their cell manufacturing technology themselves made clear that it was not for sale. Second, as already mentioned, interviews revealed that foreign companies did enter the market through joint ventures with Indian firms when the market in India was projected to expand significantly at the beginning of the 1990s. The first was a major oil company forming a joint venture for the manufacture of PV cells and modules with a large Indian company. A second company was founded at the same time as a joint venture with an Italian PV company. Interviewees from both companies stated that they had benefited, and still benefit from, knowledge transferred from abroad through the joint venture. Third, production technology was imported during the PVMTI's project duration as demand abroad was growing. There seems to be no reason why additional PV production technology and other forms of knowledge should not be imported if demand in the Indian market were growing.

However, based on this research it seems unlikely that any technology which would have been imported as a result of PMVTI would have significantly improved the innovative capabilities of Indian manufacturers. This conclusion is based on several observations. First, most Indian manufacturers lagged significantly behind with regard to technological capabilities in relation to technology leaders in industrialized countries. They were neither able to produce cells of the highest efficiency or those with the latest material designs (thin film, nano, etc.). According to the theory reviewed briefly above, this gap might require a learning effort possibly making any transfer unattractive to the recipient and to the supplier.[12] Second, the Indian market required during the PVMTI project's duration mainly subsidized low-cost modules for rural applications and unsubsidized low-cost modules for urban applications, which were manufactured with existing technological capabilities. Any additional demand created through the PVMTI would have required the same technology. Suppliers to PMVTI-funded sales companies stated that SPV systems were equal to those supplied to government programmes. In other words, the Indian market offered no incentives for domestic manufacturers to transfer PV technology that would significantly add to their innovative capabilities, which they would need to successfully compete with PV manufacturers operating at the current technological frontier in the longer term. Third, most Indian cell manufacturers stated quite clearly that they saw no reason to add to their technological capabilities and did not plan to do so immediately or with any priority. Some mentioned an intention to explore the possibility of producing thin film but have not done so until 2011.[13] This was mainly because they considered their capabilities and knowledge links sufficient for their current business models. The only exception was one major cell producer, who intended to actually reach the state of the art in PV technology production. Representatives of the company underlined, however, that this was done at that time purely for export markets.

Moreover, one has to consider the aspect of access to more advanced PV technologies. Most Indian PV manufactures were not concerned about access to more sophisticated technology at the time they were interviewed, as they wanted to access it (if at all) at some point in the future. They admitted, however, that immediate access was not likely. Two held more critical positions. One stated that access was not possible at present, and one stated that Indian manufactures were disadvantaged with regard to access and had made tremendous financial and organizational efforts to access PV breakthrough technology. This underlines that even though PVMTI would have stimulated sufficient demand to justify further investment in production technology, the acquisition of more advanced technology and the involved development of further technological capabilities would not necessarily have been without difficulties.

Summary and conclusions: The GEF, market transformation and international technology transfer

We have seen that the GEF attempts to support international technology transfer for development through projects modelled according to the market transformation approach. The overall aim of the market transformation approach is to establish or facilitate markets for certain low-carbon technologies in developing countries by removing barriers to market development. Once these markets are set to work, it is assumed that private investments will serve the markets, including the transfer of technology. Any transfers of technology might then have positive effects on economic development through expanding the technological capabilities of manufacturing firms. The case study on PMVTI India found that the project did not play any role in industry development and the formation of technological capability-building. The growth of the Indian PV industry and related international technology transfer during the project duration was mainly driven by demand in Europe and elsewhere abroad.

The following conclusions on the usefulness of the GEF market transformation approach for supporting international technology transfer have to be considered with caution, as they are based on a single case study and more comparative work is needed to substantiate and refine results. Other GEF projects modelled after the same approach might have had different results. However, a number of tentative conclusions may be drawn.

The findings firstly suggest that programmes following a market transformation approach probably need to 'create' much larger and stable markets than through PVMTI to create sufficient incentives for industrial investments that might include technology transfer. This finding contradicts the logic of the GEF market transformation approach which aims to establish functioning and growing demand through dedicated, one-time intervention that catalyses further private activity. But if larger demand was created through more rather than less GEF project based activity, possible investments in production capacities (and the possible accompanying international technology transfer) could, leaving the problems of project organisation and fund raising aside, represent significant contributions to economic development.

The findings suggest secondly that the GEF market transformation approach must not only create large demand but also sophisticated demand in order to give incentives to companies to invest in innovative capabilities to assimilate advanced technology. From the perspective of evolutionary economics and related theories on the importance of technological capabilities, investment in the production capacity needs to entail innovative capabilities in order to improve competitiveness and secure it in the long term. Indian manufacturers face e.g. competition from Chinese manufactures, which are expanding production capacities and gaining an increasing share of the world market. Also, demand from industrialised countries might shrink in the near future, as it still depends foremost on political initiative. It is therefore uncertain how long Indian PV manufactures may be successful with 'ignoring' the technological frontier.

The Indian government has, shortly after this case study was undertaken, increased its support of PV by drawing up feed-in tariffs for grid-connected PV power plants. It remains to be seen, however, how far this policy will affect the development of technological capabilities of Indian PV manufacturers. But even if we assume that the GEF market transformation approach was able to create large and sophisticated demand, the problem of access to advanced technology is not addressed within a market-based approach such as the GEF's.

In order to compensate for these shortcomings, the market transformation approach could support international technology transfer through stronger knowledge supply, as originally envisioned by the initial GEF operational strategy. Possible areas of support could be providing finance for technology acquisition or support of the various mechanisms of technological learning (international R&D, technology networks, etc.). As stated at the beginning of the chapter, it is not the aim of the GEF to support any form of economic goals since its main mission is the solution of environmental problems. It might therefore seem futile to reason about modifications of the GEF market transformation approach, which works relatively well with regard to achieving environmental goals (Haum, 2011). But even from an environmental point of view, the increase of innovative capabilities of low-carbon technology manufacturers would be beneficial. As one interviewee put it: 'We need breakthroughs on many frontiers, we need new technologies but we also need them to be cheaper in order to have rapid diffusion and to alleviate climate change.' There is no reason why those breakthroughs should be achieved in developed countries only.

Notes

1 It has to be pointed out that Joint Implementation and the Clean Development Mechanism do not explicitly aim to promote international technology transfer. They hold the potential to support it as both mechanisms aim to increase investment in low-carbon technologies in developing countries.

2 Note that this differs from the use of the term 'innovation capacities' by some other authors, including in this book, who tend to use this term to refer to the capacity to undertake incremental and adaptive innovation, and innovation at the frontier. The meaning is, however, similar.

3 Two of the most common critiques relate to the measurement of technological capabilities and their contribution to the economic performance of the recipient. In practice, there are difficulties in precisely identifying capabilities on a firm, regional and country level and estimating their exact contribution to technological change within a firm. Also, technological capabilities are only one of many factors influencing the growth and competitiveness of a firm. While acquiring technological capabilities is necessary to achieve the innovative assimilation of a technology, the relation to economic performance is less evident (Jonker et al., 2006). Technological capabilities contribute to the increased productivity of the firm as they interact with capital accumulation, but they seem not to be, in all cases, a necessity as industries might grow without upgrading their capabilities or increasing their technological learning (Dijk and Bell, 2007).

4 The idea of a GEF set of technologies did not relate to a list, but those technologies that 'have not yet been technically proven and shown to be economically viable, and for which the market mechanism is not conducive' and which support abatement of greenhouse gas emission (Gan, 1993, p259). A more recent document lists photovoltaics for grid-connected and distributed power applications, biomass gasification and gas turbines, biomass feedstock to liquid fuel conversion, solar thermal-electric applications, grid-connected wind power, fuel cells for mass transportation and distributed heat and power applications, and advanced fossil fuel gasification and power generation technologies as technologies to be 'emphasized' (GEF, 2003).

5 The reasons for narrowing down their approach are explained further below.

6 Interviewees from the World Bank and the IFC explained that the foundations of the concept of market transformation were originally developed by Word Bank staff, the GEF and the GEF STAP.

7 Key to market transformation is the idea of barrier removal through projects. Interviewees explained that projects aimed at supporting one, some or all of the mentioned 'pillars of markets'. Conditions that hindered the installation of the pillars were considered barriers which in their view differed from country to country.

8 Although the PVMTI started in 1998, the PVMTI-funded company began selling SHS in 2003. A detailed discussion of the PVMTI project in India, its success and its effects on CO_2 reduction can be found in Haum (2011).

9 As no figures were available for the years 2005 and 2007, I assumed that the average number of systems sold in the years 2003, 2004 and 2006 were installed during each year.

10 By the time the research was undertaken, this particular company had not started manufacturing thin-film cells in India (it began thin-film cell production in 2009).

11 As of July 2011, none of the four companies had upgraded their technological capabilities and started producing thin-film cells.

12 The effort undertaken by the only manufacturer that has acquired the capabilities to produce thin-film cells supports this assumption. It would, however, need substantial further research into the Indian PV industry.

13 Websites of respective companies were visited in August 2011.

References

Aboufirass, M. (2006) *Photovoltaic Market Transformation Initiative*, Presentation, Atelier, Fem Dialouge National

Aggarwal, A. (2001) 'Technology policies and acquisition of technological capabilities in the industrial sector: A comparative analysis of the Indian and Korean experiences', *Science Technology and Society*, vol 6, pp255–303

Amsden, A. (2001) *The Rise of 'The Rest': Challenges to the West from Late-Industrializing Economies*, Oxford University Press, Oxford

Archibugi, D. and Iammarino, S. (1999) 'The policy implications of the globalisation of innovation', *Research Policy*, vol 28, pp317–336

Baark, E. (1991) 'The accumulation of technology: Capital goods production in developing countries revisited', *World Development*, vol 19, pp903–913

Bell, M. (2009) *Innovation Capabilities and Directions of Development*, STEPS Working Paper 33, STEPS Centre, Brighton, UK

Bhattacharyya, S. and Srivastava, L. (2009) 'Emerging regulatory challenges facing the Rggvy Programme', *Energy Policy*, vol 37, pp68–79

Blumstein, C., Goldstone, S. and Lutzenhiser, L. (2000) 'A theory-based approach to market transformation', *Energy Policy*, vol 28, pp137–144

Bozeman, B. (2000) 'Technology transfer and public policy: A review of research and theory', *Research Policy*, vol 29, pp627–655

Bruce, A. (2007) *Capability Building for the Manufacture of Photovoltaic System Components in Developing Countries*, Ph.D. thesis, University of New South Wales, NSW

Bruun, P. and Bennett, D. (2002) 'Transfer of technology to China: A Scandinavian and European perspective', *European Management Journal*, vol 20, pp98–106

Cannice, M., Chen, R. and Daniels, J. (2003) 'Managing international technology transfer risk: A case analysis of US high-technology firms in Asia', *Journal of High Technology Management Research*, vol 14, pp171–187

Cantwell, J. (1993) 'Corporate technological specialisation in international industries', in Casson, M. and Creedy, J. (eds) *Industrial Concentration and Economic Inequality*, Edward Elgar, Aldershot, UK

Dahlmann, C. and Westphal, L. (1981) 'The meaning of technological mastery in relation to the transfer of technology', *Annals of the American Academy of Political and Social Sciences*, vol 459, pp12–26

Dechezleprêtre, A., Glachant, M., Ivan Hascic, Johnstone, N. and Ménière, Y. (2011) 'Invention and transfer of climate change mitigation technologies on a global scale: A study drawing on patent data', *Review of Environmental Economics and Policy*, vol 5, pp109–130

Derrick, A. (1998) 'Financing mechanisms for renewable energy', *Renewable Energy*, vol 15, pp211–214

Dijk, M. van and Bell, M. (2007) 'Rapid growth with limited learning: Industrial policy and Indonesia's pulp and paper industry', *Oxford Development Studies*, vol 35, no 2, June

Dosi, G. (1988) 'The nature of the innovative process', in G. Dosi, C. Freeman, R. Nelson, G. Silverberg and L. Soete (eds) *Technical Change and Economic Theory*, Pinter, London

Drennen, T., Erickson, J. and Champan, D. (1996) 'Solar power and climate change policies in developing countries', *Energy Policy*, vol 24, pp9–16

Duke, R. and Kammen, D. (2003) 'Energy for development: Solar Home Systems in Africa and global carbon emissions', in P. S. Low (ed) *Climate Change and Africa*, Cambridge University Press, Cambridge

Eberhard, A., Tolke, S., Vigh, A., Monaco, A. D., Winkler, H. and Danyo, S. (2004) 'GEF Climate Change Program Study', in GEFOOM (ed) *Evaluation*, Washington, DC

Figueiredo, P. N. (2002) 'Does technological learning pay off? Inter-firm differences in technological capability: Accumulation paths and operational performance improvement', *Research Policy*, vol 31, pp73–94

Freeman, C. and Soete, L. (2007) *Developing Science, Technology and Innovation Indicators: What We Can Learn from the Past*, UNU Merit Working Paper no 2007-001, Maastricht

Gan, L. (1993) 'The making of the Global Environmental Facility', *Global Environmental Change*, September

German Advisory Council on Global Change (2011) *World in Transition: A Social Contract for Sustainability*, Berlin

GEF (Global Environment Facility) (1995) *Revised Draft, GEF Operational Strategy*, GEF, Washington, DC

GEF (2003) *Operational Program No 7: Reducing the Long-Term Costs of Low Greenhouse Gas Emitting Technologies*, GEF, Washington, DC

GEF (2005) *OPS3: Progressing Towards Environmental Results, Third Overall Performance Study of the Global Environmental Facility*, GEF, Washington, DC

GEF (2006a) *Catalyzing Technology Transfer*, GEF, Washington, DC

GEF (2006b) *Revised Programming Document*, GEF, Washington, DC

GEF (2009a) *Investing in Energy Efficiency: The GEF Experience*, GEF, Washington, DC

GEF (2009b) *Investing in Renewable Energy: The GEF Experience*, GEF, Washington, DC

GEF (2010) *OPS4: Progress towards Impact*, GEF Evaluation Office, GEF, Washington, DC

GEF (2011) *GEF: 5 Focal Area Strategies*, GEF, Washington, DC

Goulet, D. (1989) *The Uncertain Promise: Value Conflict in Technology Transfer*, Apex, New York, NY

Greenpeace and EPIA (2011) *Solar Generation 6*, Greenpeace and EPIA, Brussels

Gunning, R. (2003) 'The photovoltaic market transformation initiative', in I. E. Agency (ed) *16 Case Studies on the Deployment of Photovoltaic Technologies in Developing Countries*, IEA, Paris

Haites, E., Duan, M. and Seres, S. (2006) 'Technology transfer by CDM projects', *Climate Policy*, vol 6, pp327–344

Haum, R. (2004) *Technology Transfer under the Clean Development Mechanism: A Case Study of Conflict and Cooperation between Germany and China in Wind Energy*, Schriftenreihe Des Iöw Berlin, Institut für Ökologische Wirtschaftsforschung, Berlin

Haum, R. (2011) *Transfer of Low-Carbon Technology under the United Nations Framework Convention on Climate Change: The Case of the Global Environment Facility and Its Market Transformation Approach in India*, SPRU – Science and Technology Policy, Brighton, UK

Heggelund, G., Andresen, S. and Ying, S. (2005) 'Performance of the Global Environmental Facility (GEF) in China: Achievements and challenges as seen by the Chinese', *International Environmental Agreements*, vol 5, pp323–348

Hoekman, B. M., Maskus, K. E. and Saggi, K. (2005) 'Transfer of technology to developing countries: Unilateral and multilateral policy options', *World Development*, vol 33, pp1587–1602

IFC (International Finance Corporation) (1998) *India, Kenya, and Morocco: Photovoltaic Market Transformation Initiative (PVMTI), Project Document*, IFC, Washington, DC

ISA (2010) *Solar PV Industry 2010*, ISA, Bangalore

Ivarson, I. and Alvstam, C. (2005) 'Technology transfer from TNCs to local suppliers in developing countries: A study of AB Volvo's truck and bus plants in Brazil, China, India, and Mexico', *World Development*, vol 33, pp1325–1344

Jäger-Waldau, A. (2011) *PV Status Report 2010*, ISPRA, Italy

Jonker, M., Romijn, H. and Szirmai, A. (2006) 'Technological effort, technological capabilities and economic performance A case study of the paper manufacturing sector in West Java', *Technovation*, vol 26, pp121–134

Kapadia, K. (2004) *Productive Uses of Renewable Energy: A Review of Four Bank–GEF Projects, Draft Report for the World Bank*, World Bank, Washington, DC

Kaufman, S., Duke, R., Hansen, R., John Rogers, Schwartz, R. and Trexler, M. (2002) *Rural Electrification with Solar Energy as a Climate Protection Strategy, REPP Research Report no 9*, Washington, DC

Kim, L. (1991) 'Pros and cons of international technology transfer: A developing country's view', in T. Agmon and M. A. Y. V. Glinow (eds) *Technology Transfer in International Business*, Oxford University Press, Oxford

Kogut, B. and Zander, U. (1993) 'Knowledge of the firm and the evolutionary theory of the multinational corporation', *Journal of International Business Studies*, vol 24, pp625–645

Krause, F. (1996) 'The costs of mitigating carbon emissions: A review of methods and findings from European Studies', *Energy Policy*, vol 24, pp899–915

Lall, S. (1992) 'Technological capabilities and industrialisation', *World Development*, vol 20, pp165–186

Liebeskind, J. P. (1996) 'Knowledge, strategy and the theory of the firm', *Strategic Management Journal*, vol 17, pp93–107

Mahlia, T. M. I. (2004) 'Methodology for predicting market transformation due to implementation of energy efficiency standards and labels', *Energy Conservation and Management*, vol 45, pp1785–1793

Martinot, E. (2002) 'Ten years of GEF-supported climate change projects: Learning from experience', in *Climate Change and Its Relevance to Bank Operations*, GEF, Washington, DC

Martinot, E., Ramankutty, R. and Rittner, F. (2000) *The GEF Solar PV Portfolio: Emerging Experience and Lessons, Monitoring and Evaluation Working Paper No 2*, GEF, Washington, DC

MFNARE (Ministry for New and Renewable Energy) (2008) *Annual Report 2007–2008*, MFNARE, New Delhi

Morrison, A., Pietrobelli, C. and Rabellotti, R. (2008) 'Global value chains and technological capabilities: A framework to study learning and innovation in developing countries', *Oxford Development Studies*, vol 36, pp39–58

Mulugetta, Y., Nhete, T. and Jackson, T. (2000) 'Photovoltaics in Zimbabwe: Lessons from the GEF Solar Project', *Energy Policy*, vol 28, pp1069–1080

Narula, R. and Dunning, J. (2000) *Industrial Development, Globalization and Multinational Enterprises: New Realities for Developing Countries, Oxford Development Studies*, no 28, Oxford

Nelson, R. (1991) 'Why do firms differ?', *Strategic Management Journal*, vol 12, pp61–74

Ockwell, D., Ely, A., Mallett, A., Johnson, O. and Watson, J. (2009) *Low Carbon Development: The Role of Local Innovative Capabilities, STEPS Working Paper 31*, University of Sussex, Brighton, UK

Ockwell, D., Haum, R., Mallet, A. and Watson, J. (2010) 'Intellectual property rights and low-carbon technology transfer', *Global Environmental Change*, vol 20, pp729–738

Peterson, S. (2008) 'Greenhouse mitigation in developing countries through technology transfer? A survey of empirical evidence', *Mitigation and Adaptation Strategies for Global Change*, vol 13, pp238–305

Porter, M. E. (1985) 'Technology and competitive advantage', *Journal of Business Strategy*, vol 5, pp60–77

Radosevic, S. (1999) *International Technology Transfer and Catch-Up in Economic Development*, Edward Elgar, Cheltenham, UK

Reddy, M. and Zhao, L. (1990) 'International technology transfer: A review', *Research Policy*, vol 19, pp285–307

Rennings, K. (2000) 'Redefining innovation: Eco innovation research and the contribution from ecological economics', *Ecological Economics*, vol 32, pp319–332

Rose, R. C., Uli, J. and Abdullah, H. (2009) 'Relationships between knowledge, technology recipient, technology supplier, relationship characteristics and degree of inter-firm technology transfer', *European Journal of Social Sciences*, vol 11, pp86–102

Rosenberg, N. and Fritschak, C. (1985) 'The nature of technology transfer', in N. Rosenberg and C. Fritschak (eds) *International Technology Transfer: Concepts, Measures and Comparisons*, Präger, New York, NY

Schlegel, J., Goldberg, M., Raab, J. and Prahl, R. (1997) *Evaluating Energy Efficiency Programs in a Restructured Industry Environment*, National Association of Regulatory Utility Commissioners, Washington, DC

Schumpeter, J. (1961) *Theory of Economic Growth*, Oxford University Press, Oxford, UK

Scott-Kemmis, D. and Bell, M. (1988) 'Technological dynamism and the technological content of collaboration: Are Indian firms missing opportunities?', in A. Desai (ed) *Technology Absorption in Indian Industry*, Wiley and Sons, New Delhi

Smith, A. (2009) 'Energy governance: The challenges of sustainability', in J. I. Scrase and G. Mackerron (eds) *Energy for the Future: A New Agenda*, Palgrave, London

Srinivasan, S. (2007) 'The Indian solar photovoltaic industry: A life cycle analysis', *Renewable and Sustainable Energy Reviews*, vol 11, pp133–147

Srivastava, L. and Rehman, I. H. (2006) 'Energy for sustainable development In India: Linkages and strategic directions', *Energy Policy*, vol 34, pp643–654

Taele, B. M., Gopinathan, K. K. and Mokhuts'oane, L. (2007) 'The potential of renewable energy technologies for rural development in Lesotho', *Renewable Energy*, vol 35, pp609–622

Teece, D. (1981) 'The market for know-how and the efficient international transfer of technology', *Annals of the American Academy of Political and Social Sciences*, vol 458, pp81–96

TERI (The Energy and Resources Institute) (2001) *Survey of Renewable Energy in India*, TERI, New Delhi

TERI (2006) *Report on Solar Photovoltaic Sector Review and Market Potential in India*, TERI, Delhi

Wei, L. (1995) 'International technology transfer and development of technological capabilities: A theoretical framework', *Technology in Society*, vol 17, pp103–120

Low-carbon Technology Transfer and Poverty Alleviation

PART V

11

STAGNATION OR REGENERATION

Technology transfer in the United Nations Framework Convention on Climate Change (UNFCCC)

Merylyn Hedger

Introduction

This chapter examines developments in the framing of technology transfer in the United Nations Framework Convention on Climate Change (UNFCCC), one of the 1992 Rio Earth Summit treaties, with a particular focus on finance and the poor.

The UNFCCC Conference process is making progress on the post 2012 climate change deal, and there is also momentum developing around the Rio+20 Conference.[1] Analyses are linking climate and the Millennium Development Goals (MDGs) in multidimensional strategies for a green economy linked to poverty eradication. Technology transfer in the UNFCCC was conceived when developing countries were a more homogenous group; there is now differentiation within this group, comprising major economies and global leaders, as well as the least developed countries (LDCs). Issues of the broadening context around technology transfer, and their significance for delivery of technology transfer for the poor, are explored to show the outstanding challenges (particularly financial) and the need for a clearer vision as to how, what and where technology transfer issues for the poor are addressed.

The chapter gives overviews of two sets of issues. First, the developments in technology transfer (TT) within the UNFCCC are outlined to show how the UNFCCC addresses poverty dimensions, limited to the special treatment of the LDCs. This group has formal status within the UNFCCC but operates within the G77 and China negotiation group of developing countries.[2] Second, the rapidly evolving and heavily contested challenges around climate finance in the UNFCCC are considered, which are important given that climate finance is widely recognized as crucial for achieving a long-term climate deal and delivery of TT. The role of the current formal financial mechanism of the UNFCCC, the Global Environment Facility (GEF), has always been controversial for the G77 and China group of developing countries. The scale of the funding challenge to deliver low carbon technologies needs to be faster than anticipated when the UNFCCC first emerged, as well as the programmed deployment of adaptation technologies due to unavoidable climate change.

In the final section, links are made to the 'green economy' agenda that emerged out of the global financial crisis and is currently being linked to energy development issues, which are considered to be the major development deficit, with over 1.6 billion people without adequate access to electricity and those dependent upon wood fuels. While these energy issues are longstanding, the Rio+20 Conference in 2011, has reinvigorated efforts to solve them in a separate track from the climate change technology transfer issues. Will this recharge negotiations within the UNFCCC or will TT issues for the poor remain out of focus?

Technology transfer in the UNFCCC

In this section the overall developments of TT in the UNFCCC are identified and their changing significance explained in relation to the emergence of new policy strategies. Subsequently, the treatment of the LDC group is considered; since the UNFCCC is a government-to-government negotiation, this is the way in which poverty issues are addressed.

Pace and approach on technology transfer

Looking back at the development of the technology transfer framework within the UNFCCC since 1992, it is clear that it has been slow, laborious and incremental. Every major climate conference has produced some ostensible policy progress in a decision; but project action on the ground has not been undertaken. Technology transfer in the past has been a critical negotiating issue for the G77 and China group, and discussions in the UNFCCC have been fraught at times, but as the negotiations have become more complex, the TT issue is now only one of several contested areas. Finally, however, if governance and funding dimensions can be resolved, a funded and structured Technology Mechanism (TM) is to be operational in 2012, following the Durban Climate Conference, some 20 years on from Rio.

Over the years a complex institutional architecture has been developed through many workshops and then reformed. Currently, the focus is on reforming this architecture through the development of the TM. Initially, the Expert Group on TT (EGTT) was established, then priority needs were identified through the Technology Needs Assessments (TNAs). These required funding organized via the Global Environment Facility – the formal financial mechanism of the convention.

TNAs, the first building block, emerged during 1998 when the countries were urged to submit their prioritized technology needs (UNFCCC, 1999, Decision 4/CP4) and the GEF was directed to provide funding for them (UNFCCC, 1999, Decision 2/CP4). Since then the GEF has financed 92 TNAs. Most effort since then within the UNFCCC has been invested in the TNA process, piece by piece. Workshops were convened to prepare handbooks, then syntheses of progress were compiled, and tools prepared for project development (UNFCCC and UNDP, 2010, p30). A major gap was then identified as the lack of capacity for project preparation, so this triggered another round of workshops and training (2008 to 2010). Subsequently, a guidebook was prepared on preparing TT projects for financing, supported by further training programmes and trainers' training (UNFCCC, 2006). An updated handbook on TNAs was produced in 2010 (UNFCCC and UNDP, 2010). Finally, the TNA process itself has also been found deficient in providing clear road maps and integration with national planning, so that National Technology Action Plans are now being developed.

All of this work, however, essentially acted as a preliminary assessment. Actual implementation linked to the UNFCCCC was finally launched through the Poznan Strategic Programme on technology transfer (UNFCCC, 2009a), which is implemented by the GEF. Only 14 pilot projects have received assistance and have only reached the project preparation stage (UNFCCC, 2010).

Since the Bali Conference, efforts have been made to develop a new institutional architecture: the *Bali Action Plan* included TT as an important area for negotiation within any post-2012 agreement (UNFCCC, 2008, p3). The emerging outcome so far was the Cancun Agreement (at COP16 in 2010), which includes the creation of a Technology Mechanism comprising the Technology Executive Committee (TEC) and the Climate Technology Centre and Network (CTCN) (UNFCCC, 2011a). The TEC will strengthen the development and deployment of new technologies and the CTCN will be the mobilization network providing support and collaboration to develop and transfer technologies (UNFCCC, 2011b).

Confusingly, changes to the current architecture of the convention are still on-going, with issues and problems, as well as discussions on the future framework. Therefore, at the June 2011 meetings in Bonn, there were three different sets of negotiations on technology transfer: the first under the Ad Hoc Working Group on Long-Term Cooperative Action (AWGLCA) looking at the new technology mechanism (UNFCCC, 2011c); another under the Subsidiary Body for Scientific and Technological Advice (SBSTA) covering the TNAs, emphasizing the need for near-term implementation of projects devised through the TNAs and calling for training workshops on preparing TT projects for financing (UNFCCC, 2011d); and the third under the Subsidiary Body for Implementation (SBI), which pointed out the need for on-going support for updating and developing TNAs, and recommended that more support be given by the GEF (UNFCCC, 2011e).

So progress on technology transfer has been slow. Looking back, it is possible to see that the importance of technology transfer as a make or break negotiation issue has declined in significance over the past 20 years. Within the UNFCCC, for the LDCs, National Adaptation Plans of Action (NAPAs) have been developed since 2001, and Nationally Appropriate Mitigation Actions (NAMAs) are now emerging: both of these activity areas have their own technology and capacity-building elements. Since the climate conference in Cancun in December 2010, the concept of National Adaptation Plans has emerged, and the UNFCCC Bonn 2011 meetings planned a workshop to elaborate guidelines and modalities for these. An Adaptation Fund is now operational and an overriding Adaptation Framework has been agreed in principle, and most of the detail, except funding, is not contentious (UNFCCC, 2011f).

On the mitigation side, NAMAs were formally established as a voluntary mechanism in the Cancun Agreement (UNFCCC, 2011a). NAMAs will outline national mitigation options in line with domestic policies. Different tiers of action are likely to be established contingent upon the level of external finance and technology available for developing countries. Around the dynamic low-carbon agenda, however, other types of integrated plans and more strategic national planning processes are also being promoted by United Nations agencies, some governments and policy institutes: the Low Emission Development Strategies (LEDS, through the Organisation for Economic Co-operation and Development (OECD) and the US); the Low Carbon Development Strategies (LCDS, through the EU); the Low Carbon Development Plans (LCDPs – Project Catalyst); the Low Emission Climate Resilient Development Strategies (LECREDs, through the United Nations Development Programme (UNDP)); the Low Carbon Growth Strategies (included in the Copenhagen Accord text of 2009); and the Low Carbon, Climate Resilient Development Programmes (LCCREDs, through the UK). Questions are already emerging about what will be the relationship between the TNAs, the TEC and NAMAs and national policy processes (UNFCCC, 2011g). Clearly, this plethora of effort is confusing when capacities within countries are limited, and could also work against the integration of climate change within national development planning processes.

Technology transfer in the UNFCCC for the poor

The UNFCCC has not delivered for LDCs and the poor. According to a UNFCCC Secretariat report on the LDCs, produced for the Fourth United Nations Conference on LDCs in May 2011, the UNFCCC has failed to create notable measures to provide for technology transfer or low-carbon development in the LDCs (UNFCCC, 2011h).

This is despite the fact that the LDCs generally receive special treatment under the UNFCCC (in Article 4, paragraph 9) as having specific needs and special situations concerning funding and transfer of technology (UNFCCC, 1992). Article 3, paragraph 14 of the Kyoto Protocol requires parties to take measures to minimize the adverse effects on developing countries and the LDCs (UNFCCC, 1998). In the Marrakech Accords (UNFCCC, 2002), the Least Developed Countries Fund (LDCF), the LDC work programme and the Least Developed Countries Expert

Group (LEG) were all established, leading to the development of the NAPAs. In the *Bali Action Plan*, LDCs were recognized as having special needs and given a seat on the Adaptation Fund Board. The Copenhagen Accord terms the LDCs, small island developing states (SIDS) and the African countries as the most vulnerable developing countries. Donors such as the European Union give special treatment to LDCs in Fast Start Funds and possibly, post-2012, in the EU Emissions Trading Scheme (ETS) by only permitting offsets from them to be credited (EC, 2009, p16).

NAPAs were intended to identify priority projects for funding, including technology dimensions; but implementation has been very weak. This is due, in part, to the complexity of GEF procedures, but also to the insignificant scale of finance for priority project implementation (COWI and IIED, 2009). At the international level, the LEG has provided technical leadership for analytical work and planning of NAPAs. Only recently has it included the preparation of work on a technical paper in its next work plan on the role and application of technology in implementing NAPAs in LDCs (UNFCCC, 2011i, pi).

Priority TT needs in LDCs are likely to be significantly different from middle-income countries (MICs) in the G77 and China negotiation group. Chronically poor people often rely heavily on climate-sensitive sectors such as agriculture and fisheries, and are less able to respond to the direct and indirect effects of climate change, due to limited assets and capacity. They tend to be located geographically in marginal areas that are more exposed to climatic hazards, such as floodplains, or on nutrient-poor soils (Tanner and Mitchell, 2008). Significantly, there are overlaps between Collier's 'bottom billion' and areas and sectors identified as being especially vulnerable to climate change by the Intergovernmental Panel on Climate Change (IPCC) (Collier, 2007; IPCC, 2007) and identified as requiring special treatment under the United Nations Climate Change Convention (Article 4, paragraph 8, UNFCCC). Geographically, these include small islands, countries with low-lying coastal areas, countries prone to natural disasters, countries liable to drought and desertification, and countries with fragile ecosystems, including mountainous countries. Critical regions of Africa, small islands and Asian and African mega-deltas contain the vast majority of the world's poorest people.[3]

It is therefore significant that analysis of the TNAs undertaken in 24 of the LDCs shows that their main technology needs were in agriculture, land use, livestock, forestry, energy waste management, and the transport and industry (UNEP, 2011, p22). The energy sector needs addressed by 87 per cent of these countries related to improved cooking stoves, where 50 per cent of LDC energy use is for cooking. The majority of LDCs highlighted their urgent needs to modernize the agriculture and forestry sectors (70 per cent) (UNEP, 2011, p22). And around 70 per cent identified water-related needs. The TNAs match the NAPAs, where 60 per cent of the priority projects were for food security, terrestrial ecosystems, water resources and coastal zones (UNFCCC, 2009). This analysis suggests that scaling up and diffusion for LDCs are the key elements, and not the intellectual property right (IPR) issues, which often get considerable exposure within the UNFCCC. The IPR agenda has been given profile by some major economies and is still raised at the highest levels

– for example, at the Major Economies Forum in April 2011 (Major Economies Forum, 2011).

Climate finance in the UNFCCC

Context

Delivery of technology transfer through the UNFCCC in its broadest sense at programmatic scale over long timeframes depends upon new, additional and innovative funding being generated, whether under the formal Technology Mechanism, Adaptation Framework or through supported NAMAs or LEDS. Financing of a sustainable climate deal is still a vision, not a reality, although pieces of the structure of a future package are now in place. Delivery of new and additional sources of finance under the UNFCCC appears problematic at the scale required and committed to (US$100 billion a year by 2030) compared to the US$30 billion in total over 2010 to 2012 for Fast Start Funds, although the UNAGF thought it 'challenging but feasible' (UNAGF, 2010, p4). There is also confusion, not only about whether the scale of the funds needed can be generated from innovative sources of funding, but also because there is increased pressure on public funds with evidence of switches from official development assistance (ODA) into climate rather than the creation of new and additional flows (Hedger, 2011, p41). There are also pressures on policy and regulatory frameworks which can generate other flows, such as the EU ETS. While there is a proliferation of approaches at the international level, increased pressure is placed on the implementation capacity of the national level. Here the formal TT agenda is likely to be marginal and embedded in many sectors. Encouragingly, climate finance is beginning to be dealt with through mechanisms established under the aid effectiveness agenda and made central to development planning in some countries (Hedger, 2011, p34).

Negotiations and discussions on TT within the UNFCCC do not have the energy that is generated around the finance issues. Indeed, in many ways, issues on TT have been subsumed within the broader finance agenda and the work on mitigation and adaptation. For example, the G77 and China tabled a draft decision for the Durban Conference of the Parties (COP) in December 2011, on the financial mechanism of the convention which reiterates 'the need for enhanced and urgent action on the provision of financing resources and investment to support action on mitigation, adaptation and technology cooperation to developing country parties' (UNFCCC, 2011j, p2).

This section examines three dimensions of the climate finance issue: developments within the UNFCCC; specific funding challenges relating to public and private sources, and early experience through the Fast Start Funds (FSFs); and the evolving relationship between development and climate finance.

Finance developments within the UNFCCC

Providing finance for vulnerable countries was a fundamental part of the UNFCCC's mandate in 1992. Once the reality of climate change became clear following the IPCC's *Fourth Assessment Report* and *The Inconvenient Truth* film by Al Gore, delivery became an overwhelming necessity. The LDCF was established in 2001 through the Marrakech Accords to fund urgent and immediate adaptation actions in LDCs and is operated by the GEF (UNFCCC, 2002). The Adaptation Fund was also established in 2001, under the Kyoto Protocol, but this did not become operational until the Bali Climate Conference. With a lack of trust in the GEF by developing country partners (Mohner and Klein, 2007) and with donors wanting to scale up funding in ways that suited their objectives, the Climate Investment Funds (CIFs) were set up through the World Bank in 2007/2008. Contentiously for the G77 and China group, these are outside the UNFCCC and, thus, its direct control (TWN, 2008). With the publication of the Stern Review in 2006 and the UNFCCC finance report (UNFCCC, 2007) the Bali Action Plan put finance at the heart of the negotiations. The failure of developed countries to deliver their GHG emission cuts and obligations on climate finance has contributed significantly to the lack of trust and failure to achieve a post-Kyoto 2012 deal (Stewart et al., 2009, p5). However, it should be noted that work on climate finance costs, sources and delivery is just beginning (Haites, 2011).

The GEF and World Bank-related funding mechanisms have procedural and accessibility problems according to the G77 and China developing country group. Issues relate to delays in accessing funds (due to the stringent procedures under which the World Bank operates) and to the inadequacy of funds (Persson et al., 2009, p91). With a long running history of problems and antipathy to the GEF, the developing countries fought hard at Bali to get the Adaptation Fund developed in a different way so that they had a majority vote and direct access (ENB, 2007). Indeed, these arguments are surfacing again around the emerging Technology Mechanism (UNFCCC, 2011b). After some delays in getting procedures and modalities agreed, the Adaptation Fund is now making allocations for funding; but again its funds are limited, chiefly from 2 per cent of proceeds of Certified Emissions Reductions from the Clean Development Mechanism (CDM), with some limited additional funds from donors.

Agreement on the Green Climate Fund (GCF) emerged in the first instance from the package in the Copenhagen Accord, and was formalized in the Cancun Agreements. In the build-up to Copenhagen, the FSFs emerged driven by EU leaders. Before the Cancun Conference, the important BASIC group of countries (Brazil, South Africa, India and China) emphasized that both finance features of the Copenhagen Accord – short-term funds (US$30 billion 'fast track' 2010 to 2012) and medium-term funds (US$100 billion annually by 2020) – must be operationalized and provided by developed countries if there was to be any chance of a deal (BASIC Ministers, 2010). Reporting on provision of, and access to, these resources (US$30 billion during 2010 to 2012) was formalized in Cancun to improve transparency. Little was agreed on how the level of long-term funding would be generated and accessed: according to the *Bali Action Plan* (UNFCCC, 2008), this needed to be adequate, predictable, new and additional. Reference was made to the report of the High-Level Advisory Group (UNAGF, 2010) which had identified that it was challenging, but feasible, to meet the goal of mobilizing US$100 billion a year by 2020. The report also recognized that grants and highly concessional loans were crucial for adaptation in the most vulnerable developing countries, such as LDCs, SIDS and Africa.

Modifying procedures and approaches, the GEF and the World Bank are positioning themselves to be able to take advantage of the possibility of disbursing the increased funds from the GCF (UNFCCC, 2010; IISD, 2011, p2). However, there are contested issues around the GCF concerning institutional architecture and governance, and problems are already visible around flows of both public and private sources of funding and access to these. In general, donors are more comfortable with working with multilateral funding arrangements, and resist the establishment of new mechanisms.

It is expected that the Technology Mechanism will have a funding window under the GCF. However, the facilitator's note of the discussions in Bonn makes it clear that there are significant areas to be decided about financing arrangements for the Climate Technology Centre and Network (CTCN). Emergent texts suggest that it is expected that the CTCN will be funded from a variety of sources, including the financial mechanisms of the convention, bilateral, multilateral and private-sector channels, and in-kind contributions from the host organization and participants of the network (UNFCCC, 2011b). In the first meetings of the Transitional Committee of the GCF, funding windows have been mentioned for adaptation and mitigation, and possibly SIDS and/or LDCs, but not a separate technology window (Harmeling, 2011a, 2011b).

Funding challenges

Apart from the scale of funding and contested governance issues, another issue has been the balance between public and private sources of funding, differences which were reflected in the UNAGF report. Developed countries see the private sector as playing an important role in supplying low-carbon technologies; by contrast, the developing countries argue that the provision of financial support is a responsibility of developed country nations. Since the collapse of the international financial system in 2008/2009, these countries have been able to imply that funds from the private sector are not 'reliable and predictable' as stipulated in the *Bali Action Plan*. Developed countries think that innovative funding, linked to the private sector, will deliver over the long term (EC, 2011).

There are some specific issues for poor countries within these debates. The financial and economic fiscal crisis left many LDCs in a fragile fiscal condition. Current account deficits have widened for many non-oil exporting countries and especially African LDCs. Overseas development assistance (ODA) will remain a critical source of external financing and this can complement and catalyse private investment from both domestic and private sources. ODA to LDCs was US$37 billion in 2009, less than in 2008 – well below the official targets such as MDG8, which commits to increases (UNEP, 2011, p19). ODA is a catalyst for foreign direct investment – on the rise in LDCs since 1990, but it mainly flows in to the primary sector (oil, gas and minerals) and less to manufacturing and infrastructure services that are crucial for development. Furthermore, it is geographically concentrated in a few LDCs (UNEP, 2011, p19).

Experience of the key market mechanism currently operating under the Kyoto Protocol, the CDM, has not been uniformly successful, particularly for LDCs as most projects have been focused in a small number of countries. The top four host countries of CDM projects (Brazil, China, India and Mexico) have received approximately 76 per cent of the projects. Registered CDM projects in the LDCs accounted for only 25 out of more than 2800 projects worldwide (UNFCCC, 2011h). Obstacles include issues around country risk, the small size of the economy and low profitability of projects. These figures can explain why there has been such a keen interest in the development of a Reduced Emissions from Deforestation and Forest Degradation mechanism (REDD), and also for a soil carbon land-use mechanism in some afforested LDCs and parts of Africa (Ellis and Kamel, 2007; World Agroforestry Centre, 2009; Wolemberg et al., 2011). Additionally, it seems that low-carbon technologies will need to be enabled by massive public investment. The EU has stated that in the absence of an ambitious international agreement, only credits from new CDM projects registered in LDCs will be permitted (EC, 2009, p16). An interesting new development is that the World Bank has announced in August 2011 that it will start a UK£130 million CDM fund for LDCs at Durban on the back of this policy (Pointcarbon, 2011).

In general, the financial crisis reduced credit availability and the confidence of private capital markets in investment. While fossil fuels are widely available and cheaper, if not subsidized, public and private funds are unable to deliver the necessary technologies without major changes in investment frameworks stimulated by strong policy support mechanism and incentives. Business leaders have been asking for long-term policy signals:

> Private investment will only flow at the scale and pace necessary if it is supported by clear, credible and long-term policy frameworks that shift the risk-reward balance in favour of less carbon-intensive investment.
>
> (IIGCC et al., 2010, p1)

The financial crisis slowed down spending in infrastructure in almost all emerging countries, highlighting the difficulty of obtaining financing in the new environment of limited credit, including the energy sector. Private capital has undertaken a 'flight to quality' and is unlikely to be present in the poorest countries. More than 75 per cent of infrastructure spending in developing countries, closer to 90 per cent in many International Development Association (IDA) countries, is financed entirely by the public sector, which in turn is partly financed by governments accessing private markets (Sierra, 2009). The implication here is that ODA from public sources is vital, especially to protect investment in low-carbon climate-resilient pathways.

Financing the investments necessary for low-carbon energy in the future is challenging. After five consecutive years of robust growth, the value of the global carbon market stalled in 2010 (World Bank, 2011, p9). Prices have declined in both the European Union Emissions Trading Scheme and the trade generated by the Certified Emissions Reductions (CERs), and the value of each trading allowance has fallen substantially. In addition, the theft in January 2011 of trading allowances reignited controversy in the carbon market's effectiveness (World Bank, 2011, p42). For new and emerging renewable technologies, the challenges have been greater. The lack of certainty around a post-2012 agreement has significantly impeded private financing of emissions reductions in developing countries (Griffith-Jones et al., 2009).

Fast Start Funds

Analysis of the EU's FSFs reveals some issues. Collectively, the EU (the European Commission (EC) and its member states) is a major player in the FSFs and is giving US$7.2 billion for the period of 2010 to 2012, with around 33 per cent in adaptation funding with priority to most vulnerable and least developed countries. Around 40 per cent will go to mitigation, for NAMAs and LEDS, to promote the deployment of clean energy technologies, and 13 per cent for REDD (EC, 2011). Member states and the EC interpret what can be designated FSFs in different ways, and this relates to the fundamental 'new and additional' issue about funding within the UNFCCC. The terminology about 'new and additional' was used within Article 4.3 of the UNFCCC and Article 11-2 of the Kyoto Protocol, with the intention of ensuring that no ODA funds would be diverted by Annex I (developed countries) to meet their obligations under the convention (Yamin and Depledge, 2004). The term has never been defined, but generally is considered to refer to above the formal ODA target of 0.7 per cent of gross national income (GNI). Only five countries have reached the 40-year-old 0.7 per cent target, but the EU 15 countries in 2005 did commit to reach this by 2015. It is already clear that it is difficult to define what is actually 'new and additional' above ODA when that is increasing (Fallasch and De Marez, 2010; Stadelmann et al., 2010). This means, for example, that as the UK is increasing its aid budget, this allows for some ambiguity about the increase in climate funding.

Most of the EU's FSFs will be deployed through existing and already operational cooperation instruments and initiatives (the bilateral and multilateral arrangements) and not directly to the UNFCCC's Adaptation Fund or the new GCF. The EC's FSF comes on top of preliminarily programmed support for climate-relevant actions in developing countries in the period of 2010 to 2012 in the order of 900 million Euros. However, there is no room for complacency; at the BASIC ministerial meeting on climate change (26 to 27 February 2010), the ministers noted that, despite declarations at Copenhagen and Cancun, actual disbursement of funds is lacking even to SIDS, Africa and the LDCs (who need it most) and that a sizeable flow of funds should begin before discussions on the GCF gain momentum (BASIC Ministers, 2010).

Relationship of development and climate finance

Analyses around the UNFCCC system inevitably have a narrow vision of climate, and focus on the international and global scale. However, when a broader development cooperation lens is put on the picture, some different issues arise, especially when these are linked to studies at the national level. Climate change needs to be seen as part of the bigger development picture, as it enhances the need for development assistance and requires increased focus on major questions such as energy access, water scarcity, and energy and food security, and presents new challenges on coordination and capacity. Furthermore, as attention is moving into the potential monitoring, reporting and verification (MRV) systems which are required as an integral part of the Cancun agreements, it is evident that climate finance is challenging to track, with a multiplicity of climate projects and initiatives being implemented in countries (OECD, 2010, 2011a). Technology dimensions and capacity building are embedded within these areas (such as water, agriculture and energy) in many ways.

Climate change finance is now part of the EU's larger development assistance agenda, focused on poverty alleviation and the fulfilment of the United Nations Millennium Development Goals (MDGs). The EU – the world largest donor – has made climate change an increasingly important component of its development cooperation effort. It has identified that a significant part of its portfolio contributes to climate change intervention strategies. EU cooperation has been promoting adaptation and mitigation synergies across all development sectors (water, agriculture, forests, fisheries, rural development, health, the promotion of energy efficiency, and renewable energies) alongside poverty alleviation (EC, 2009, p9). An analysis of the EC aid portfolio shows that commitments for climate-relevant interventions have been globally increasing since 2002, totalling today around 2.5 billion Euros. Overall, climate change commitments in the development portfolio increased between 2002 and 2008 to 1.7 billion Euros. This is marked up as 'demonstrating that a significant amount of climate change integration has already taken place in development cooperation' (EC, 2010). Projects branded as climate change include water efficiency, renewable energy and infrastructure.

It seems probable that there will be continuing strong drivers from the development finance framework for climate change from the Busan round of the international aid effectiveness process (the *Paris Declaration*), which took place in December 2011, and from the MDG process for more impact upon policy alleviation. Due to the financial problems of many European governments, there is an anticipation of increased pressures on aid budgets. In 2007, EU-27 increased contributions as a share of GNI from 0.08 to 0.42 per cent, but decreased in volume to 49 billion

Euros. A total of 12 EU member states maintained or increased their development cooperation budgets, but others (Germany, Italy, Austria, Greece and The Netherlands) cut theirs (Scholze, 2010).

While it is evident that there is a need to scale up climate finance on the adaptation side, budgets of national governments are likely to dwarf international public climate finance, even if the US$100 billion a year mark is achieved. It is vital that a clear additional role be identified for climate finance for it to be effective, lever additional funds, and be capable of MRV. On the mitigation side, country case studies suggest that the pressures to create a new institutional architecture to handle climate change funds is diverting effort from working out exactly what needs to be funded. At present, analysis at the country level suggests that the funding modalities are guiding activities, rather than the other way around (Brown and Peskett, 2011). In each country, there is a plethora of funding mechanisms, resulting from donor governments changing policies, all of which leave an on-going legacy and result in effort in countries being disjointed (Hedger, 2011). With new models emerging under the UNFCCC, first with the Adaptation Fund and now the Global Climate Fund, care needs to be taken that international finance remains responsive to country needs.

A new push or the *coup de grâce*? Rio 1992 to Rio 2012

Within the UNFCCC, there has been little space made to focus on the climate technology needs of the poorest. To an extent, this is due to the fact that the UNFCCC is only one of the UNCED agreements made at the Earth Summit in Rio in 1992. Its core objective is stabilization of atmospheric greenhouse concentrations. It had been seen to be one of the more dynamic and effective of the Rio conventions, which led to the explosion of interest in the Copenhagen climate conference in 2008. The central process on development launched at the United Nations Conference on Environment and Development (UNCED) Rio 1992 was around Agenda 21, an action plan for implementing sustainable development. With clear road blocks ahead in the UNFCCC regarding the prospect of a new legal post-2012 framework, much attention is now focused on the Rio+20 process, which culminates in Rio in June 2012. The two themes are the green economy in the context of sustainable development and poverty eradication, and the institutional framework for sustainable development (UNGA, 2011). The concept of a green economy has become a focus of policy debate during recent years, triggered by the opportunity which arose from the global financial crisis (Ocampo, 2010).

Little has actually happened as a result of the other Rio agreements since 1992, but the energy needs of the poor have had considerable attention in international conferences over the past two decades. Included in Agenda 21 was Chapter 9, which covered the need for environmentally sound energy systems, particularly new and renewable sources of energy. Within the broader UN context, these are recognized as meeting the essential livelihood needs of the poor, who are often not only economically impoverished, but energy poor as well. In fact, the first major

UN event on 'New and Non-Conventional Sources of Energy' was actually held in 1981. Alleviating energy poverty is now seen as a prerequisite to achieving the UN MDGs by 2015, and renewable energies are seen to contain opportunities for small-scale, decentralized energy production, ideal for reaching rural and remote areas not serviced by existing energy grids.

The energy agenda has been directly negotiated since the Rio 1992 Conference under the aegis of the Commission for Sustainable Development (CSD) and was its focus theme in 2001 and 2007. Indeed, while CSD 9 in 2001 did come to an all-embracing decision, with recommendations on energy for sustainable development, CSD 15 in 2007, although addressing energy issues, did not reach any consensus in its final outcome documents. This was despite the fact that the World Summit on Sustainable Development (WSSD) in 2002 addressed renewable energy in several of the chapters of the *Johannesburg Plan of Implementation*. The CSD and the sustainable development framing have not been successful, not only in delivery but also as a way of developing international consensus, so several other initiatives have been spawned. These all encompass elements of technology.

A bewildering number of global and regional initiatives have been established, all seemingly looking at clean energy, sustainable and renewable energy and their technology dimensions. One of the first international-sponsored events on energy issues to be established was the Global Forum on Sustainable Energy in 2000, which has met almost every year since. Under the auspices of the German government, a separate process, the International Renewable Energy Conference, was launched in 2004 to follow up the WSSD. The Global Renewable Energy Forum was also started by Brazil in 2008. In Asia, the Clean Energy Forum was established by the Asian Development Bank (ADB) in 2006, with meetings annually since then to share knowledge. From 2009, the United Nations Industrial Development Organization (UNIDO) has sponsored the Vienna Energy Conference, looking at energy and development debates more specifically. In addition, there is the World Future Energy Summit series and the Global Energy Assessment, to be published later in 2011, as well as the First Global Green Growth Forum, which is to be held in October 2011 and which will look at the new model of economic growth in relation to poverty reduction and social inclusion.

One new institution has emerged – the International Renewable Energy Agency (IRENA) – which had its first formal Assembly in April 2011. But its precise focus and status has yet to be resolved. In its first assembly, one of the themes pursued was that IRENA should promote development, demonstration and deployment of renewable energy technologies, including facilitating knowledge dissemination on the development and transfer of technologies. It could facilitate partnerships and access to funding or become a source of funds. In the Ministerial Round Table, some countries (India, Nicaragua and Australia) said that IRENA should concentrate on facilitating technology transfer to assist in energy poverty reduction. Similar bodies include the Renewable Energy Policy Network for the 21st Century (REN21) and the Renewable Energy and Energy Efficiency Partnership (REEEP).

It is therefore widely recognized outside the UNFCCC that there is an ever more urgent need for technologies which reduce carbon energy consumption, and also to ensure that development pathways in the future are based on fundamentally different technologies. Some new initiatives were started by the G20 group in the immediate wake of the financial crisis in 2008, and some governments did establish green investments in their stimulus packages to reduce carbon dependency (e.g. China, Korea and the US). A key document for Rio+20 states that some changes have begun: 'growing number of countries are experimenting with a more comprehensive reframing of their national development strategies and polices along green economy lines, including as "low-carbon green growth strategies"' (UNGA, 2011, p2). Despite these encouraging efforts, there is also concern: the *World Economic and Social Survey 2011* argues for a fundamental technological overhaul, otherwise there is a risk of failure in fulfilling global commitments to end poverty and averting the catastrophic impacts of climate change and environmental degradation (UNDESA, 2011).

There is now much riding on the Rio+20 process, with its green economy and the institutional framework themes. The hope is that it will trigger global, economic, social and economic renewal (Evans and Steven, 2011). Ban Ki Moon has expressed the hope that governments will adopt the three 2030 energy goals of universal access to electricity, 40 per cent energy intensity reduction and a global mix of 30 per cent renewables during the Rio+20 Conference. According to Barbier (2010), the expanded vision offered by a Global Green New Deal (GGND) to confront multiple crises (fuel, food and financial) could ensure financial stability, reviving growth and creating jobs, while addressing other global challenges, such as reducing carbon dependency, protecting ecosystems and water resources and alleviating poverty. Definitions and debates about the precise nature and definitions of green growth, green economy and its relationship to sustainable development are not our concern here.[4] The point is that issues pertinent to the debates around delivering technologies for adaptation and mitigation which meet the needs of the poor are being fully debated and explored within another set of UN meetings. Within the report of the UN Secretary General for the Preparatory Committee for the CSD, green technology issues and various policy tracks towards a green economy are examined.

What is novel is that the green growth framing focused on a broader economic approach to technology issues, not just related to energy directly but also to water, waste, agriculture and infrastructure. A sector-wide assessment has emerged so that it is more evident that the challenge is about scaling up delivery of effective technologies, rather than transplants of specialized IPR-protected industrial technologies. The importance of nationally coherent enabling environments, with plans and systems of regulations and incentives, is vital. The green economy perspective also adds in the need for the restoration and enhancement of natural capital, notably agriculture (UNGA, 2011, p15).

The LDCs currently present a low-carbon profile due to their low levels of carbon emissions. As their economies rely on natural capital (agriculture, forest resources, biodiversity, tourism, mineral and extraction) there exists a large potential for renewable energies. Furthermore, refocusing policies and investments that are most relevant to the livelihoods of the poor would be more conducive to inclusive growth and jobs, and make a significant contribution to the MDGs (UNEP, 2011). The argument is that international sources of funding are needed to support clean energy technology adoption and trade-related capacity to catalyse and sustain LDC transition to a green economy, and they are well suited to benefit from this integrated approach. There are known technology and institutional packages which can help delivery in key sectors such as Climate Smart Agriculture (Wolemberg et al., 2011) and integrated water resources management.[5] So far, delivery of technology transfer to meet the needs of LDCs have not been effective and more systemic and integrated discussions under the Rio+20 agenda may work better.

Discussion and conclusions

The UNFCCC has been one of the more successful of the Rio treaties and has a wide reach and influence. But no notable measures have provided technology transfer or low-carbon development in the LDCs. After many years, each with many weeks of negotiations, it seems that finally there is a prospect of its Technology Mechanism being established. However, there is a real danger that this will be too little too late. The Technology Mechanism is now only one of several strands of activity linked to the UNFCCC on action on mitigation and adaptation. In respect of the poor, the UNFCCC is a country-to-country negotiation, and it is through the LDC group that special attention is given to the most vulnerable. Needs assessments (TNAs) of LDCs show that agriculture, water, forestry and household energy needs are priorities; however, these are not being given specific attention within the UNFCCC in relation to technology transfer. Other parts of the UN system are now regenerating their activity on energy and development for the poorest 1.6 billion, coalescing for the Rio+20 Conference. Efforts delivered under the UNFCCC may be established in a small silo of isolated projects at international level unless new cross-linking mechanisms are devised.

Initial reactions to the global economic crisis provided some scope for optimism that transformative green growth pathways might emerge, delivering on employment, energy security and climate change. While some of the initial optimism has dissipated, several new reports have emerged, and there is evidence of a coherent reframing of issues, which may affect the delivery and types of low-carbon technologies for the poorest. The 'green economy' does provide a broader approach to technology needs which could work for the poorest countries, if backed up.

Financing of a sustainable climate deal is still a vision, not a reality, although pieces of the structure of a future package are now in place. Delivery of the new and additional sources of finance at the scale required under the UNFCCC is looking problematic (US$100 billion a year by 2030). In addition, in relation to the TT needs of the poorest, it is not clear if they will be given priority through the UNFCCC route.

There are tantalizing opportunities for change. There are known solutions which would improve livelihoods, provide climate resilience and energy for development. There have been innumerable international conferences over the years, and yet more are planned. Actions could finally get under way, and within a broader framing than the UNFCCC TT framework. The big unknown is whether global leaders in the mature and emerging economies can rise to the challenges. Can they lead to new pathways fast enough?

Notes

1 This chapter was written in August 2011. The key relevant developments at Durban were that while the Green Climate Fund (GCF) was launched, there was no agreement on how the long-term financing would be mobilized; and the relationship of the Technology Mechanism to the GCF has not yet been clarified.
2 The Group of 77 and China is the name of the key bloc of developing nations within the UNFCCC process.
3 Although China and India have many millions of poor people who are also vulnerable to climate change, especially in coastal areas, they have not prioritized their needs from a technology transfer perspective. In fact, a recent analysis has shown that three-quarters of the world's 1.3 billion poor live in middle-income countries India, China, Nigeria, Pakistan and Indonesia – the 'new bottom billion' (Sumner, 2010).
4 But see, for example, OECD (2011a, 2011b) and UNEP (2011).
5 There is a considerable literature on IWRM (see, for example, UNFCCC, 2011k).

References

Barbier, E. (2010) *A Global Green New Deal: Rethinking the Economic Recovery*, UNEP and Cambridge University Press, Cambridge
BASIC Ministers (2010) Joint Statement issued at the conclusion of the Third Meeting of BASIC Ministers, Cape Town, 25 April 2010, http://moef.nic.in/downloads/public-information/BASIC-statement.pdf
BASIC Ministers (2011) Text of Joint Statement issued at the Conclusion of the Sixth Basic Ministerial meeting on Climate Change, New Delhi, India, 26–27 February 2011
Brown, J. and Peskett, L. (2011) *EDC2020 Working Paper: Climate Finance in Indonesia: Lessons for the Future of Public Finance for Climate Change Mitigation*, http://www.edc2020.eu/fileadmin/publications/EDC_2020_-_Working_Paper_No_11_-_Climate_Finance_in_Indonesia.pdf
Collier, P (2007) *The Bottom Billion: Why the Poorest Countries Are Failing and What Can Be Done about It*, Oxford University Press, Oxford, UK
COWI and IIED (2009) *Evaluation of the Operation of the Least Developed Countries Fund for Adaptation to Climate Change*, GEF Evaluation Office and Ministry of Foreign Affairs, Evaluation Department Government of Denmark, Denmark, www.evaluation.dk
EC (2009) *Supporting a Climate for Change: The EU and Developing Countries Working Together*, Luxemburg Publications Office of the European Union, doi: 10.2773/82318
EC (2010) *European Commission Climate Action*, http://ec.europa.eu/clima/policies/finance/index_en.htm, accessed 24 August 2011
EC (2011) *EU Fast Start Finance Report to the UNFCCC Secretariat*, 6 May 2011, 9888/11 Ecofin 249. ENV 343, Council of the European Union, Brussels
Ellis, J. and Kamel, S. (2007) *Overcoming Barriers to Clean Development Mechanism Projects*, OECD COM/ENV/EPOC/IEA/SLT, Paris
ENB (2007) Summary of the Thirteenth COP, UNFCCC, 3–15 December, vol 12, no 354, Tuesday, 18 December, p20, http://www.iisd.ca/download/pdf/enb12354e.pdf

ENB (2011) vol 30, no 4, p7, http://www.iisd.ca/irena/irenaa1/

Evans, A. and Steven, D. (2011) *Making Rio 2012 Work: Setting the Stage for Global Economic, Social and Ecological Renewal*, Centre on International Cooperation, New York University, New York, NY, www.nyu.edu

Fallasch, F. and De Marez, L. (2010) *New and Additional? A Discussion Paper of Fast-Start Finance Commitments of the Copenhagen Accord*, Climate Analytics, www.climateanalystics.org

Griffith-Jones, S., Hedger, M. and Stokes, L. (2009) *The Role of Private Investment in Increasing Climate Friendly Technologies in Developing Countries*, Background Paper for UNDESA for World Economic and Social Survey 2009, http://www.un.org/en/development/desa/policy/wess/wess_bg_papers.shtml

Haites, E. (2011) 'Climate change finance', in *Climate Policy*, vol 11, pp963–969

Harmeling, S. (2011a) *Successful Start for the Design of the Green Fund*, Report on the first meeting of the Transitional Committee to design the Green Climate Fund, May 2011, http://www.germanwatch.org/kliko/ks48e.pdf

Harmeling, S. (2011b) *Countdown to Capetown*, Report on the second meeting of the Transitional Committee to design the Green Climate Fund, August 2011, Germanwatch, www.germanwatch.org

Hedger, M. (2011) *Climate Finance in Bangladesh: Lessons for Development Cooperation and Climate Finance at the National Level*, Working Paper no 12, European Development Cooperation 2020 project, http://www.edc2020.eu/fileadmin/publications/EDC_2020_-_Working_Paper_No_12_-_

IIGCC, INCR, IGCC and UNEPFI (2010) *Global Investor Statement on Climate Change: Reducing Risks, Seizing Opportunities and Closing the Climate Investment Gap* (Statement supported by 268 investors), Institutional Investors Group on Climate Change, November 2010, London, http://www.iigcc.org/__data/assets/pdf_file/0015/15153/Global-Investor-Statement.pdf

IISD (2011) *CIF Partnerships Forum Bulletin*, issue 1, vol 172, no 3

IPCC (Intergovernmental Panel on Climate Change) (2007) *Climate Change Impacts, Adaptation and Vulnerability: Contribution of Working Group 11 to the Fourth Assessment Report*, IPCC, Cambridge University Press, Cambridge

Major Economies Forum (2011) http://www.majoreconomiesforum.org/past-meetings/tenth-meeting-of-the-leaders-representatives-of-the-major-economies-forum-on-energy-and-climate.html, accessed 21 August 2011

Mohner, A. and Klein, R. (2007) *The Global Environment Facility: Funding for Adaptation or Adapting to Funds?*, Climate and Energy Working Paper, Stockholm Environment Institute, Stockholm

Ocampo, J. (2010) 'The transition to a green economy: Benefits, challenges and risks from a sustainable development perspective. Summary of background papers', in *The Transition to a Green Economy: Benefits, Challenges and Risks from a Sustainable Development Perspective*, Report by a Panel of Experts to Second Preparatory Committee Meeting for UNCS, Prepared under the direction of UNDESA, UNEP and UNCTAD, http://www.uncsd2012.org/rio20/content/documents/Green%20Economy_full%20report.pdf

OECD (Organisation for Economic Co-operation and Development) (2010) *Development Perspectives for a Post 2012 Climate Financing Architecture*, OECD, Development Assistance Committee, Draft

OECD (2011a) *Monitoring and Tracking Long-Term Finance to Support Climate Action*, May 2011 (B. Buchner, J. Brown and J. Corfee-Morlot), OECD/IEA project for the CC Expert Group on the UNFCCC, www.oecd.org/env/cc/ccxg

OECD (2011b) *Towards Green Growth*, OECD Publishing, Paris, http://dx.doi.org/10.1787/9789264111318-en

Persson, A., Klein, R., Siebert, C., Atteridge, A., Muller, J., Hoffmaister, L. M. and Takama, T. (2009) *Adaptation Finance under a Copenhagen Agreed Outcome*, Research report, Stockholm Environment Institute, Stockholm

Pointcarbon (2011) 'World Bank to launch $130m CDM fund for LDCs', http://www.pointcarbon.com/news/1.1568527

Scholze, I. (2010) *European Climate and Development Financing before Cancun*, Opinion no
    7, December 2010, EDC 2020, http://www.edc2020.eu/fileadmin/publications/
    EDC_2020_Opinion_No_7_European_Climate_and_Development_Financing_
    Before_Cancun_v2.pdf
Sierra, K. (2009) Statement to Opening Plenary Energy Week, 31 March–2 April 2009,
    World Bank, Washington, DC, http://siteresources.worldbank.org/INTENERGY/
    Resources/335544-1232567547944/5755469-1239633250635/Kathy_Sierra.pdf
Stadelmann, M., Roberts, T. and Amchaelowa, A. (2010*) Keeping a Big Promise: Options for
    Baselines to Assess 'New and Additional' Climate Finance*, CIS Working Paper no 66 2010,
    University of Zurich, http://ssrn.com/abstract=1711158
Stern, N. (2007) *The Economics of Climate Change: The Stern Review*, Cambridge University
    Press, Cambridge
Stewart, R., Kingsbury, B. and Rudyk, B. (2009) 'Climate finance for limiting emissions
    and promoting green development', in R. Stewart, B. Kingsbury and B. Rudyk (eds)
    Climate Finance, New York University, Abu Dhabi Institute, New York University Press,
    New York, NY
Sumner, A. (2010) *Global Poverty and the New Bottom Billion: Three-Quarters of the World's
    Poor Live in Middle-Income Countries*, IDS Working Paper 349, IDS, Brighton, UK
Tanner, T. and Mitchell, T. (2008) *Entrenchment or Enhancement: Could Climate Change
    Adaptation Help Reduce Chronic Poverty?*, Working Paper 106, Chronic Poverty Research
    Centre, Manchester, UK
TWN (2008) *World Bank's Funds will Undermine Global Climate Action*, Third World
    Network, Malaysia, www.twnside.org.sg
UNAGF (2010) *Report of UN Secretary General's High level Advisory Group on Climate Change
    Financing*, 5 November 2010, http://www.un.org/wcm/content/site/climatechange/
    pages/financeadvisorygroup/pid/13300
UNDESA (2011) *The Great Green Technological Transformation*, World Economic and Social
    Survey 2011, Overview, UN, New York, http://www.un.org/en/development/desa/
    policy/wess/wess_current/2011wess_overview_en.pdf
UNEP (2011) *Towards a Green Economy: Pathways to Sustainable Development and Poverty
    Eradication – A Synthesis for Policy Makers*, www.unep.org/greeneconomy, http://www.
    unep.org/greeneconomy/Portals/88/documents/ger/GER_synthesis_en.pdf
UNEP, UNCTAD and UN-OHRLLS (2011) *Why a Green Economy Matters for the Least
    Developed Countries*, UNEP, http://www.unctad.org/en/docs/unep_unctad_un-ohrlls_
    en.pdf
UNFCCC (1992) *United Nations Framework Convention on Climate Change*, UNFCCC
    Secretariat, Bonn, http://unfccc.int/resource/docs/convkp/conveng.pdf
UNFCCC (1998) *Kyoto Protocol*, UNFCCC Secretariat Bonn, http://unfccc.int/resource/
    docs/convkp/kpeng.pdf
UNFCCC (1999) *Report of the Conference of the Parties on its Fourth Session*, Buenos Aires,
    2–14 November 1998, Addendum, Part Two: Action Taken by the Conference of the
    Parties at its Fourth Session, FCCC/CP/1998/16/Add.1, http://maindb.unfccc.int/
    library/view_pdf.pl?url=http://unfccc.int/resource/docs/cop4/16a01.pdf
UNFCCC (2002) *Report of the Conference of the Parties Held at Marrakesh 2001*, Addendum,
    Part Two: Actions Taken by the Conference of the Parties, UNFCCC Secretariat, Bonn,
    http://unfccc.int/resource/docs/cop7/13a02.pdf
UNFCCC (2006) *A Guidebook on Preparing Technology Transfer Projects for Financing*,
    UNFCCC Secretariat, Bonn, http://unfccc.int/ttclear/pdf/PG/EN/UNFCCC_
    guidebook.pdf
UNFCCC (2007) *Report on Existing and Potential Investment and Financial Flows Relevant to
    the Development of an Effective and Appropriate International Response to Climate Change*,
    UNFCCC Secretariat, Bonn
UNFCCC (2008) *Report of the Conference of the Parties on its Thirteenth Session, Held in Bali
    from 3 to 15 December 2007*, FCCC/CP/2007/6/Add.1*14 March 2008, Addendum,
    Part Two, Decisions adopted, p3, Decision 1/CP.13, Bali Action Plan, and Decision

3/CP.13, Development and Transfer of Technologies under the Subsidiary Body for Scientific and Technological Advice, p12, UNFCCC Secretariat Bonn, http://unfccc. int/resource/docs/2007/cop13/eng/06a01.pdf#page=3

UNFCCC (2009a) *Poznan Strategic Programme on Technology Transfer*, FCCC/CP/2008/7/ Add.1 Page 3 Decision 2/CP.14, Development and Transfer of Technologies, UNFCCC Secretariat, Bonn, http://maindb.unfccc.int/library/view_pdf. pl?url=http://unfccc.int/resource/docs/2010/sbi/eng/25.pdf

UNFCCC (2009b) *Least Developed Countries under the UNFCCC*, UNFCCC Secretariat, Bonn, Figure 3

UNFCCC (2010) *Report of the Global Environment Facility on the Progress Made in Carrying out the Poznan Strategic Programme on Technology Transfer*, FCCC/SBI/2010/25, UNFCCC Secretariat, Bonn, http://maindb.unfccc.int/library/view_pdf. pl?url=http://unfccc.int/resource/docs/2010/sbi/eng/25.pdf

UNFCCC (2011a) *March Report of the Conference*, Held in Cancun from 29 November to 10 December 2010, Addendum, Part Two: Action Taken by the Conference of the Parties at Its Sixteenth Session, Decisions adopted by the Conference of the Parties, FCCC/CP/2010/7/Add.1 UNFCCC Secretariat, Bonn, http://unfccc.int/resource/ docs/2010/cop16/eng/07a01.pdf#page=2

UNFCCC (2011b) *Report by the Chair of the Expert Workshop on the Technology Mechanism*, Bangkok, 5–8 April, 2011 FCCC/AWGLCA/2011/INF.2, UNFCCC Secretariat, Bonn, http://unfccc.int/ttclear/jsp/TechnologyMechanism.jsp

UNFCCC (2011c) *Work of the AWG–LCA Contact Group on Technology Development and Transfer*, Note by facilitator, UNFCCC Secretariat, Bonn, http://unfccc.int/files/ meetings/ad_hoc_working_groups/lca/application/pdf/draft_text_16.06.2011@ 1700_edited.pdf and G77 and China Draft decision under the agenda item on Development and Transfer of Technologies for AWG–LCA: http://unfccc.int/files/ meetings/ad_hoc_working_groups/lca/application/pdf/2011-06-16_g77-china__ text_on_ctc-n-1700-1.pdf

UNFCCC (2011d) *Development and Transfer of Technologies*, SBSTA FCCC/SBSTA/2011 L10, UNFCCC Secretariat, Bonn, http://unfccc.int/resource/docs/2011/sbsta/eng/ l10.pdf

UNFCCC (2011e) *Development and Transfer of Technologies*, UNFCCC Secretariat, Bonn, SBI FCCC/SBI/2011/L.10

UNFCCC (2011f) *Work of the AWG_LCA Contact Group, Enhanced Action on Adaptation*, Note by facilitator, Draft decision text, Version 17-06-11, UNFCCC Secretariat, Bonn, http://unfccc.int/files/meetings/ad_hoc_working_groups/lca/application/pdf/ lca_agenda_item_3.3_adaptation_17_june.pdf

UNFCCC (2011g) *Interlinkages between Technology Needs Assessments and National and International Climate Policy Making Processes*, Background Paper, III UNFCCC Workshop on Technology Needs Assessments, Wissenschaftszentrum, Bonn, Germany, 1–2 June 2011, http://unfccc.int/ttclear/jsp/TrnDetails.jsp?EN=TNAWshpBonn

UNFCCC (2011h) *The LDCs: Reducing Vulnerability to Climate Change, Climate Variability and Extremes, Land Degradation and Loss of Biodiversity*, Contribution to Fourth UN Conference on LDCs, May 2011, UNFCCC Secretariat, Bonn

UNFCCC (2011i) *Report of the Nineteenth Meeting (14–17 March 2011) of the Least Developed Countries Expert Group*, FCCC/SBI/2011/4, UNFCCC Secretariat, Bonn

UNFCCC (2011j) *Work of the AWG_LCA Contact Group Finance*, Note by facilitator, Version 17-06-11, p2, UNFCCC Secretariat, Bonn, http://unfccc.int/files/meetings/ad_hoc_ working_groups/lca/application/pdf/facilitators_note_finance_17june_09.50.pdf

UNFCCC (2011k) *Climate Change and Freshwater Resources*, A synthesis of adaptation actions undertaken by Nairobi Work Programme partner organizations, UNFCCC Secretariat, Bonn

UNFCCC and UNDP (2010) *Handbook for Conducting Technology Needs Assessments*, UNDP, November, New York, http://unfccc.int/ttclear/pdf/TNA%20HANDBOOK%20 EN%2020101115.pdf

UNGA (2011) UNGA A/CONF.216/PC/7 22-12-10, Preparatory Committee for the UN Conference on Sustainable Development (UNCSD), Second Session, 7–8 March 2011, Objectives and themes of the UNCSD, Report of the Secretary General, UN, New York, NY

Wolemberg, E., Campbell, B., Holmgren, P., Seymour, F., Sibanda, L. and von Braun, J. (2011) *Actions Needed to Halt Deforestation and Promote Climate-Smart Agriculture*, CCAFS Policy Brief no 4, CGIAR Research Program on Climate Change, Agriculture and Food Security (CCAFS), Copenhagen, www.ccafs.cgiar.org

World Agroforestry Centre (2009) *Africas's Biocarbon Experience: Lessons for Improving Performance in the African Carbon Markets*, Policy brief no 06 2009, COMESA, ASB, SADC, EAC, World Agroforestry Centre, Nairobi, Kenya

World Bank (2011) *State and Trends of the Carbon Market in 2011*, Environment Department, World Bank, Washington, DC, www.carbonfinance.com

Yamin, F. and Depledge, J. (2004) *The International Climate Change Regime: A Guide to Rules, Institutions and Procedure*, Cambridge University Press, Cambridge

12

PRO-POOR LOW-CARBON DEVELOPMENT

Implications for Technology Transfer

Frauke Urban and Andy Sumner

Introduction

Global climate change is considered one of the greatest threats to international development efforts. It poses risks to humans, the environment and the economy. The Intergovernmental Panel on Climate Change (IPCC) reports in its *Fourth Assessment Report* (4AR) that the global mean surface temperature has risen by 0.74°C ±0.18°C during the last century. This increase has been particularly significant over the last 50 years (IPCC, 2007). However, more recent research indicates that climatic changes occur at a much faster rate than assumed a few years ago in the IPCC's 4AR (e.g. Richardson et al., 2009). Today most climate scientists agree that the possibility of staying below the 2°C threshold between 'acceptable' and 'dangerous' climate change becomes less likely as no serious global action on climate change is taken (Richardson et al., 2009; Tyndall Centre, 2009). A rise above 2°C is likely to lead to abrupt and irreversible changes (IPCC, 2007). These changes are expected to make it difficult for contemporary societies to cope with, and they could cause severe societal, economic and environmental disruptions which could severely threaten international development throughout the 21st century and beyond (Richardson et al., 2009).

Developing countries, and especially the poor in low-income countries, have historically contributed very little to climate change. However, they are often the most vulnerable to climate change due to their limited resources, high population growth and limited capacity to adapt. Poor people and poor communities in developing countries are also the least likely to have access to low-carbon technologies or access to sufficient financial means for purchasing low-carbon technologies. Nevertheless, low-carbon technologies can satisfy basic human needs, such as energy provision for cooking and heating.

Low-carbon development (LCD) and technology transfer (TT) debates, to date, have been mainly about high- and middle-income countries. However, there are good reasons why even the poorest countries with low emissions might be interested in pursuing low-carbon development. Indeed, LCD can be an opportunity for low-income countries to pursue pro-poor development in a carbon-constrained world. This chapter argues that there is a need to link up pro-poor policy debates with the low-carbon and technology transfer debates as part of a post-Millennium Development Goals (MDGs) agenda. The chapter explores several policy responses to low-carbon development and technology transfer, and analyses how pro-poor these policy responses are. The authors distinguish between different approaches to low-carbon development, discuss how low-carbon development can be pro-poor, and elaborate upon what this means for the technology transfer debate.

The chapter is structured as follows: the second section links climate change and the poverty domains of the MDGs. The third section then seeks to define different approaches to low-carbon development and addresses the link to technology transfer. Section four focuses on pro-poor LCD policy responses for low-carbon development. The final section concludes the chapter.

The Millennium Development Goals and climate change

As livelihoods shift in response to more extreme climatic conditions, issues of climate change adaptation and mitigation need to cut across all poverty reduction efforts, including any post-MDGs architecture. Despite a goal dedicated to environmental sustainability, some of the fundamental criticisms of the MDGs have been based on issues of sustainability and the lack of attention to tackling climate change – the impact of which is likely to affect poor people more than others.

Climate change is directly related to the poverty concerns of the MDGs (see Table 12.1).

The observed impacts of climate change include melting glaciers, increases in global surface temperatures, heavier precipitation, increases in tropical cyclone activity and higher frequency of droughts, especially in the (sub)tropics (IPCC, 2007). This will have severe effects on people living in vulnerable areas, such as people living in drought-prone areas where climate change will deteriorate the already harsh living conditions, which can result in famines and malnutrition.

Recent concerns about the achievability of the MDGs and what to do after 2015 in a post-MDG era have focused on the fact that today 960 million people, or 72 per cent, of the world's poor live in middle-income countries (Sumner, 2010). In 1990 most of the world's poor people (93 per cent) lived in poor countries – meaning low-income countries (LICs). Two decades on, the world's poor – 72 per cent, or almost 1 billion poor people – now live in middle-income countries (MICs). This requires further exploration. It points towards the role of inequality as

TABLE 12.1 Millennium Development Goals (MDGs) 1 to 7 and climate change-relevant poverty impacts

Millennium Development Goals (MDGs)	Climate change-relevant poverty impacts
Goal 1: Eradicate extreme poverty and hunger Goal 2: Achieve universal primary education	Climate change is likely to impact upon poor people's livelihoods and food security.
Goal 3: Promote gender equality and empower women	The climate-induced destruction of infrastructure, loss of livelihoods and disaster-related migration could be a barrier to achieving universal primary education and gender equality in education.
Goal 4: Reduce child mortality Goal 5: Improve maternal health Goal 6: Combat HIV/AIDS, malaria and other diseases	Climate change-induced extreme weather events are likely to result in higher prevalence of vector- and water-borne diseases, declining food security and decreased availability of potable water.
Goal 7: Ensure environmental sustainability	Climate change will directly impact upon natural resources, ecosystems and the Earth's natural cycles. This is predicted to reduce the quality and quantity of natural resources and ecosystems.

a cause of persistent poverty and it raises a whole range of questions for the future of international development (Sumner, 2010). The rise of emerging economies such as China and India and their rapid population growth play a role in this shift; however, the percentage of global poverty in the MICs minus China and India has risen from 7 to 22 per cent over the last two decades, while the proportion of the world's poor accounted for by China and India has fallen from 75 to 50 per cent. This means that inequality in other MICs has significantly risen and has caused a growing divide between the rich and the poor. The situation is fairly similar taking into account education, nutrition and the new United Nations Development Programme's (UNDP's) multidimensional poverty index.

This has implications for international development and the MDGs, including implications for aid architecture, global governance and other key issues, such as climate change and technology transfer. Development pathways which aim at tackling climate change, while at the same time focusing on social and economic development and achieving the MDGs, are urgently needed. Low-carbon development can be one way to achieve this.

What is low-carbon development and how is it linked to technology transfer?

Defining low-carbon development

There is currently no internationally agreed definition of LCD. Definitions that do exist mainly focus on mitigation, which neglects the importance that LCD can play in low-income countries and adaptation. Low-carbon development is defined by the UK's Department for International Development (DFID) as using less carbon for growth by switching to low-carbon energy and improving energy efficiency and energy savings; protecting and promoting carbon sinks; promoting low- or zero-carbon technologies and business models; and introducing policies which discourage carbon-intensive practices (DFID, 2009, p58).

The definition is low-carbon growth-centred (see below discussion). LCD has been defined more broadly by Skea and Nishioka (2008, p6) as 'compatible with the principles of sustainable development, ensuring that the development needs of all groups within society are met, making an equitable contribution towards the global effort to stabilize the atmospheric concentration of CO_2 and other greenhouse gases at a level that will avoid dangerous climate change, through deep cuts in global emissions'. However, definitions such as this focus on mitigation, which neglects the importance that adaptation can play in low-income countries. This definition is also focused on developed countries as it calls for 'deep cuts in [global] emissions'. At a global level this is crucial, but at a national level this is not an option for low-income countries as they (generally) have very low emissions and the main issue is how to achieve development in times of climate change.

Low-carbon development is not only important for developed countries that caused the bulk of climate change, but it is also important for developing countries, particularly for the poor in low-income countries. The poor in low-income countries have contributed least to climate change. For them, low-carbon development is not about cutting greenhouse gas (GHG) emissions, but about the benefits and opportunities that LCD can bring.

Low-carbon development can be beneficial to the poor as it can: provide climate-friendly modern energy for electrification as an alternative to traditional fuels and fossil fuels; provide 'green jobs'; promote sustainable use of forest and land-use resources; increase energy security; improve environmental quality; increase community participation; and contribute to capacity-building. Low-carbon development can also be beneficial from a cost perspective. Fossil fuel resources such as oil are costly and can lead to a 'carbon lock-in', with infrastructure and investments bound to a carbon-intensive economy for decades. Thus, relying on them can mean greater costs in the long run. The Emissions Trading Scheme (ETS) under the United Nations Framework Convention on Climate Change (UNFCCC) has also introduced a price for carbon. Having a high price attached to carbon could mean a competitive disadvantage for low-income countries in relation to global markets (Urban and Sumner, 2009; Urban, 2009, 2010).

Low-carbon development and technology transfer

Low-carbon development can be achieved by technology changes, sectoral changes and behavioural changes (Urban and Sumner, 2009; Urban, 2010). Technology changes will require access to low-carbon technologies. Low-carbon technology includes renewable energy technology such as biomass, solar, wind and hydropower technology; and 'cleaner' fossil fuel technology such as natural gas, carbon capture and storage, and nuclear energy and energy-efficient technology (IPCC, 2007).

Many developing countries, and especially low-income countries, do not have access to modern low-carbon technology or cannot afford it. This is particularly the case for poor people and poor communities in these countries. Middle-income countries such as China, India, Brazil and Mexico have profited to some extent from access to low-carbon technology through the Clean Development Mechanism (CDM); however, very few low-income countries have profited so far. For instance, while China received about 31 per cent of all registered CDM projects worldwide, India 28 per cent, Brazil 11 per cent and Mexico 8 per cent, all registered CDM projects in Sub-Saharan Africa combined (excluding South Africa) accounted for less than 0.5 per cent of global projects by 2010 (IGES, 2010). The percentage of projects which actually benefitted poor people is estimated even lower.

Technology transfer is therefore crucial as it aims to increase access to low-carbon technology for developing countries, enhance innovation and absorptive capacity, and might reduce barriers created by intellectual property rights (IPRs). Neverthe-less, the debate on technology transfer is complex, let alone the challenges associ-ated with enabling low-carbon technology transfer that benefits the poor: there are many actors, different actions and proposals for action, and conflicting interests and motives (Ockwell et al., 2010). The debates tend to be mainly from a techni-cal and economic angle and a top-down perspective; the interests of the poor in low-income countries are often only marginally represented. This chapter aims to shed some light on the policies and actions required to make low-carbon develop-ment pro-poor, which also has implications for the technology transfer debate.

Different approaches to low-carbon development

LCD can be thought of in terms of changes in production, such as changes in supply or economic growth, and/or consumption such as demand, consumption patterns or lifestyles. Figure 12.1 and Table 12.2 provide four contrasting inter-pretations, resulting from where policy-makers place themselves on two different dimensions of response: their approach to growth; and their focus on production- or consumption-related policy measures.

The first two types of LCD (here labelled 'low-carbon growth' and 'low-carbon lifestyles') assume that economic growth is compatible with significant reduc-tions in carbon emissions – the latter two (here labelled 'equilibrium economy' and 'coexistence with nature') assume that it is not. The 'low-carbon growth' and 'equilibrium economy' approaches both put the emphasis on reducing the produc-tion of carbon through technological changes (from inefficient to more efficient, from polluting to less polluting) and/or sectoral changes (structural changes taking place in the economy). For example, in China, many inefficient older coal-fired power plants are being replaced with more efficient, less polluting new plants; and in India, the service economy has been rapidly growing over recent years, while the share of agricultural value added has declined (Van Ruijven et al., 2008).

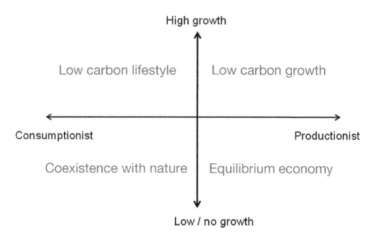

FIGURE 12.1 Types of low-carbon development (LCD)

TABLE 12.2 Types of low-carbon development (LCD)

Type of LCD	Focus and approach
Low-carbon growth. Focuses on the production side of an economy and on how goods and services can be produced with lower emissions. It aims at decoupling economic growth from carbon emissions (e.g. halving emissions, but doubling GDP).	Focus mainly on mitigation, though adaptation also plays a role. Approach: technological change, sectoral change.
Low-carbon lifestyles. Focuses on the consumption side of a growing economy and on the consumer's ability to reduce emissions by consuming climate-friendly products. It implies lifestyle changes and behavioural changes and also leads to a decoupling of carbon emissions (e.g. halving emissions, but doubling GDP).	Focus equally on mitigation and adaptation. Approach: behavioural changes, sectoral change, technological change.
Equilibrium economy. Focuses on the production side of an economy and aims at social development rather than growth. No decoupling is necessary as growth is neutral (e.g. halving emissions, but keeping GDP stable).	Focus mainly on mitigation, though adaptation also plays a role. Approach: technological change, sectoral change.
Coexistence with nature. Focuses on the consumption side of an economy and aims at social development rather than growth. No decoupling is necessary as growth is neutral (e.g. halving emissions, but keeping GDP stable).	Focus equally on mitigation and adaptation. Approach: behavioural change, sectoral change, technological change.

The 'low-carbon lifestyles' and 'coexistence with nature' approaches focus on reducing demand through lifestyle and behavioural changes, as well as through sectoral and technological changes. Behavioural change refers to changes in behaviour and lifestyle: using public transport instead of travelling by car; switching to 'green' electricity instead of fossil fuel-powered electricity; buying local products instead of imported 'air mile products'. Not all options are equally valid for developing countries and specific groups, as discussed in table 12.2.

Of course, the options presented in Table 12.2 are not mutually exclusive. Many policy-makers will favour a mix of production-side and consumption-side approaches to LCD. The debate about the appropriate mix is part of the wider discussion.

While general discussions about the limits of decoupling growth from emissions are fraught (see Barrett et al., 2008; Ockwell, 2008; Sustainable Development Commission, 2009), many case studies argue that low-carbon growth is possible – for example, for China (IEA, 2007), India (World Bank, 2008), South Africa (Government of South Africa, 2008) and Mexico (Project Catalyst, 2008).

How can low-carbon development be pro-poor?

Low-income countries have contributed least to climate change. For them, LCD is not about cutting emissions, but about the benefits and opportunities that LCD can bring for achieving a higher development status. As mentioned above, there are good reasons why LCD can be beneficial for low-income countries (such as access to modern low-carbon energy as an alternative to traditional fuels and fossil fuels; increased energy security from a quality and supply perspective; 'green jobs'; payments for sustainable forest and land-use management; improved environmental quality; increased competitiveness in a world that has a price attached to carbon; etc.). Some of the issues, such as access to low-carbon technologies and energy security, relate directly to the technology transfer debate.

The types of appropriate policy measures will differ for different country income groups, resource availability and the LCD definition taken. LCD pathways can differ between countries having high fossil fuel resources and those not having abundant fossil resources. Countries with high fossil fuel resources usually tend to promote primarily so-called 'cleaner' fossil energy which emits less greenhouse gas emissions than conventional coal and oil (such as natural gas or fossil fuel power plants with carbon capture and storage). Countries with low fossil fuel resources usually tend to promote primarily renewable energy. Forest resource availability can also be important: countries with large forest resources tend to aim to achieve LCD through climate-friendly forest and land-use management. There are several UNFCCC mechanisms for LCD, as indicated in Table 12.3.

The first commitment period of the Kyoto Protocol will end in 2012 and a new climate change agreement will be needed for the post-2012 era. Until recently, only the Clean Development Mechanism (CDM) was accessible for developing countries. Since COP15 and COP16 in Copenhagen and Cancun, there has been significant progress in the mechanism relating to Reducing Emissions from

Deforestation and Forest Degradation (REDD). REDD+, which includes emissions from deforestation, forest degradation and stresses the role of conservation, sustainable management of forests and enhancement of forest carbon stocks, was established as a mechanism under the Copenhagen Accord (UNFCCC, 2010). A year later, the Cancun Agreements further extended the role of forestry and land-use conservation (UNFCCC, 2011a) and outlined guidelines for emission baselines from the forestry and land-use sectors (UNFCCC, 2011b). Both the CDM and REDD mechanisms, in essence, pay developing countries for low-carbon development. For the CDM, developed countries implement projects leading to emission reductions in developing countries. Developing countries gain access to climate-friendly technology, while developed countries gain emission reduction credits to offset their emissions. Similar approaches are in place for REDD where developed countries pay developing countries for sustainable forest and land-use management. Countries such as Brazil, Indonesia and the Democratic Republic of Congo have extensive forest resources which are threatened by deforestation and degradation and fall under the remit of REDD. Another mechanism is currently in discussion and might be part of a future climate change agreement – Land Use, Land-Use Change and Forestry (LULUCF); but it needs to be clarified how the mechanism will work and in how far it overlaps with REDD approaches (UNFCCC, 2011b).

Mechanisms for LCD are some of the key issues for a post-2012 climate agreement. The CDM is currently on its way to being reformed, including approaches at programmatic and sectoral level rather than only at project level, as has been the case in the past. The Cancun Agreements also outline plans to include certain types of carbon capture and storage (CCS) projects in the reformed CDM (UNFCCC, 2011c).

A further outcome of COP15 and COP16 comprised Nationally Appropriate Mitigation Actions (NAMAs) for developing countries. The purpose of NAMAs is to create national mitigation options which are in line with domestic policies and which are developed in 'the context of sustainable development, supported and enabled by technology, financing and capacity building, in a measurable, reportable and verifiable manner' (IEA/OECD, 2009, p7). Each NAMA will depend upon the nationally appropriate mitigation actions and the financial support available. As of February 2012, 44 developing countries have submitted their NAMAs, including a number of least developed countries such as Afghanistan, Bhutan, Ethiopia and Sierra Leone, and emerging economies such as China, India, Mexico, Brazil and South Africa.

Besides the UNFCCC mechanisms, there are many other options of how to achieve LCD depending upon each country's national and local priorities and plans, and the funding and technologies that are available. It is important to have policies and practices in place which are suited for the national circumstances and the local needs. The meaning, scope and scale for LCD differ within different groups of countries. Upper- and middle-income groups in developing countries, particularly in middle-income countries, may have consumption patterns which are in some ways similar to developed countries. Reducing excessive consumption and making

TABLE 12.3 Low-carbon development mechanisms

LCD mechanisms	What is it?
Clean Development Mechanism (CDM)	Developed countries implement projects leading to GHG emissions reductions in developing countries. Developing countries gain access to climate-friendly technology, while developed countries gain emission reduction credits to offset their emissions.
Emission trading (EM)	Mechanism that sets a cap on greenhouse gas emissions and introduces a trading system. Once emission allowances are exceeded, emission credits must be bought from those who have emitted less. Emission trading is currently in place for developed countries only, but might be extended to a global level in the future.
Joint Implementation (JI)	Developed countries can invest in emission reduction projects in other developed countries as an alternative to reducing emissions domestically. JI is currently in place for developed countries only.
Reducing Emissions from Deforestation and Forest Degradation (REDD) and Land Use, Land-Use Change and Forestry (LULUCF)	REDD was mentioned as a viable mechanism in the Copenhagen Accord (UNFCCC, 2010) and the Cancun Agreements aimed to operationalize both REDD and LULUCF (UNFCCC, 2011a, 2011b). The key principle is that developing countries can be paid for climate-friendly forest and land-use management, while developed countries can gain emission reduction credits to offset their emission obligations.
Nationally Appropriate Mitigation Actions (NAMAs)	The purpose of Nationally Appropriate Mitigation Actions is to outline national mitigation options which are in line with domestic policies and which are developed in 'the context of sustainable development, supported and enabled by technology, financing and capacity building, in a measurable, reportable and verifiable manner' (IEA/OECD, 2009). More than 40 developing countries have submitted their NAMAs as a response to the Copenhagen Accord as of early 2011 (UNFCCC, 2010).

'greener' choices can therefore be an important issue. The poor in developing countries have, however, contributed very little to climate change and their main priority is social and economic development. For poor and vulnerable groups and low-income countries, the main issue is how to achieve development in times of climate change. The benefits and opportunities of LCD, such as provision of 'green' jobs, increased access to modern technology and access to electricity, contributions to energy security and improved environmental quality, can be valuable.

What is missing so far in the debate is an explicit pro-poor concern or concern for distributional issues (i.e. how do different types of LCD impact upon the poor?). Since the main goal of the UNFCCC mechanisms is to reduce greenhouse gas emissions, we need to link up pro-poor policy debates with the low-carbon debates as part of a post-MDG agenda or paradigm.

LCD can be beneficial to the poor as it can open up new opportunities. LCD can create new 'green' jobs and new 'green' industries. The need for green jobs for the poor has been stressed by a recent UNDP report (UNDP, 2009). Green jobs are defined by the UNDP (2009, p2) as:

> ... involving the implementation of measures that reduce carbon emissions or help realise alternative sources of energy use ... [to] align poverty reduction and employment creation in developing countries with a broader set of investments in environmental conservation and rehabilitation to also preserve bio diversity, restore degraded land, combat erosion, and remove invasive aliens etc. ... well designed interventions can contribute directly to the poverty-environment nexus by allowing income generated from environmental activities to ease the pressure on generating income through exploiting the environment. Environmental sector targeted public employment programmes can also be deployed to specifically address environmental concerns and create employment for the poor at the same time.

In addition to green jobs for the poor, LCD can provide climate-friendly energy for electrification and can increase the energy-generating capacity which is often scarce in developing countries. This can be achieved, for example, by renewable energy, which already plays a prominent role at the community level in some poor communities in developing countries. LCD can result in the introduction of energy-efficient technologies which use fewer resources and reduce costs. LCD can lead to an enhanced energy security due to relying on fewer fossil fuel imports and instead using more locally abundant energy sources, such as solar and hydropower. It may also provide a better quality of energy supply.

This is directly linked to technology transfer issues. The key recipients of technology transfer and technology cooperation are emerging economies such as China and India. While these countries are middle-income countries, they have millions of poor people who live without basic services, such as access to electricity. In India alone, 380 million people did not have access to electricity in 2005 (IEA, 2007). Technology transfer which can enable providing access to electricity and/ or access to modern energy should be a key propriety. It needs to be ensured that poor people and poor communities can benefit from technology transfer and cooperation mechanism, such as the CDM.

At the same time, the CDM needs to be reformed to enable low-income countries and least developed countries, particularly in Sub-Saharan Africa, to get access to CDM projects. These projects should ideally aim at increasing access to electricity and modern energy for the poor. Nevertheless, a reforming of the CDM alone will not be sufficient as investments in CDM depend upon the ease of doing business and the associated investment risks of each country.

Key policies for pro-poor LCD can be drawn by linking up pro-poor growth debates (see discussion in McKay and Sumner, 2008; Sumner and Tiwari, 2009) and LCD debates (NIES, 2006, Barrett et al., 2008; Ockwell, 2008; Urban, 2009).

The following examples indicate policies for pro-poor LCD:

- *Redistributive policies and public expenditure:* for example, this can take place when the government revenues made by 'green' industries are distributed to pro-poor sectors such as health and education.
- *Support for specific sectors which are crucial for the poor, such as agriculture and forestry:* this requires specific sectoral investments, market development and infrastructure for pro-poor productive sectors.
- *Social protection for adaptation and combining the synergies between mitigation and adaptation:* for example, social protection measures to reduce vulnerability to climate change.
- *Community participation:* LCD provides opportunities to involve communities on a small-scale local level, such as rural electrification with renewable energy. This can enable sharing the profits from LCD on a community level.
- *Development to foster capacity for the legislative, economic and technical frameworks needed to achieve low-carbon pathways:* for example, capacity-building to ensure that local policy-makers can develop the legislative frameworks needed for LCD.
- *Increasing the rate of 'green' job creation:* this will require investments, development of the finance sector and increased investments in small-scale infrastructure.
- *Pro-poor biofuel policies:* biofuels are considered climate-friendly fuels, but can lead to competition between land for food and land for biofuels, which can result in an increase in food prices. This usually hits the poorest the hardest. Pro-poor biofuel policies therefore need to be introduced which promote the growing of biofuels by the poor, create local employment opportunities, and enable the investments to go to low-income countries (Peskett and Prowse, 2008).
- *Pro-poor forest and land-use policies:* REDD and LULUCF are two possible mechanisms where payments from developed countries are directed to developing countries for climate-friendly forest and land-use management, and thus could benefit the poor by ensuring that smaller farmers and foresters can engage in the carbon market (Peskett and Prowse, 2008).

In relation to technology transfer, the following actions might contribute to making it pro-poor:

- Reform the CDM to ensure that a higher number of CDM projects are located in low-income and least developed countries.
- Make it a prerequisite that a certain share of CDM projects needs to be linked to providing low-carbon energy to poor people and communities who do not have access to electricity and/or modern energy.
- Establish incentives for more investment in low-carbon technologies in low-income and least developed countries. This needs to involve national-level and international-level financiers, authorities and institutions.
- Increase the share of technology cooperation between developed countries and developing countries by promoting joint ventures, joint research and development (R&D) projects, etc.

- Establish a mechanism which raises funds for low-carbon technologies in low-income and least developed countries.
- Facilitate access to intellectual property rights (IPRs).

IPRs may be the most fiercely discussed issue in the technology transfer debate. IPRs divide the main players into two groups: those who want to strengthen the protection of IPRs and those who want to loosen them. On the one hand, it is argued that developing countries need to tighten their legal frameworks for increasing IPR protection to cut out cheap imitation; on the other, it is argued that low-carbon technology should be a public good and that IPR protection needs to be loosened (Ockwell et al., 2010; UNFCCC, 20011d). Scientifically, there is high uncertainty both ways: it is not clear at all whether IPRs are a catalyst or a barrier to technology transfer since the different stages of technology development are poorly understood (Ockwell et al., 2010). What is certain is that industrialized countries hold over 97 per cent of all patents, while developing countries – and particularly low-income countries – hold a marginal share of patents (CAFOD, 2001). The process of registering patents is complicated, lengthy and expensive. This seems to be a major reason for many small- and medium-sized enterprises (SMEs) and companies in developing countries for not registering their patents.

A new mechanism for how to deal with IPRs for low-carbon technologies should be established under the framework of the UNFCCC. This could include the pooling of patents for use in developing countries at low-cost rates. This could also include low cost or free access to publicly funded technologies and concessional rates for privately funded technologies in accordance with the current development status of countries. Intellectual property (IP) rules need to be shortened and facilitated to make the process easier for low-income countries with regard to patenting. At the same time, developing countries need to respect IPRs. Low-income countries should have privileged access and privileged pricing for IP, enabling them to access climate-relevant technologies both for adaptation and mitigation.

Conclusion

The development model of 'pollute first, clean up later' is not viable any longer due to climate change and resource scarcity. New development pathways are needed in times of climate change. LCD is a development pathway which can achieve economic and social development while tackling global climate change.

LCD needs to be pro-poor. LCD should therefore be accompanied by mechanisms, incentives and institutions to support a pro-poor low-carbon economy, such as improved access to low-carbon technology for the poor and targeting support to those groups who are the most vulnerable to the impacts of climate change. In relation to low-carbon technology transfer, reforms are needed. Suggestions for reforms include ensuring that a higher number of CDM projects are located in low-income and least developed countries; making it a prerequisite that a certain share of CDM projects needs to be linked to providing low-carbon energy to poor people; establishing a mechanism which raises funds for low-carbon technologies in low-income and least developed countries; and facilitating the access to intellectual property rights for low-carbon technologies.

The IPCC's (2007) *Fourth Assessment Report* warns that global greenhouse gas emissions need to be cut by 80 per cent in 2030 compared to 2000 levels to avoid 'dangerous climate change' (defined as a global temperature rise above 2 degrees Celsius). Leading scientists suggest that even more drastic cuts are needed and warn that a temperature rise above 2 degrees is becoming more and more likely (Richardson et al., 2009). Mitigation of greenhouse gases is therefore becoming increasingly important for global development. High-income countries need to make drastic emission cuts now. In the long run, emerging economies and middle-income countries with high emissions will also need to reduce their emissions. Low-income countries have very low emissions. LCD for low-income countries will primarily be about achieving development in times of climate change. This means that equitable pro-poor low-carbon development pathways are increasingly needed.

References

Barrett, M., Lowe, R., Oreszczyn, T. and Steadman, P. (2008) 'How to support growth with less energy', *Energy Policy*, vol 36, no 12, pp4592–4599
CAFOD (2001) *Intellectual Property Rights and Development*, CAFOD, London
COP15 (2009) 'Paradise lost?', United Nations Climate Change Conference, Copenhagen 2009, http://en.cop15.dk/news/view+news?newsid=881
DFID (UK Department for International Development) (2009) *Eliminating World Poverty: Building our Common Future*, DFID White Paper, DFID, London
Government of South Africa (2008) *South Africa Long-Term Mitigation Scenarios*, http://www.environment.gov.za/HotIssues/2008/LTMS/A%20LTMS% 20Scenarios%20for%20SA.pdf
IEA (International Energy Agency) (2002) *World Energy Outlook 2002: Energy and Poverty*, IEA/OECD, Paris
IEA (2007) *World Energy Outlook 2007*, IEA/OECD, Paris
IEA/OECD (2009) *Linking Mitigation Actions in Developing Countries with Mitigation Support: A Conceptual Framework*, http://www.oecd.org/dataoecd/27/24/42474721.pdf
IGES (Institute for Global Environmental Strategies) (2010) *CDM Project Data Analysis*, http://www.iges.or.jp/en/cdm/report_cdm.html#cdm_a
IISD (2005) *Vulnerability and Adaptation in Developing Countries*, IISD, http://www.iisd.org/ pdf/2007/climate_bg_vulnerability_adap.pdf
IPCC (Intergovernmental Panel on Climate Change) (2007) *Fourth Assessment Report on Climate Change*, http://www.ipcc.ch/ipccreports/assessments-reports.htm
McKay, A. and Sumner, A. (2008) *Economic Growth, Inequality and Poverty Reduction: Does Pro-Poor Growth Matter*, IDS In Focus, 3.2, IDS, Brighton, UK
NIES (National Institute for Environmental Studies) (2006) *Developing Visions for a Low Carbon Society (LCS) through Sustainable Development: Executive Summary*, NIES, Ibaraki
Ockwell, D. G. (2008) 'Energy and economic growth: Grounding our understanding in physical reality', *Energy Policy*, vol 36, no 12, pp4600–4604
Ockwell, D. G., Mallett, A., Haum, R. and Watson, J. (2010) 'Intellectual property rights and low carbon technology transfer: The two polarities of diffusion and development', *Global Environmental Change*, vol 20, pp729–738
Peskett, M. and Prowse, L. (2008) 'Mitigating climate change: What impact on the poor?', *ODI Opinion* 97, April 2008, Overseas Development Institute, London

Project Catalyst (2008) *Low-Carbon Growth: A Potential Path for Mexico*, Project Catalyst, Centro Mario Molina, Mexico

Richardson, K., Steffen, W., Schellnhuber, H. J., Alcamo, J., Barker, T., Kammen, D. M., Leemans, R., Liverman, D., Munasinghe, M., Osman-Elasha, B., Stern, N. and Wæver, O. (2009) *Climate Change: Global Risks, Challenges and Decisions, Synthesis Report*, http://climatecongress.ku.dk/pdf/synthesisreport/

Skea, J. and Nishioka, S. (2008) 'Policies and practices for a low-carbon society', *Climate Policy*, vol 8, Supplement: Modelling Long-Term Scenarios for Low-Carbon Societies, pp5–16

Sumner, A. (2010) *Global Poverty and the New Bottom Billion: Three-Quarters of the World's Poor Live in Middle-Income Countries*, IDS Working Paper 349, IDS, Brighton, UK

Sumner, A. and Tiwari, M. (2009) After 2015: International Development Policy at a Crossroads, Palgrave Macmillan, Chippenham, UK

Sustainable Development Commission (2009) *Prosperity without Growth: The Transition to a Sustainable Economy*, Sustainable Development Commission, London

Tyndall Centre (2009) *Climate Change in a Myopic World*, Tyndall Briefing Note no 36, http://www.tyndall.ac.uk/publications/briefing_notes/bn36.pdf

UNDP (United Nations Development Programme) (2009) *Green Jobs for the Poor: A Public Employment Approach*, Poverty Reduction Discussion Paper PG/2009/002, http://www.undp.org.gy/documents/bk/PG-2009-002-discussion-paper-green-jobs.pdf

UNFCCC (United Nations Framework Convention on Climate Change) (2010) *Copenhagen Accord*, http://unfccc.int/resource/docs/2009/cop15/eng/11a01.pdf

UNFCCC (2011a) *Cancun Agreements*, Outcome of the work of the Ad Hoc Working Group on Long-Term Cooperative Action under the Convention, http://unfccc.int/files/meetings/cop_16/application/pdf/cop16_lca.pdf

UNFCCC (2011b) *Cancun Agreements: Land Use, Land-Use Change and Forestry*, http://unfccc.int/files/meetings/cop_16/application/pdf/cop16_lulucf.pdf

UNFCCC (2011c) *Further Guidance Relating to the Clean Development Mechanism*, http://unfccc.int/files/meetings/cop_16/conference_documents/application/pdf/20101204_cop16_cmp_guidance_cdm.pdf

UNFCCC (2011d) *Enabling Environment*, http://unfccc.int/ttclear/jsp/EEnvironment.jsp

Urban, F. (2009) *Enabling Environments for Low Carbon Economies in Low Income Countries*, CAFOD, London

Urban, F. (2010) 'Pro-poor low carbon development and the role of growth', *International Journal of Green Economics*, vol 4, no 1, pp82–93

Urban, F. and Sumner, A. (2009) *After 2015: Pro-Poor Low Carbon Development*, IDS in Focus Policy Briefing 9.4, IDS, Brighton, http://www.ids.ac.uk/go/bookshop/ids-series-titles/ids-in-focus-policy-briefings/ids-in-focus-policy-briefing-9

Van Ruijven, B., Urban, F., Benders, R. M. J., Moll, H. C., Van der Sluijs, J., De Vries, B. and Van Vuuren, D. P. (2008) 'Modeling energy and development: An evaluation of models and concepts', *World Development*, vol 36, no 12, pp2801–2821

World Bank (2008) *Low Carbon Growth in India*, World Bank, Washington, DC

WRI (World Resources Institute) (2005) *Navigating the Numbers: Greenhouse Gas Data and International Climate Policy*, http://pdf.wri.org/navigating_numbers_chapter6.pdf

13

CLIMATE CHANGE MITIGATION TECHNOLOGY AND POVERTY REDUCTION THROUGH SMALL-SCALE ENTERPRISES

Annemarije L. Kooijman-van Dijk

Climate change mitigation technologies: Potential links to poverty reduction

Small-scale enterprises in developing countries have an important role to play in linking climate change mitigation to poverty reduction strategies through their crucial role as a source of income for the poor. The energy consumption in small-scale enterprises is significant in the total energy consumption by the poor, and targeting this group of enterprises may therefore provide an appropriate entrance to reach the poor with climate change-induced interventions. Small-scale enterprises contribute to a more equitable distribution of income and accessibility to the poor through their wide geographic dispersion and their flexible functioning as fall-back or diversification of household incomes beyond agriculture (Mead and Liedholm 1998; Ellis 2000; ILO 2005). Although there are few national-level statistics on small-scale enterprises in developing countries, partly due to the fact that a large majority of small-scale enterprises in these countries operate in the informal sector, it is known that small-scale enterprises dominate by their numbers. National-level data from India indicates that over 90 per cent of the 1.5 million registered micro-, small- and medium-scale enterprises in 2006 to 2007 were run by only the owner. Even for factories (with powered appliances having a minimum of 10 employees, or without powered appliances with a minimum of 20 employees), 72 per cent of the total number in India still has fewer than 50 employees. A cross-country comparison of data by Davis et al., (2010) also indicates a high relevance of off farm resources in rural areas across developing regions, accounting for 50 per cent of total income in sampled countries from Eastern Europe and Latin America and for all but Vietnam among Asian countries, and between 22 and 41 per cent in African countries.

Considering this role of small-scale enterprises in providing incomes to the poor, the energy consumption patterns of small-scale enterprises therefore links climate change mitigation to poverty reduction in two directions. First, the impact upon the sustainability of global production is significant following the sheer numbers of small-scale enterprises. Second, it is clear that policies and technologies that affect the viability of this sector, whether positive or negative, will have economic and social impacts.

Two routes to the reduction of greenhouse gases by small-scale enterprises are through the use of renewable energy sources and increasing energy efficiency. The availability and adoption of technologies for renewable energy supply and energy efficiency links to poverty reduction, as their use influences many aspects of enterprise operation which, in their turn, affect the income and work circumstances of the poor. For both renewable energy supply and energy efficiency technologies, the potential size of beneficiary groups among small-scale enterprises is high. Renewable energy can contribute to increasing access to modern energy services, especially in rural areas at a distance from energy infrastructure where decentralized renewable energy supply can offer services at lowest cost. The scale of lacking energy infrastructure with the global population lacking access to electricity, estimated at 1.4 billion, or 21.1 per cent (IEA, 2010), shows that there is a substantial role for renewable energy options to play in creating access. Lacking electricity supply infrastructure is most prevalent in rural areas, with, respectively, only 14.3 and 51.2 per cent of rural populations in sub-Saharan Africa and South Asia having electricity access (IEA, 2010).

For energy efficiency, the potential demand is not as much defined by lack of energy access, but rather by the appliances used for energy services. The most substantial quantities of energy are used in manufacturing and processing enterprises, where energy efficiency gains can often be made in upgrading production technologies and processes. Taking the example of India again, it is known that the potential for energy efficiency technologies is high, as many Indian small-scale industries are working with technologies at least 50 years old, such as in energy-intensive industries in foundry and forging (World Bank, 2010). The data on India show that the SMEs involved in manufacturing account for over 80 per cent of the total number of industrial enterprises in the country (MSME, 2010). Enterprises in areas with low access to electricity or fossil fuels may also benefit from energy efficiency improvements, saving on energy use from biomass. A survey of energy consumption in Kenya (Kamfor, 2002) indicated that 97.5 per cent of the energy (in joules) in the surveyed energy-intensive cottage enterprises was provided by firewood or wood for charcoal. It is estimated that currently 2.7 billion people, or nearly 40 per cent of the global population, rely on traditional use of biomass for cooking (OECD/IEA, 2010), where energy efficiency gains can be made by improving the efficiency of stoves or fuel switching.

Although both renewable energy supply and energy efficiency have a clear potential to provide other and more direct benefits to small-scale enterprises than climate change mitigation, the understanding of actual impacts and influencing factors is limited. This often leads to assumptions that potential benefits follow naturally once the technologies are in place. Such assumptions may be positive for policy and project justification, but they are detrimental to achieving actual impacts upon climate change and poverty reduction.

For entrepreneurs, many potential benefits of the availability of renewable energy supply or energy efficiency technologies occur if their uptake in the enterprise leads to innovations that are much broader and much more complex than the uptake of the technology, such as in changes in production and accessing new markets. Only if the steps in the chain between (renewable) energy supply or supply of energy efficiency technologies and impacts are established, can any statements be made on true attribution of changes in enterprise to the technological inputs (Kooijman-van Dijk, 2008).

This chapter addresses the problem that although small-scale enterprises are an important means for poverty reduction and reducing vulnerability, even renewable energy and energy efficiency projects and policies that target poverty reduction often exclude or do not reach this sector. Rather than taking the mainstream approach of analysing opportunities and barriers for diffusion of technologies from the perspective of climate change objectives, here technologies for climate change mitigation will be viewed from the perspective of small-scale enterprises. Taking this approach leads us to consider these technologies as innovations no different from any other changes that potentially bring benefits to the enterprise. Making use of theory in the field of diffusion of innovations described in the following section, the core of this chapter is formed by analysis of empirical data on renewable energy and energy efficiency. The empirical evidence is taken from studies specific for small-scale enterprises, which is analysed with regards to diffusion for use in small-scale enterprises and their potential role in contributing to poverty reduction. By taking such a demand and user-oriented approach, this chapter aims to contribute to improving the positive poverty reduction impacts and reducing negative impacts upon small-scale enterprises of climate change technology projects.

Diffusion of innovations for small-scale enterprises

The diffusion of innovations, such as the uptake of renewable energy or energy efficiency in small-scale enterprises, does not necessarily bring the enterprises to the cutting edge of the global market, but they can be crucial for (or contribute to) the viable operation of the enterprise. It is important to realize that it is true in all parts of the world that many entrepreneurs do not aspire for their enterprise to innovate or grow, but rather the enterprise serves to maintain a traditional way of life or work (Nooteboom, 1994). Research in The Netherlands indicates that only about 20 per cent of small-scale enterprises could be characterized as 'dynamic' as a measure of innovativeness (Nooteboom, 1994). In developing countries, the small-scale sector is dominated by artisans working with established, if not traditional, production technologies and methods, and producing traditional products. Many small-scale enterprises remain a (additional) source of income for mainly the owner (Grosh and Somolekae, 1996; Liedholm and Mead, 1999; Ellis 2000). Where enterprises' main function is to increase stability of income and reduce vulnerability to poverty, the interest to venture into innovations that entail risks is naturally low.

In a broad categorization of factors influencing adoption of innovations, Rogers (2003) distinguishes three aspects: the characteristics of the entrepreneurs as potential adopters; the characteristics of the innovations (in this case, climate change mitigation technologies); and the policy and institutional context for the diffusion to take place.

Where most research on innovations tends to focus on high-tech innovations, research specific for developing countries tends to focus on the organization of diffusion and on the characteristics of entrepreneurs that facilitate diffusion. Lall (1992) and Romijn (1996) identify 'technological capabilities' of entrepreneurs or enterprises as key. Such technological capabilities describe the substantial demands on skills, awareness and prioritization that are posed on entrepreneurs for successful adoption of technologies, even if the development efforts have been done elsewhere.

The characteristics of entrepreneurs influencing adoption choices of energy innovations are studied in this chapter in relation to categories of an entrepreneur's financial, physical, and natural, human and social assets as described in the Sustainable Livelihoods approach. The Sustainable Livelihoods approach (Bebbington, 1999; DFID, 2001) can be used to describe the factors influencing a person's strategies towards personal goals, taking as a starting point one's assets and access to assets in terms of finance, human assets, natural assets, physical assets and social assets.

Financial assets and access to finance form a condition for the investments necessary to realize enterprise innovations. Substantial profits are rare in the small-scale sector: these niches are typically limited to relatively well-off entrepreneurs (Barrett et al., 2001). In addition, access to credits and willingness to use credits is strongly related to financial independence and social structures that limit the personal consequences of risks involved with investments (Kooijman-van Dijk, 2008). Physical and natural characteristics such as availability and location of a physical site and building for enterprise operation, and access to natural resources are relevant for the ability to innovate, especially through access to customers. Financial assets can provide opportunities to overcome issues of location, but not of availability of resources, as will be discussed in the following section. Human assets have been partly discussed above as capabilities, but in the case of adoption of simple technologies, this can be interpreted as skills to use (rather than adapt) the technologies. However, the skills and social assets to be able to translate innovations into profit through accessing new markets or new customers are crucial to all innovations, regardless of level of complexity.

Within the category of factors influencing the diffusion related to the characteristics of the innovations themselves, Rogers (2003) identifies five types of variables that determine rates of adoption of innovations: the perceived attributes of the innovation; the type of innovation decision; communication channels; the nature of the social system; and the extent of change agents' promotion efforts. Promotion of renewable energy supply tends to focus on two of these variables: the communication channels and the promotion efforts of the change agents, such as local sales persons.

Rogers (2003) argues strongly that the first of the variables, taking the point of view of the potential accepter or rejecter of an innovation, is key to diffusion taking place. Five conceptually distinct characteristics are used to describe the factors that play a role in defining this perception:

1. relative advantage as perceived by the individual as better than the idea it supersedes;
2. compatibility with existing values, experiences and needs of potential adopters;
3. complexity of the innovation;
4. trialability, or the degree to which an innovation can be tried out;
5. 'observability' or visibility of results from earlier adopters to potential adopters of an innovation.

It is the first of these factors (the perceived advantage to the entrepreneur) that entails the link to poverty reduction of the discussed technologies, as their impacts upon enterprise operation are the key to changing entrepreneurs' livelihoods. Taking the above user perspective, increasing a focus on improving the relative advantage of innovations for potential users can therefore be expected to simultaneously increase the diffusion of technologies and have positive impacts upon poverty reduction.

Innovations for small-scale enterprise: Renewable energy and energy efficiency

Moving from general mechanisms and potential links between climate change mitigation technologies and poverty reduction through small-scale enterprise to understanding practice, the analysis below is based on empirical evidence specific to renewable energy and energy efficiency technologies. The characteristics of the innovations and the characteristics of the entrepreneurs that influence the adoption process are discussed.

Analysing the characteristics of innovations as perceived by potential users is a complex matter. The characteristics are related not only to the characteristics of the renewable energy technology, the alternatives available and the characteristics of the entrepreneur and his or her assets, but also on the enterprise sector and scale, the market for enterprise products, and the policy and technology supply context. Nevertheless, the analysis below is able to distil general findings, making use of empirical data on renewable energy and energy efficiency in small-scale enterprises. The findings on renewable energy supply below are based on the study by Kooijman-van Dijk (2008), which provides detailed analysis based on empirical data from the Indian Himalayas. The findings on energy efficiency in this chapter are based on a literature review of empirical studies on energy efficiency in small-scale enterprises in developing countries.

General characteristics of renewable energy supply as an innovation for small-scale enterprises

As the characteristics of renewable energy depend strongly upon the energy sources, energy forms and their supply mode, these will first be elaborated upon. Useful energy forms that can be provided by renewable sources are electricity, thermal energy and mechanical energy, and this is what is of interest to the end-user. The main renewable energy sources applicable to developing country contexts are wind, solar and hydro, as well as biomass and geothermal sources.

First we take a closer look at electricity supply from renewable energy sources. For electricity supply, the mode of supply organization is a large distinction. If the supply mode is through a centralized grid connection, the electricity supplied is typically from a range of sources. Without great dependency upon one energy source, the impact of renewable energy supply is then through its impact upon the electricity market, which may influence cost and reliability of supply. If the electricity supply mode is in decentralized form, though, the user does notice the impacts of the source through potential limits in volume of supply or reliability of supply. Although technically it is possible to adapt technical design (having larger systems) and storage capacity (adding batteries or water storage) to achieve required volumes and reliability of supply, the costs related to doing so are largely inhibitive. This is why, in practice, there are still large differences in supply patterns and therefore also appropriate end-uses depending upon the resource.

Solar energy is a relatively reliable energy source, with a high predictability of the resource through daily and seasonal irradiation patterns. The high certainty of solar irradiation, across regions, is a major reason for the popularity of photovoltaics (PV) as an electricity supply technology. This does not make PV automatically appropriate for use in small-scale enterprises, as the costs per kilowatt hour (kWh) of electricity production by PV leads to applications being focused on small demands. For rural electrification, typical products are solar lanterns and Solar Home Systems based on PV, which can supply a few hours of light, radio, small television or laptop or mobile telephone charging. In innovative organizational forms such as village systems, the potential of PV to meet other enterprise energy demands is becoming greater, depending largely upon costs of electricity storage.

The renewable energy sources of hydro and wind are less predictable than solar, and the availability of these resources is highly specific to the exact location. This locational characteristic of most renewable energy technologies has implications for the energy services and locations at which energy is provided, and therefore also on the potential benefits for enterprises. Where electricity generation from wind or hydro are feasible, typical system designs are at the community level, except for the smallest hydro systems, which are designed for individual household use similar to Solar Home Systems.

From both hydro and wind, common applications are for mechanical energy. Mechanical energy is a form of energy that typically finds its demand in the enterprise and agricultural sectors. These renewable energy sources are well proven to meet small-scale enterprise energy demands in milling or carpentry, and the diffusion of technologies in this field typically consists of upgraded traditional systems.

The third form of energy supply is that of thermal energy. By far the most common renewable energy source for thermal energy is biomass. Biomass is used at a large range of scales. At the top end, biomass is used in large-scale industry for process heat. At small scales, biomass is the most common resource for cooking, which is also an energy service demanded by enterprises such as restaurants or food processing. Technologies for the promotion of small-scale biomass applications are typically regarded as energy efficiency measures. These are discussed after the discussion related to renewable energy technologies below.

Renewable energy supply as an innovation: Relative advantage to the entrepreneur

This sub-section focuses on the first of the characteristics of the innovation influencing diffusion of innovations, according to Rogers (2003): that of relative advantage to the entrepreneur. Relative advantage is crucial to the understanding of the diffusion of innovations and of impacts upon poverty reduction.

For energy supply, location is related to the viability of the renewable energy supply option and to the alternatives available (is there a grid or access to fuels, or not?). The assets of the entrepreneur define a large part of the freedom to choose between alternatives. The financial and physical assets (defining, for instance, ownership of a workshop for the enterprise) and social assets (especially household and childcare tasks for women) of an entrepreneur have a high impact upon whether the location of an enterprise is close to markets or to energy access. If there is an electricity grid in the area, the demand, prices of alternatives and policies regarding connection define whether renewable energy technologies have a role to play. In the case of renewable energy being promoted as a means of rural electrification where there is no grid nearby, having electricity from renewable energy sources is often the only realistic supply option to meet small electricity demands locally. For larger demands, diesel generators form the default supply in un-electrified areas, and whether these can be substituted depends upon the renewable energy supply. This implies that renewable energy technologies provide additional options for entrepreneurs to access modern energy supplies.

Small-scale energy supply from hydro is very much bound to the location of the natural resource, especially in the case of direct mechanical energy, which is generated in direct proximity to the river. This inflexibility of location has a large impact upon the advantages of using this energy source for enterprises. In the Indian Himalayas, the location of the existing water mills in valleys and gorges along streams often does not coincide with the location of villages and roads which are located on hillsides and hilltops. Water mills are part of the tradition of this region; but only in those cases where the location of enterprises formed no barrier to customers were the businesses doing well. The upgrading of mechanical mills to the production of electricity for enterprise use is often promoted as a way of reviving traditional mills; but in the studied sites, such upgrading was found not to have any significant impact upon enterprise operation. Instead, a trend was perceived of new milling enterprises being established at locations close to customers, along roads and in newly electrified off-road villages, following access to diesel or grid electricity. In the case of direct use of hydro for enterprise activities, it appears therefore that the crucial context factor for benefits to small-scale enterprises is the location relative to customers.

In the Indian Himalayas, PV systems and decentralized hydro power stations were located in remote hill villages, typically several hours away by car and an additional several hours walk to rural towns. Less remote villages are, or are planned to be, electrified from the central grid. It is exactly the remoteness of the location that increases the relative advantage of using renewable energy sources, and at the same time impedes growth or the introduction of new products or services by an enterprise. Typical enterprise sectors even in the most remote villages are small grocery shops, tailors, millers, and carpenters, blacksmiths and masons. Energy services are lighting, mechanical energy and heating. The small volumes of energy supply available from PV cater only for the first of these services. Mechanical energy can be provided from hydro power, as discussed above. The impacts of having access to these energy services were found to be limited in terms of enterprise innovations, growth or incomes. This is related to the lack of demand from local customers in these remote locations in combination with the entrepreneurs' lacking financial, physical and social assets, while such assets are crucial to identify business ideas and implement them. The poverty impacts of the renewable energy services in remote locations are therefore limited to well-being rather than income improvement aspects of poverty reduction unless market links for enterprise products are established. This appears to be a phenomenon related to the remoteness of villages that are the most appropriate for energy supply through renewable energy, and therefore it is a factor that should be taken into consideration.

Nevertheless, the alternative provided from conventional energy supply through the grid or fuel distribution systems often also brings problems in terms of accessibility and reliability that may outweigh those that accompany renewable energy supply. For grid electricity, power shedding and scheduled and unscheduled down times can typically be experienced on a weekly, if not daily, basis. Such problems may contribute to a perceived advantage of energy from renewable sources even when electricity from the grid is available.

Renewable energy supply as an innovation: Compatibility, complexity, 'trialability' and 'observability'

Next to the relative advantage of the innovation, the other characteristics mentioned by Rogers (2003) related to the perception of the innovation are also relevant for the diffusion of renewable energy technologies: compatibility, complexity, 'trialability' and 'observability'. The compatibility of the design of most decentralized energy-supply systems is an issue, as enterprise energy demands other than lighting are low. Many examples of subsidized programmes for standardized PV systems have been shown to effectively rule out the diversification of available technical designs according to customer demands. The complexity of renewable energy supply is not an issue for end-users, but rather for installation and operation and maintenance. The complexity for the entrepreneur is not so much related to the functioning of the technology as how to optimize the impacts of changes in the enterprise. The relevance of the visibility of results from early adopters is already widely recognized in project development as a crucial factor in influencing potential customers' attitudes – for example, through providing initial systems at highly visible locations to enter new markets.

Energy efficiency as an innovation: Relative advantage, compatibility, complexity, trialability and observability

For energy efficiency, the characteristics of an innovation that influences adoption decisions are different than those of energy supply. First, the relative advantage of investing in energy efficiency is not obvious. Energy-efficient technologies and innovations may require significant investments in skills, time or money; but the benefits are difficult to see as they can only be measured as reduced consumption of energy not necessarily in absolute terms, but in reference to a scenario without these innovations. In the case of fluctuating manufacturing volumes, energy savings can hardly be perceived if a baseline has not been established. The recognition of the role of perception shows that one should be cautious about the opinions or stated requirements of entrepreneurs when assessing the potential of energy efficiency and cleaner production, as risk avoidance may lead to negative perception. What is measured, then, has a closer link to the levels of knowledge, awareness and expectations of entrepreneurs regarding market and policy developments than with actual potential benefits. Brown (2001), Jaffe and Stavins (1994) and Kounetas and Tsekouras (2008) refer to the 'energy efficiency gap' or 'paradox' as the situation in which cost-effective technologies exist but remain unadopted by many firms. Brown (2001) explains that obstacles to clean energy technologies include a low priority of energy issues among consumers, capital market imperfections, and incomplete markets for energy-efficient features and products, meaning that the energy efficiency aspects cannot be considered separate from the appliance. The benefits do exist, but they are not perceived strongly enough for the diffusion of innovations in this field to take place.

Second, energy efficiency is often not compatible with the existing values, experiences and needs of potential adopters, where (especially for small-scale enterprises) priority is given to survival in the short term and long-term strategies often do not exist.

Rogers's (2003) third characteristic is the complexity of the innovation. There are many different types of innovations in energy efficiency. Innovations in the energy efficiency of production processes include not only complex adaptations for which technological capabilities are obviously of great significance, but also innovations such as in lighting appliances that can be interpreted as changes in consumer goods, and innovations that can be interpreted as operational management (Kooijman-van Dijk, 2011). Trialability depends upon the type of energy efficiency innovation. Whereas operations management improvements have a low barrier, improvements requiring investments typically affect the whole production process, which cannot be tried without risk of influencing production and products.

Finally, observability, or visibility of energy efficiency impacts, is low. The experience in the case of energy efficiency improvements in small-scale enterprises in the foundry sector described by Pal et al., (2008) was that unit owners with successful innovations do not share their positive experiences with other small enterprises in the same sector in the cluster; in fact, they were found to even present a negative impression of the technology. Research by Soni (2007) also encountered this phenomenon of negative promotion of energy efficiency due to enterprises within the same cluster viewing one another as competitors. Whether this lack of willingness to spread positive experience also holds between enterprises of different scales is not known. Pal et al., (2008) do support the idea that large-scale enterprises will be willing to function as showcases for small-scale enterprises.

A way of overcoming barriers to spreading positive experiences with energy efficiency would be cluster-wide cooperation through associations to strengthen the market position of the cluster as a whole, focusing on the co-benefits of energy efficiency innovations in reaching new markets, rather than increasing competition between cluster members by focusing only on the benefits of energy efficiency in cost reduction. Increasing energy efficiency and modernizing production with regard to labour circumstances can be an effective instrument in improving such a market position, especially for the global market.

Conclusions and discussion

Although renewable energy and energy efficiency technologies can potentially contribute to socio-economic development through small-scale enterprises, this potential is not fully realized. Assumptions that potential benefits follow naturally once the technologies are in place are detrimental to achieving actual impacts upon climate change and poverty reduction, as the problems for entrepreneurs are not identified. Looking at the promotion of climate change mitigation technologies from the perspective of diffusion of innovations stresses the need to focus on the potential users, not only for the diffusion of the technology which is relevant from the technology-supply perspective in the long term, but also to increase the impacts upon climate change mitigation and poverty reduction.

This chapter shows that the perception of the innovation of potential end-users is indeed relevant to the diffusion of renewable energy and energy efficiency technologies, and that knowledge of the characteristics of the innovation is beneficial for understanding barriers to the diffusion of these innovations. It shows that not only are the 'objective' financial benefits relevant, but that for small-scale industries in developing countries in which entrepreneurs are typically lacking access to knowledge and awareness, and views on future market and policy development are frequently based on experiences in the past, rather than national or global trends, subjective perceptions play a strong role in influencing acceptance of innovations (Kooijman-van Dijk, 2008).

Concluding from the above, the empirical evidence on adopting climate change mitigation technologies in small-scale enterprise shows that the often cited risk adversity of entrepreneurs appears to be not so much a character trait, but at least partially related to the financial assets and social networks of an entrepreneur. Because taking risks may endanger the sustainability of a poor entrepreneur's livelihood, their livelihood assets are highly relevant for the diffusion of innovations even if there is a good chance of making profits.

The above leads to a discussion of implications for policies for climate change mitigation. First, the need to include the demands of small-scale enterprises is not met by current technology transfer agreements, which focus on high-tech and institution building for technological innovations. Second, small-scale enterprises are rarely targeted, and impacts upon income generation are limited if there is no special targeting of small-scale enterprises. Such special targeting should include a focus on the entrepreneur's perceptions, including opportunities for learning by doing and creating access to tacit knowledge, as stressed by Romijn (1996). For example, on-the-job training in enterprises that renewable energy or energy efficiency technologies would be a way of contributing to realistic expectations on potential advantages and to skills to operate or work with these technologies.

However, it is clear from the above analysis that support in linking climate change mitigation technologies to poverty reduction should move beyond the scope of the technologies. For energy supply, business support may lead to increasing the energy demand for new products and services; therefore, basing demand on current demand patterns could be short sighted and limit the development of the small-scale enterprises. On the other hand, small-scale enterprises may move from remote to more centralized locations once financial assets allow for this, so that energy planning for the inclusion of small-scale enterprise demand in rural areas may not require uniform design of supply.

Popular policy instruments in the field of renewable energy and energy efficiency are co-financing of hardware and awareness-raising of environmental and cost benefits. Such policy instruments may not be the most appropriate to reach small-scale enterprises. Awareness-raising activities, rather than focusing on changing the attitude of the entrepreneur, would be productive in altering social assets if improvements of links to the value chain are sought. For both renewable energy and energy efficiency, this means links to downstream markets for enterprise products; and for energy-efficient production technologies, links to upstream technology suppliers are required. Initiatives for steering the CO_2 emissions of production through the demands on enterprise products have a potential to function mainly in the most extreme polluting enterprise sectors, where customers will find it worthwhile to monitor these issues. Small-scale enterprises can be influenced in their energy choices by markets with a high demand for quality products, while only very few access niche markets with a willingness to pay for sustainable energy in the production process.

Schemes specifically supporting access to markets related to the impacts of climate change mitigation technologies are scarce. An exception is the recently established Indian Technology and Quality Upgradation Support scheme (DCSME, 2010) in which the strategy to improve the competitiveness of small-scale industry includes, first, sensitizing small-scale manufacturing industry to save costs by improving energy efficiency, and, second, quality improvement of products, for which meeting international standards and improving energy efficiency are considered positive with regard to consumer preference. Hopefully this is a sign of recognizing the need to pay specific attention to small-scale enterprises in climate change mitigation policy, especially by emphasizing and supporting the benefits for entrepreneurs. Taking the perspective of entrepreneurs, finding and promoting additional advantages to technical innovations apart from the environmental sustainability aspect can be a highly beneficial strategy, not only for the entrepreneurs, but for climate change impacts.

References

Barrett, C. B., Reardon, T. and Webb, P. (2001) 'Nonfarm income diversification and household livelihood strategies in rural Africa: Concepts, dynamics, and policy implications', *Food Policy*, vol 26, no 4, pp315–331

Bebbington, A. (1999) 'Capitals and capabilities: A framework for analysing peasant viability, rural livelihoods and poverty', *World Development*, vol 27, no 12, pp2021–2044

Brown, M. A. (2001) 'Market failures and barriers as a basis for clean energy policies', *Energy Policy*, vol 29, pp1197–1207

Davis, B., Winters, P., Carletto, G., Covarrubias, K., Quinones, E., Zezza, A., Stamoulis, K., Azzarri, C. and Di Giuseppe, S. (2010) 'A cross-country comparison of rural income generating activities', *World Development*, vol 38, no 1, pp48–63

DCSME (2010) *Scheme Technology and Quality Upgradation Support*, http://www.dcmsme. gov.in/schemes/technology&quality10.pdf, accessed May 2010

DFID (UK Department for International Development) (2001) *Sustainable Livelihoods Guidance Sheets*, http://www.eldis.org/go/livelihoods/

Ellis, F. (2000) *Rural Livelihoods and Diversity in Developing Countries*, Oxford University Press, Oxford

Grosh, B. and Somolekae, G. (1996) 'Mighty oaks from little acorns: Can micro-enterprise serve as a seedbed of industrialisation?', *World Development*, vol 24, no 12, pp1879–1890

IEA (International Energy Agency) (2010) *World Energy Outlook 2010*, http://www. worldenergyoutlook.org/database_electricity10/electricity_database_web_2010.htm

ILO (International Labour Organization) (2005) *World Employment Report 2004–05: Employment, Productivity and Poverty Reduction*, ILO, Geneva

Jaffe, A. B. and Stavins, R. N. (1994) 'The energy efficiency gap: What does it mean?', *Energy Policy*, vol 22, pp804–810

Kamfor Ltd (2002) *Study On Kenya's Energy Demand, Supply and Policy Strategy for Households, Small Scale Industries and Service Establishments*, Final Report for the Ministry of Energy, Nairobi, Kenya, September 2002

Kooijman-van Dijk, A. L. (2008) *The Power to Produce: The Role of Energy in Poverty Reduction through Small-Scale Enterprises in the Indian Himalayas*, Ph.D. thesis, University of Twente, Enschede, The Netherlands

Kooijman-van Dijk, A. L. (2011) 'Climate change mitigation: Beware of negative social impacts on small-scale enterprise', Paper presented at the International Sustainable Development Research Conference, 8–10 May, New York, NY

Kounetas, K. and Tsekouras, K. (2008) 'The energy efficiency paradox revisited through a partial observability approach', *Energy Economics*, vol 30, pp2517–2536

Lall, S. (1992) 'Technological capabilities and industrialization', *World Development*, vol 20, pp165–186

Liedholm, C. and Mead, D. C. (1999) *Small Enterprises and Economic Development*, Routledge, Oxon, UK

Mead, D. C. and Liedholm, C. (1998) 'The dynamics of micro and small enterprises in developing countries', *World Development*, vol 26, no 1, pp61–74

MSME (Ministry of Micro, Small and Medium Enterprises), Government of India (2010) *NMCP Results Framework Document for R.F.D. Ministry of Micro, Small and Medium Enterprises (2010–2011)*, MSME, New Delhi

Nooteboom, B. (1994) 'Innovation and diffusion in small firms: Theory and evidence', *Small Business Economics*, vol 6, pp327–347

OECD/IEA (2010) *Energy Poverty: How to Make Modern Energy Access Universal?*, IEA, Paris

Pal, P., Sethi, G., Nath, A. and Swami, S. (2008) 'Towards cleaner technologies in small and micro enterprises: A process-based case study of foundry industry in India', *Journal of Cleaner Production*, vol 16, pp1264–1274

Rogers, E. M. (2003) *The Diffusion of Innovations*, Free Press, New York, NY

Romijn, H. (1996) *Acquisition of Technological Capabilities in Small Firms*, Ph.D. thesis, KUB, Tilburg

Soni, P. (2007) *Global Solutions Meeting Local Needs: Climate Change Policy Instruments for Diffusion of Cleaner Technologies in the Small-Scale Industries of India*, Ph.D. thesis, Vrije Universiteit, Amsterdam

World Bank (2010) *India: Financing Energy Efficiency at Micro, Small and Medium Enterprise (MSMEs)*, Project document, World Bank, Washington, DC

PART VI

Low-Carbon Technology Transfer in the Context of Other Global Concerns

14

THE ROLE OF TRADE AND INVESTMENT IN ACCELERATING CLEAN ENERGY DIFFUSION

Private-sector Views from South Asia

Mahesh Sugathan and Muthukumara S. Mani

Technology issues have always been at the forefront of the global climate change debate. We need nothing short of a technology revolution to deal with climate change. However, if we look over the horizon, there does not appear to be enough of a sense of urgency being exhibited when it comes to technology, either internationally or at the country level. Trade and investment frameworks will play a major role in enabling a critical framework for private-sector investment. As this chapter will highlight, private-sector views emerging from South Asia in the course of World Bank consultations make clear that governments need to respond not only in terms of attractive and conducive policies, but also effective and meaningful implementation on the ground.

Technology transfer in the climate change context

Reducing greenhouse gas (GHG) emissions, while accommodating both economic growth and population growth, entails focusing on decarbonizing energy supply as well as increasing energy efficiency.

Both strategies require developing new and improved technologies, and enabling their greater diffusion through various channels, particularly to developing countries where GHG emissions can be reduced at a much lower cost. Technology issues have always been at the forefront of the global climate change debate. It is even suggested that just as the food crises of 1960 triggered the pursuit of a 'green' revolution, we need nothing short of a technology revolution to deal with climate change, especially if one were to get rapidly growing developing countries on a low-carbon path.

However, if we look over the horizon, there is not enough of a sense of urgency being exhibited when it comes to technology, either internationally or at the country level. Although an Expert Group on Technology Transfer was created by the United Nations Framework Convention on Climate Change (UNFCCC) to identify ways to facilitate and advance technology transfer activities, Cancun also resulted in an agreement to establish a new Technology Mechanism. This would comprise a Technology Executive Committee and Climate Technology Network and Centre. The main goal of the mechanism is to accelerate the development and transfer of climate-friendly technologies, particularly to developing countries, to support action on climate mitigation and adaptation. The precise roles and relationships between these entities still need to be clarified as these will determine their effectiveness in facilitating technology diffusion. However, the decisions to create the fund as well as the Technology Mechanism represent significant milestones in multilateral efforts to facilitate technology diffusion under the UNFCCC, and will create opportunities for reducing the cost burden for these technologies, particularly for developing countries.

Energy technologies, both currently in use and under development, have the potential to reduce carbon emissions substantially. Such options include renewable energies, carbon capture and storage, more efficient power generation from fossil fuels, nuclear power, and improved efficiency of end-use technologies, industry and transport. While clean energy-generating technologies and energy efficiency technologies have equal abatement potential, it is often difficult to say where the biggest bang for the buck is regarding taxation/regulation/investment. It is often cited that energy efficiency measures are somewhat 'low-hanging fruits' as far as costs of mitigation are concerned (as popularized in McKinsey marginal abatement cost curves), yet measures to create incentives and to scale up energy efficiency measures are still demonstrably lacking, and other barriers to diffusion of energy-efficient technologies continue to persist (Brown, 2001). On the clean energy side, some of the key barriers, including market failure, both in the innovation as well as diffusion sides of development, deployment and diffusion have been well documented (see, for instance, Kofoed-Wiuff et al., 2006; Barton, 2007; and Mallett et al., 2010). A recent World Bank study (Avato and Coony, 2009) suggests that the research, development and deployment (RD&D) activities needed to commercialize clean energy and energy-efficient technologies have finally – after a period of significantly reduced activity – increased over the last two to three years. From the mid-1980s to the early 2000s, energy research and development (R&D) spending was well below historic highs. By 2003, public energy R&D spending in the Organisation for Economic Co-operation and Development (OECD) had fallen to 60 per cent from its peak in 1980, and private-sector spending had fallen from US$8.5 billion in the late 1980s to US$4.5 billion in 2003. While absolute investments in energy innovation continue to lag behind historical levels, the trend appears to be reversing as concerns about climate change, energy security and high oil prices are prompting intensified private and public R&D activities.

The study adds, however, that these renewed efforts will face significant barriers that affect the ability to develop and deploy promising clean energy options, such as uncertain future value of CO_2 emissions abatement; provision of a public good being hampered by free-riding across space – countries that free-ride on the mitigation efforts of others; the 'Valley of Death' phenomenon, which occurs when promising technologies languish between public- and private-sector RD&D efforts in innovation; intellectual property rights (IPRs) issues, where the large RD&D investments needed for technical advances in certain clean energy technologies will be undermined by uncertain global IPR protection; challenges including developing and transferring technology to developing countries (a substantial source of incremental GHG emissions growth, requiring foreign and domestic resources and expertise, such as from OECD countries to deploy the needed clean energy technologies); subsidies for conventional energy products at both the retail and production levels reduce to below-cost price with which new energy technologies must compete; and deployment of clean energy and energy-efficient technologies that is often hampered by trade barriers.

All this suggests that an 'activist' public policy should support development, deployment and diffusion of clean technologies. Technology transfer occurs through a number of channels such as trade, investment, international joint ventures, licences and international development assistance. Successful absorption of technology, however, depends upon the technological absorptive capacity of the economy, as well as firms and organizations. Cohen and Levinthal (1990) coined the term absorptive capacity and defined it as the ability of innovating firms to assimilate and replicate new knowledge gained from external sources. The ability to recognize, value and exploit external sources of knowledge, in general, and technology, in particular, is crucial in explaining organizations' innovative capabilities. Torodova and Durisin (2007) have refined the concept of absorptive capacity and conclude that acquisition, assimilation, transformation and exploitation of external knowledge are four distinct dimensions of absorptive capacity.

Absorptive capacity is often governed by the overall macroeconomic and governance environment, which influences the willingness of entrepreneurs to take risks on new and new-to-the-market technologies; and the level of basic technological literacy and advanced skills in the population, which determines a country's capacity to undertake the research necessary to understand, implement and adapt them. In addition, because firms are the key mechanism by which technology spreads within an economy's private sector, the extent to which financing for innovative firms is available – through the banking system, remittances or government-support schemes – also influence the extent to (and speed with) which technologies are absorbed (Hoekman et al., 2005; World Bank, 2008a).

Another World Bank study (2008a) suggests that liberalizing trade can significantly increase the diffusion of clean technologies in developing countries. While liberalization policies may help in gaining access to international technology, the success of technology diffusion, in general, depends upon a range of other enabling factors, particularly the capacity to absorb and improve technologies in the host countries. Beyond trade, foreign direct investment (FDI) can be an important means of transferring technology given the fulfilment of certain conditions such as the presence of absorptive capacity, as mentioned above. Various barriers are often cited as reasons that inhibit diffusion of specific technologies. These range from weak environmental regulations, fiscal feasibility, financial and credit policies, economic and regulatory reforms, and the viability of technology to local conditions (including availability of local skills and know-how); in the case of some cutting-edge technologies, they could include access to IPRs (see Watson et al., 2010). Therefore, governments need to put in place appropriate policy infrastructure, governance and competition systems, as well as a conducive regulatory framework on IPRs in order to be effective conduits for technology transfer and diffusion.

Policies to ease barriers to diffusion of clean technologies in developing countries

Technology transfer and diffusion is essentially motivated by economic incentives and largely takes place in the private sector. Thus, policies that influence market conditions (such as current and potential market demand, market access and expected price) in the host country play a critical role in technology transfer and diffusion.

While the private sector will be key for driving diffusion and deployment of clean and climate-friendly technologies, investors, both foreign and domestic, will consider a number of factors when making decisions on low-carbon energy investment, as well as investments in energy-efficient products and services, a large number of which can be rolled together under the broad heading of investment climate. Investors look for such things as political and macroeconomic stability, an educated workforce, adequate infrastructure (transportation, communications, energy), a functioning bureaucracy, the rule of law and a strong finance sector, as well as ready markets for their products and services. In doing so, they assess how risky or difficult it will be to make an investment in a given country using a given technology, and add this to the expected costs.

On the other hand, overarching challenges in this area from a climate change perspective include, first, government commitment to changing the trajectory of carbon emissions; and, second, government incentives for low-carbon investments. The types of policy barriers will differ fundamentally from country to country, and diagnostic studies could help to identify the full range of potential actions that are needed to help make clean energy investment more attractive to both domestic and foreign investors.

There are three broad ways in which a government can encourage the growth of a market for clean energy:

1. Implement policies that *indirectly encourage* the purchase of clean energy (examples: pollution reduction targets; carbon cap-and-trade programme). These are policies which broadly encourage an overall outcome (e.g. reduced

carbon emissions) for which increasing the purchase of renewable energy is one of a number of potential means to the end (other means include reducing use, increasing efficiency, etc.).

2. Implement policies that *directly encourage* the purchase of clean energy (examples: Renewable Portfolio Standards and Renewable Electricity Standards). These are policies which directly require an increase in the purchase of renewable energy (e.g. by utilities).

3. Purchase renewable energy (examples: buying and installing solar panels on rooftops; installing wind turbines on military bases) and energy-efficient appliances (*compact fluorescent lamp* (CFL) bulbs and other equipment) directly. In this approach, government entities act as consumers, buying directly from renewable energy companies.

Any government policy which ensures or encourages (or discourages) growth in market demand for clean energy and energy efficiency will also encourage (or discourage) capital investment in clean energy generation and consumption. In addition to these policy interventions, governments can directly intervene in the markets, providing guarantees to encourage private credit and financing of renewable and energy-efficient technology. The year 2010 was, for instance, marked by significant milestones in the renewable energy landscape in India, such as the government's new Solar Mission to develop 1GW of grid-connected capacity by 2013; and the launch of Renewable Energy Certificates and Renewable Purchase Obligation schemes. This was also a year that saw a strong 25 per cent growth in renewables investment in India over 2009 to US$3.8 billion, ranking eighth in the world. Wind projects were the biggest single item, at US$2.3 billion, followed by US$400 million each for solar, and biomass and waste-to-energy. Similarly, in Pakistan, investment in renewables tripled to US$1.5 billion as the country financed 850MW of new wind capacity across 16 projects. The Pakistan government has announced upfront preferential tariffs for wind power and has other generous incentives, such as exemptions from customs duty or sales tax for machinery imports, exemption from income taxes and withholding tax on imports, a guaranteed rate of return of 17 per cent in US dollar terms on equity (rather than the 15 per cent allowed for thermal power projects), repatriation of equity along with dividends without incurring any additional penalties, and flexibility in the mode and currency of financing (ADB, 2010). There are also other players involved in low-carbon investments which may be as important as the private sector, such as multilateral development banks and other agencies. The Asian Development Bank (ADB), for instance, has announced support for the first private-sector-led wind power project in Pakistan and has also approved a US$21.6 million grant for the development of a rural renewable energy project in Bhutan. Several hydro projects in Nepal similarly have been supported by assistance from development agencies.

The South Asian experience

We outline below some key ingredients necessary for promoting clean energy technologies based on extensive consultations with private-sector firms in India, Pakistan, Bangladesh and Sri Lanka. The findings are based on a pilot project recently launched by the World Bank in South Asia to assess the Investment Climate for Doing Climate Business. The project will eventually aim at developing a tool – the Climate Investment Readiness Index (CIRI) – that will evaluate and compare enabling environments in countries for supporting private-sector investment in climate mitigation technologies. Once developed, CIRI will assess progress made by countries in moving towards a low-carbon growth path and will help inform private-sector investment in 'climate mitigation' technologies.

While the project has initially focused on policies to promote renewable energy, as well as policies required to expand sales of energy-efficient appliances in the residential and commercial sectors, such as lighting, heating and cooling technologies, in the future it may also expand its scope to include other low-carbon technologies, such as carbon capture and storage (CCS) and super-critical technologies, etc.

Policy environments: The distinction between paper and practice

The project initially undertook a comparative assessment of the environment for policy, regulation and incentives (henceforth PRIs, or PRI frameworks) for grid-connected renewable energy existing in a select group of high greenhouse gas-emitting countries in South Asia. Also included were a select group of high GHG-emitting developing countries in Asia, Africa and Latin America. The objective was to find out whether these countries had the key PRIs that are normally required to attract investments in renewable energy generation, at least on paper. There was no distinction made between domestic and foreign investments, and both types of incentives were covered. The indicators selected were based on an extensive survey of literature on some of the main PRIs introduced by countries to attract private-sector investments in renewable energy. Scoring for renewable energy PRIs was conducted on a scale of 10. For the presence of a PRI indicator, the country was assigned a score of 1 for that indicator, whereas the absence of an indicator meant the assigned score was 0. Certain PRIs for renewable energy (RE), such as existence of law and policy on RE and trade-related incentives were split into two sub-components, each carrying a score of 0.5. The presence of both sub-components meant a country would get a score of 1 for the category as whole, whereas the presence of only one sub-component meant a score of 0.5. Currently, each PRI indicator is assigned a weight of 1. The tool is designed in a flexible manner so that weights can be changed if the importance of a particular PRI indicator changes in the future and a new score is arrived at.

The selected indicators are illustrated in Table 14.1. The list is by no means exhaustive or detailed; but it does provide a fair indication of some of the key elements that a country needs to have, at least on paper, in order to provide a clear signal for private investors. Grid-connected electric power generation was focused

TABLE 14.1 Renewable energy generation policy, regulation and incentives

Cross-cutting policy, regulation and incentive (PRI) indicators
1   Existence of law and policy on renewable energy (RE)
a Existence of RE policy
b Existence of RE law
2   Existence of RE target
3   Obligation for designated entities to purchase/off-take RE
4   Availability of tradable instruments for RE generation
Grid-connected indicators (solar PV, onshore wind, small hydro and biomass)
5   Availability of designated preferential tariffs
6   Grants, subsidies and incentives related to capital/investment tax credits
7   Incentives linked to generation/production tax credits
8   Income tax holidays/exemptions
9   Trade-related incentives
a Customs duty exemptions
b Absence of 'local content' requirements for power producers
10   Other tax exemptions or concessions (sales, VAT, energy tax)

on as this was a key activity in both developed as well as developing countries and would also ensure an easier degree of comparability in later stages of the project. Certain indicators, such as renewable energy targets and tax incentives on equipment sales, are relevant for the expansion of both grid-connected and off-grid renewable energy; but otherwise the two are quite different in terms of indicators that matter.

An examination of the presence or absence of these PRIs across countries is quite revealing. In sectors such as onshore wind, countries with a very good record in installed wind power capacity as of 2010, such as China (ranking first in total installed capacity with 44.7GW) and India (ranking with 13.2GW), do come out with high scores. However, a number of European countries with a good installed wind power presence by 2010, such as Spain (20.7 GW), the UK (above 5.5GW) and The Netherlands (above 2GW), come out with much lower scores compared to certain countries such as the Philippines (33MW).[1]

This is also seen in other renewable energy sectors, such as solar PV. One obvious reason, of course, is that a number of countries are very well endowed with geographical advantages for certain renewable energy resources, such as solar and wind, as well as an attractive market in terms of population numbers, with a certain purchasing power capacity. Countries that may not have these advantages or are early entrants may need to offer more incentives to make it worthwhile for the private sector to establish a presence. As the market develops and costs of RE deployment come down, countries can then make do with fewer incentives.

Consultations with the private sector in South Asia, as well as existing literature on the subject, reveal that two elements critical for private-sector investment in renewable energy are the attractiveness of the power purchase tariff and its stability (i.e. the duration for which the preferential tariff will be made available). This enables the private-sector player to plan for stable returns and eventually break even. The level and duration of a power purchase tariff that will be attractive will, of course, vary from country to country depending upon the local cost conditions of investment. In addition to stability of the power purchase tariff, other critical ingredients include some assurance of purchasing the renewable energy that is produced, as well as policies that reduce upfront investment costs, such as accelerated depreciation and other investment-related incentives. Policy stability, in general, ensures predictability in terms of calculating returns on investment. An unpredictable policy environment (where PRIs are altered or dropped without warning) will not be attractive for private investors. Studies on the US Production Tax Credit (such as Barradale, 2010) indicate that the constant uncertainty surrounding whether or not it would be renewed was frustrating for investors.[2]

The PRI score tables do not attempt to measure the *degree of PRI stability*. In this regard, whether the private sector feels a certain PRI environment also ensures that the desired level of stability can be better gauged through private-sector interactions and detailed surveys as each country context may be different, and what may be attractive for some countries may not be to others. Country contexts, however, are important to keep in mind. For instance, rather than introducing uncertainty, the biennial review of renewable energy policies in Germany was viewed by many actors as an opportunity to debate the future of the renewable energy sector.[3]

As a general rule of thumb, however, longer timeframes for maintaining PRIs (for preferential tariffs, firms often cite at least 20 years) is a good indicator of policy stability. As many countries (particularly in the developing world) have only recently started introducing PRIs, it may take a while before the degree of stability can be properly ascertained. Table 14.2 indicates the existence or absence of grid-connected PRIs in three South Asian countries (India, Pakistan and Sri Lanka), as well as the presence or absence of four renewable power sectors (solar PV, onshore wind, small hydro and biomass). A detailed overview of the role of trade and investment linkages, as well as private-sector perceptions, follows.

Table 14.2 (a–e) reveals that South Asian countries, like many others beyond the region, have already put in place policies needed to attract clean energy technology investments over the medium term, including feed-in tariff regimes, mandatory renewable energy targets, and tax incentives. But more such policies may be necessary and existing policy distortions may need to be removed if one is to envision massive scale-up in investments.

Discussions with the private sector suggest that having well-developed regulatory frameworks and policies for clean energy on paper may not suffice to attract private-sector investments if, among other things, the private sector perceives that the implementation of these policies is weak and if a number of other supportive policies – such as transparency, access to electricity grids and information on key aspects (e.g. mapping of available sites for establishing renewable energy plants), as well as data relating to wind speeds and solar radiation – are missing. Digging deeper, we were actually able to categorize the private perception into ten Cs as prerequisites for providing a conducive environment for 'Doing Climate Business':

TABLE 14.2

(a) Cross-cutting grid-connected renewable energy (RE) policy, regulation and incentives in South Asia

Cross-cutting grid-connected PRI indicators	India	Pakistan	Sri Lanka
1 Existence of law and policy on renewable energy	X	X	X
a Existence of RE policy	X	X	X
b Existence of RE law	X	X	X
2 Existence of RE target	X	X	X
3 Obligation for designated entities to purchase/off-take RE	X	X	X
4 Availability of tradable instruments for RE generation	X		

(b) Solar PV grid-connected RE policy, regulation and incentives in South Asia

Solar PV grid-connected indicators	India	Pakistan	Sri Lanka
1 Availability of designated preferential tariffs for solar PV	X		
2 Grants, subsidies and incentives related to capital/investment tax credits	X	X	
3 Incentives linked to generation/production tax credits	X		
4 Income tax holidays/exemptions	X	X	X
5 Trade-related incentives			
Customs duty exemptions (zero duty on major components and equipment)	X	X	X
Absence of 'local content' requirements for solar PV power producers		X	X
6 Other tax exemptions (sales, VAT, energy tax, etc.)	X	X	

(c) Onshore wind grid-connected RE policy, regulation and incentives in South Asia

Onshore grid-connected indicators	India	Pakistan	Sri Lanka
1 Availability of designated preferential tariffs for onshore wind	X	X	X
2 Grants, subsidies and incentives related to capital/investment tax credits	X		
3 Incentives linked to generation/production tax credits	X		
4 Income tax holidays/exemptions		X	X
5 Trade-related incentives			
Customs duty exemptions (zero duty on major components and equipment)		X	X
Absence of 'local content' requirements for on-shore wind power producers	X	X	X
6 Other tax exemptions (sales, VAT, energy tax, etc.)	X	X	X

Continued

TABLE 14.2 (*Cont'd*)

(d) Small hydro grid–connected RE policy, regulation and incentives in South Asia

	Onshore grid-connected indicators	*India*	*Pakistan*	*Sri Lanka*
1	Availability of designated preferential tariffs for small hydro	X	X	X
2	Grants, subsidies and incentives related to capital/investment tax credits	X		
3	Incentives linked to generation/production tax credits			
4	Income tax holidays/exemptions	X	X	X
5	Trade-related incentives			
	Customs duty exemptions (zero duty on major components and equipment)		X	
	Absence of 'local content' requirements for small-hydro power producers	X	X	X
6	Other tax exemptions (sales, VAT, energy tax, etc.)	X		

(e) Biomass grid-connected RE policy, regulation and incentives in South Asia

	Onshore grid-connected indicators	*India*	*Pakistan*	*Sri Lanka*
1	Availability of designated preferential tariffs for biomass	X		X
2	Grants, subsidies and incentives related to capital/investment tax credits	X		
3	Incentives linked to generation/production tax credits			
4	Income tax holidays/exemptions	X	X	X
5	Trade-related incentives			
	Customs duty exemptions (zero duty on major components and equipment)			
	Absence of 'local content' requirements for biomass power producers	X	X	X
6	Other tax exemptions (sales, VAT, energy tax, etc.)	X	X	

Clarity and coherence. Policies/laws on clean energy should be very clear and transparent, as well as coherent. They should send a strong signal on the intent of the
country to move towards cleaner/low-carbon energy options.

Consistency. Policies have to be consistently implemented across sectors and regions
within a country. For example, in a federal structure, one needs to ensure that
there is a national standard guaranteeing a minimum level of renewable energy
development, with states being allowed to be more aggressive in requiring
more renewable energy deployment if needed. This is very evident in India,
for instance, where the state-level incentive scheme for solar energy introduced
by the Gujarat government has attracted a great deal of interest, and the state
has already received proposals for setting up around 365MW of solar power.[4]
At the same time, there may be a need to avoid too many competing and/or
overlapping policies that could create confusion in the minds of investors.

Commitment/credibility. For the policies to be credible, governments should signal a long-term commitment to the RE sector, backed by a comprehensive and transparent regulatory and tariff structure.

Clearances. Investors are often bogged down by the number of clearances they have to get to set up, for example, a wind or solar farm. This can be considerably eased by setting up single-window clearance systems for specific sectors. In some South Asian countries, firms during the course of consultations have pointed out the slow-down in implementation time, particularly for small hydro projects, owing to the large number of agencies involved in clearances, as well as additional clearances required. While many of these clearances, such as environmental ones, are justifiable, the process could certainly be stream-lined, for instance, by having one nodal entity as a coordinating point for the private sector.

Capacity. As countries ambitiously expand their clean energy portfolio, the capacity of agencies should be considerably expanded to ensure that targets are met and policies adhered to and laws complied with.

Compliance. Investors often are concerned about the commitment of utilities to ensure compliance with Purchase Power Agreements (PPAs). It is therefore important to establish transparent cost-recovery rules and prudency tests for utility compliance with policies, and enforcement of contractual arrangements. It also necessary to ensure utilities' compliance with any obligations they may have to purchase renewable energy, or to ensure compliance through other means available, such as the purchase of Renewable Energy Certificates. In India, responding to private-sector concerns, the government has introduced a payment security mechanism that would enable the Ministry of Non-Renewable Energy (MNRE) to avail of budgetary support to possible payment-related risks by power producers to banks in case of defaults by distribution utilities for the bundled power.[5]

Coordination. Coordination across a multitude of agencies involved in the clean energy sector (regulatory agencies, implementing agencies, utilities, distributing companies, etc.) is critical to ensure that clean energy policies are implemented consistently and efficiently.

Collateral. Banks are often reluctant to finance clean energy projects because of concerns regarding the bankability of PPAs, which are often related to the compliance of utilities. Policies and institutional mechanisms to encourage clean energy financing through specialized vehicles or institutions (which can act as catalysts, enable effective risk mitigation and ensure adequate financing) is something that countries should consider until clean energy becomes as competitive as conventional energy.

Connectivity. Grid connectivity or access to grids provides an important aspect of the investment criteria of firms looking for investments especially in the renewable sector. Transparent rules, procedures and standards for grid connectivity are important aspects of attracting clean energy investments.

Cartography. Since the quality and availability of renewable energy (wind, solar, hydro, biomass) various across locations, it is very important to have these sources and their potential sites mapped accurately as it will have a bearing on returns to investment.

Consultations with the private sector in South Asia threw up a number of issues related further to the ten Cs. Some key ones are listed below, with the relevant countries where these issues cropped up in brackets:

- The need for *more attractive project financing terms and conditions* (whereby bank loans are disbursed on the merit of a particular project) and which are often the only option available for smaller firms, as opposed to balance-sheet financing that larger companies can also access (*India, Pakistan and Bangladesh*).
- *More attractive interest rates and longer loan tenures*, as well as *subsidies* that can help to reduce the high upfront costs associated with renewable power projects (*India, Pakistan, Bangladesh, Sri Lanka and Nepal*).
- A number of firms mentioned that the Clean Development Mechanism (CDM) funds were cumbersome to access and they wanted a more streamlined process (*India*).
- A federal system of government *creates opportunities as well as challenges.* While, on the one hand, the government creates regulatory 'space' and autonomy for sub-federal policies, on the other, it also throws up challenges, such as the variable levels of enforcement with respect to renewable purchase obligations (RPOs) and additional state-level clearances that are required (*India and Pakistan*).
- *Access to information on renewable energy-related data* (that helps in choosing the best site locations), as well as easy access to grid (to evacuate the power generated) are vital but are not always easily available (*India, Pakistan and Bangladesh*).
- *Transparency and good-governance are essential but are not always present.* This was particularly true for some project bidding and allotment processes, as well as for obtaining clearances in states where corruption was an issue. Streamlining the clearance processes would reduce the scope for corrupt practices (*India, Pakistan, Nepal and Bangladesh*).
- *Better access to equipment technology and training to set up manufacturing establishments* – in particular, drawing on greater intra-regional cooperation and collaboration. Greater awareness creation, particularly regarding equipment standards, is necessary (*Pakistan*).
- *Better availability and access to feed stock and regulation of feed-stock prices* (in the case of biomass) (*India and Pakistan*).

The relevance of trade

As countries pursue low-carbon development paths, trade and trade policy will play an important role in enabling countries' access to climate-friendly goods and

TABLE 14.3 Change in trade volumes in high GHG-emitting developing countries from liberalizing clean energy technologies

Technology option	Liberalization scenario 1 (%) Elimination tariff (only)	Liberalization scenario 1 (%) Elimination tariff and non-tariff barriers
Clean coal technology	3.6	4.6
Wind power generation	12.6	22.6
Solar power generation	6.4	13.5
Efficient lighting technology	15.4	63.6
All four technologies	7.2	13.5

Source: World Bank, 2008a, p53

services. Depending upon their manufacturing capacities and cost conditions, it will also enable them to create a competitive production base in these technologies. In the clean energy sector, few countries have the domestic capacity or know-how to produce all that they need. This is particularly true for developing countries, and although building such capacities could be their long-term goal, trade liberalization can provide rapid access to key technologies. Trade liberalization, whether 'locked in' through the World Trade Organization (WTO), regional or bilateral agreements, or undertaken autonomously, can lower clean energy equipment costs for consumers (industries or households) by enabling them to purchase these at world market prices. Within the context of the current global trade regime, a World Bank study has found that a removal of tariffs and non-tariff barriers (based on *ad valorem* equivalents of selected measures, such as quotas and technical regulations) for four basic clean energy technologies (wind, solar, clean coal and efficient lighting) in 18 of the high GHG-emitting developing countries will result in trade gains of up to 13 per cent (World Bank, 2008a). This is illustrated in Table 14.3 above.

Trade in clean energy products and equipment, as well as clean energy services, are in turn closely driven by domestic regulatory policies and incentives for clean energy. Feed-in tariffs in Spain, for instance, were a big driver for exports of solar PV panels from China and India to Spain (Jha, 2009). Even today, countries such as China and India export wind turbines and solar panels to other countries, notably in the European Union (EU) and US; but as domestic PRIs in emerging economies make them attractive destinations for RE power producers, much of their manufacturing activity may be directed internally towards domestic RE power expansion.

Again, there are differences between sectors. Transportation costs play a major role in the case of wind (given bulky parts and components).Hence, equipment production is likely to be based in countries where sufficient market incentives for wind power generation are created, unlike in the case of solar, where modules and cells are much easier to trade. This trend may be reinforced by the lower labour component required in wind energy (and, consequently, less important role of labour costs).

TABLE 14.4 Average employment generation over life-time of facility

Energy source	Manufacturing, construction and installation	Operations and maintenance/ fuel processing	Total	Project average
Solar photovoltaic	0.16–0.84	0.07–0.57	0.23–1.42	0.87
Biomass	0.01–0.03	0.16–0.21	0.19–0.22	0.21
Wind power	0.03–0.14	0.05–0.13	0.1–0.26	0.17
Coal-fired	0.03	0.08	0.11	0.11
Natural gas fired	0.01	0.1	0.11	0.11

Note: The data are average annual employment numbers distributed over the lifespan of an installation and takes into consideration the very different factor capacities of renewable (low factor capacities) and fossil fuel (high factor capacities) energy sources.
Source: Wei et al., (2009)

Cross-border investment is predicted to increasingly displace trade as a driver of global integration in the wind sector, although trade will continue to play a role in enabling firms to optimize their supply chains (Kirkegaard et al., 2009). On the other hand, solar power equipment has been more trade intensive and the shift of light manufacturing jobs to Asia looks set to continue. It is very likely, however, that services associated with solar installations would be sourced close to centres of demand (Kirkegaard et al., 2010).

Comparative advantage, as well as trade and investment policies will again determine how much and what part of the RE supply chain is sourced domestically, and how much and what part is imported and from which countries:

The dilemma of balancing objectives. Towering clean energy production costs, accessing efficient and reliable technologies and creating a domestic manufacturing base for clean energy equipment.

Countries, both developing and developed, increasingly aim at multiple objectives while promoting clean energy. Clean energy is seen as a promising sector not only in terms of its prospects for enabling a switch to lower carbon growth trajectories and improving energy security, but also as an engine to generate domestic manufacturing jobs, as well as encouraging innovation and technology deployment. One study for instance projects a greater average employment per unit energy produced over plant lifetime in the case of renewable energy plants compared to fossil-fuel plants as illustrated in Table 14.4. Policy-makers as well as financial institutions want to ensure that equipment deployed for clean energy generation perform well and are of sound quality. These objectives may often be at odds with each other unless a country is blessed with both specific manufacturing cost advantages and technological know-how. However, even these may not extend to all clean energy technologies. In trying to meet and balance these objectives, policy-makers usually rely on domestic policy instruments, some of which are trade related or can have a trade impact. Trade policies include tariffs and local content measures, as well as policies determining market access for services and regional trade agreements.

Domestic renewable energy-related policies that may have an impact upon trade include subsidies, as well as technical requirements and performance standards for renewable energy equipment. With regard to direct trade policy instruments, the following section focuses mainly on tariffs and local content measures in the South Asian context. It examines two key domestic policies designed to promote the uptake and quality of renewable energy – namely, subsidies as well as technical standards on equipment and their trade and investment impacts.

Four other policies that are of relevance to trade and investment in RE but beyond the immediate scope of this chapter deal with trade in services, intellectual property rights, government procurement and regional trade agreements. Procurement also has a link with local content measures and will be touched upon while discussing this topic.

Trade policy considerations for RE investors: Tariffs and local content requirements

Tariffs

Most renewable energy sources (with the exception of biomass) benefit from very little or no operating costs as their inputs come free. Most RE costs arise due to upfront investment expenses of equipment and capital costs. Hence, governments usually put in place schemes and measures to lower these upfront capital costs, such as grants, subsidies, investment tax credits, sales tax and VAT exemptions on equipment used in renewable energy production. Import duty concessions and exemptions are a trade-related policy instrument that can be used to lower equipment costs.

Governments may also wish to create incentives for fledgling domestic manufacturers of this equipment or induce foreign investors to set up manufacturing facilities within the country as opposed to serving the market through exports. In such cases, they may apply some level of tariffs on final products while extending reductions or duty concessions to intermediate inputs and raw materials. If these tariffs lead to domestic equipment manufacturing being costlier than imports, costs for renewable energy producers (who use this equipment) may rise and, in turn (depending upon the degree of impact), lead to higher prices for the renewable energy generated. Policy-makers in their pursuit to create a domestic manufacturing base often have to tread a fine line between ensuring that renewable energy producers are not penalized and domestic equipment producers are protected from import competition. If tariff protection is extended for too long and fails to promote competitive domestic manufacturing, it could adversely affect renewable energy deployment in the country. Hence, if tariff protection is resorted to, it is preferable that it is time limited and has sunset clauses built in while providing sufficient opportunity for domestic manufacturers to become competitive.

Tariff architecture – the way in which tariffs for final products and intermediate goods are designed – also has an impact upon the extent to which a country pursuing renewable energy development can benefit from global supply chains. Rapid deployment of renewable energy may require reduction or elimination of tariffs on final equipment. The impact of tariff protection will vary from country to country. If no domestic firms engaged in manufacturing exist, higher tariffs on intermediate inputs or raw materials will not have any impact upon local production. For countries that desire to create a manufacturing capacity on final equipment, it may not make sense to have high tariffs on the inputs that go into making them, while having zero tariffs on the final equipment. This would lead to an *inverted duty structure* and penalize domestic manufacturers of final equipment. Countries that seek to create domestic manufacturing capacities for final equipment may often maintain some degree of tariff protection on final equipment while lowering or eliminating duties on inputs. Other countries may choose to specialize in the manufacture and exports of parts and components that are usually less technology intensive compared to final equipment.

Bound tariffs represent the ceiling levels up to which a country is permitted to raise its tariffs under WTO rules. If tariffs are unbound, then a country can, in principle, raise its actual tariffs to any level. Whatever the level at which countries 'bind' their tariffs, binding increases predictability for their trading partners as they know the maximum possible tariffs that they can face at any given time. Once 'bound', ceiling tariff levels cannot be changed except in exceptional circumstances and under procedures laid down under the WTO. More commonly, exporters face tariffs that are actually 'applied' at the border and which may be significantly lower than 'bound' ceiling rates. The greater the gap between bound and applied tariff levels, the greater is said to be the 'water' in tariffs. Even if the applied tariffs are very low in a particular country, a big gap between bound and applied tariffs decreases 'predictability' for exporters to that country.

It is important to note that tariff reductions negotiated at the WTO and in regional trade agreements usually relate to bound tariffs. If bound tariff levels are very high, then depending upon the level of reduction members may negotiate, there may be very little or no impact upon the actual tariffs applied. A lower bound tariff, while not immediately increasing trade flows, will most certainly increase predictability against tariff increases beyond a certain level for a country's trading partners.

To illustrate prevailing gaps between bound and applied tariffs, Table 14.5 provides a snapshot of both 'bound' as well as the latest available data on 'applied' import tariffs for South Asian countries (except the Maldives and Bhutan) in select RE equipment and their components. The six-digit Harmonized System (HS) code for the product is indicated in brackets.

As is clear from Table 14.5 a–g, applied rates are low or even zero for solar PV cells and modules in South Asia. Sri Lanka and Bangladesh have unbound rates and Pakistan has a high bound rate of 50 per cent. Hence, the degree of tariff uncertainty for solar panel exporters to these three countries is greater despite low applied tariffs. Table 14.5 also reveals that India, Nepal and Bangladesh apply higher duties on silicon, the input material that currently accounts for more than

TABLE 14.5

(a) Bound and applied tariffs for solar PV panels and cells (HS 854140) in South Asia

Country	Bound ad valorem tariffs percentage	Applied ad valorem tariff percentage (year)
Bangladesh	Unbound	5% (2008)
India	0 (duty free)	0 (2008)
Nepal	0 (duty free)	0 (2010)
Pakistan	50%	5% (2010)
Sri Lanka	Unbound	2.5% (2009)

(b) Bound and applied tariffs for silicon used in solar panel manufacture (HS 280461) in South Asia

Country	Bound ad valorem tariffs percentage	Applied ad valorem tariff percentage (year)
Bangladesh	Unbound	12% (2008)
India	40%	5% (2008)
Nepal	20%	10% (2010)
Pakistan	30%	5% (2010)
Sri Lanka	Unbound	2.5% (2009)

(c) Bound and applied tariffs for wind-powered generating sets in South Asia

Country	Bound ad valorem tariffs percentage	Applied ad valorem tariff percentage (year)
Bangladesh	Unbound	3% (2008)
India	25%	7.5% (2008)
Nepal	30%	15% (2010)
Pakistan	55%	5% (2010)
Sri Lanka	Unbound	2.5% (2009)

(d) Bound and applied tariffs for hydraulic turbines and water wheels of a power less than 1000kW (1MW) (HS 841011) in South Asia

Country	Bound ad valorem tariffs percentage	Applied ad valorem tariff percentage (year)
Bangladesh	50%	3% (2008)
India	25%	7.5% (2008)
Nepal	20%	5% (2010)
Pakistan	50%	10% (2010)
Sri Lanka	Unbound	2.5% (2009)

(e) Bound and applied tariffs for hydraulic turbines and water wheels of a power greater than 1000kW (1MW) but less than 10,000kW (10MW) (HS 841012) in South Asia

Country	Bound ad valorem tariffs percentage	Applied ad valorem tariff percentage (year)
Bangladesh	50%	3% (2008)
India	25%	7.5% (2008)
Nepal	20%	5% (2010)
Pakistan	50%	10% (2010)
Sri Lanka	Unbound	2.5% (2009)

Continued

TABLE 14.5 (*Cont'd*)

(f) Bound and applied tariffs for hydraulic turbines and water wheels of a power greater than 10,000kW (10MW) (HS 841013) in South Asia

Country	Bound ad valorem tariffs percentage	Applied ad valorem tariff percentage (year)
Bangladesh	50%	3% (2008)
India	25%	7.5% (2008)
Nepal	20%	5% (2010)
Pakistan	50%	5% (2010)
Sri Lanka	Unbound	2.5% (2009)

(g) Parts of hydraulic turbines and water wheels (HS 841090) in South Asia

Country	Bound ad valorem tariffs percentage	Applied ad valorem tariff percentage (year)
Bangladesh	50%	3% (2008)
India	25%	7.5% (2008)
Nepal	20%	7.5% (2010)
Pakistan	50%	5% (2010)
Sri Lanka	Unbound	2.5% (2009)

Source: WTO Tariff Download Facility, www.wto.org

50 per cent of total module costs. This inverted duty structure imposes costs, even if small, on established domestic manufacturers or those planning to set up domestic manufacturing facilities (which could also include foreign investors). It may be noted that there is broad agreement amongst analysts that the falling price of polysilicon will be the single most important driver of future cost improvements (Kirkegaard et al., 2010). Cutting silicon tariffs will certainly facilitate cost reductions, as would the entry of new manufacturers and increased supplies – a trend that is already happening.

In its 2011–2012 annual budget, the Indian government announced that the basic customs duty on raw materials used in the manufacture of solar modules and cells was being reduced to zero. India already applies zero duty on the import of PV cells and modules (HS 854140).[6] This category also includes light-emitting diodes (LEDs) as well. If raw material duties were not reduced to zero, this would raise costs for domestic manufacturers.

Along with import duties, other import-related taxes and charges – even minor ones – can contribute to raising costs. The Ministry of New and Renewable Energy in India (MNRE), for instance, sought the removal in early March 2011 of the 1 per cent excise duty on silicon wafers imposed by the Finance Ministry on the grounds that it would hamper the growth of domestic industry, as these were not produced in India. The excise duty on silicon wafers, along with other local taxes are seen as raising costs by 5 to 10 per cent for domestic solar module manufacturers already facing competition from China. The measure is also expected to discourage new Indian and foreign firms from entering the solar cell and module manufacturing industry in India.[7]

In the category of wind-powered electric generating sets (HS 850231), South Asian countries show similar profiles as in solar PV cells, with high levels of bound tariffs and much lower applied tariffs (in single digits). It is interesting to note here that Nepal (which is now seeking to deploy wind power in select regions in the country), without a domestic wind-turbine manufacturing industry, applies higher tariffs (based on data available) than India (which has an established equipment manufacturing industry).

Small hydro is a renewable energy sector where South Asian countries (with the exception of Bangladesh and the Maldives) have natural advantages and prospects for further development (see Table 14.A1). The definition of small hydro varies from country to country, and upper capacity can range from 10MW to 50MW (World Bank, 2008c).

Projects with the range of 100kW and above feed power into the grid and are commercial by nature. Projects below 100kW are mostly off-grid options being harnessed for rural village electrification and come under the social sector. Tariffs provided in Table 14.5(d) are for hydraulic turbines and water wheels for various categories of hydel projects from micro-hydro (less than 1MW), small hydraulic turbines (1MW to 10MW) and large turbines (greater than 10MW). This range captures more or less all categories of small-hydro plants as defined. In addition, parts of hydraulic turbines and water wheels (HS 841090) are also included.

Most countries in South Asia can be seen to apply fairly uniform rates for all types of hydraulic turbines, as well as for parts. Small hydro uses fairly well-established technologies and it is possible that turbine costs are not significantly affected due to prevalent tariffs. The presence of turbine and parts manufacturers may also account for higher applied duties in India, although it is surprising that Nepal (with immense potential for hydro-power development), facing a serious power crisis, also applies similar duty levels. It has, however, recently announced a series of measures, including zero duties on materials required to produce solar power, with the aim of ending the crisis in a period of five years.[8] However, it appears that Nepal, in practice, applies zero duty for equipment, parts and components not produced domestically. Nevertheless, the government stipulates that the amount of imported equipment, machinery and spare parts should not exceed 20 per cent of the total cost (see www.nicci.org). This policy statement, as well as prevailing tariffs, appears reflective of the trend among countries to protect and encourage domestic equipment manufacturing while simultaneously trying to incentivize RE deployment. For smaller developing countries and least developed countries, tariff revenues may also be an important consideration.

In summary, if the market is sufficiently attractive, higher tariffs on RE equipment may induce domestic and foreign investors to set up local manufacturing facilities to evade tariff protection. In sectors such as wind where equipment is bulky and transportation costs are important, there is already a trend towards investments clustering closer to centres of demand as opposed to relying on trade. Applied tariffs on key RE equipment, as can be seen, are already in single digits in South Asia and this is the case for most countries. Having higher tariffs on raw materials

and inputs used in equipment manufacture could discourage equipment manufacturers by raising production costs. Domestic power producers, on the other hand,
may want the flexibility that a free-trade regime offers them to import equipment
from wherever is most competitive and which meets their quality requirements.
Domestic equipment manufacturers, in response, often argue that the initial protection will help them ramp up the scale of production and eventually lead to a reliable
and competitive source of supply for power producers.

Local content policies and requirements

As with tariffs, local content requirements (LCRs) that mandate the use of locally
made components or technologies in RE projects could induce a certain degree
of investment in RE markets that are otherwise attractive. Local content policies
have been widely used in the past in a number of countries, both developed as well
as developing, such as Spain, Brazil and China. In September 2010, Japan launched
dispute settlement proceedings against Canada at the World Trade Organization on
13 September by saying that the province of Ontario's Green Energy Plan unfairly
pressures its producers of clean power to buy hardware from local manufacturers
(*ICTSD Bridges Weekly*, 15 September 2010).[9]

The extent to which LCRs distort competition and affect trade may depend
upon the way in which they are designed. If policies are laid out in broad terms
(such as stipulating a certain percentage or value of the investment to be sourced
locally), it may offer more flexibility to investors as opposed to LCRs that are very
specific and detail the components and parts to be sourced locally. Of course, the
percentage of local content will also matter. LCRs may be sub-national and force
firms to locate in specific geographic locations, as well. Depending upon the size
and attractiveness of the domestic market, the availability of local suppliers with the
ability to manufacture the equipment and components required, LCRs can influence a firm's ability or inability to optimize its supply chain. In many cases, as in
wind power in China, foreign firms will establish domestic manufacturing facilities
in response to LCRs.

Local content requirements may be mandatory throughout the country for all
renewable energy projects or they may be specific to a particular programme or
incentive scheme, such as the Jawaharlal Nehru National Solar Mission (JNNSM)
for solar PV and solar thermal projects in India. The JNNSM requirements are the
only significant local content requirement currently prevailing in the RE sector in
South Asia. The US has repeatedly complained that the domestic content requirements under India's National Solar Mission hurt manufacturers in the US, as well as
American investors developing solar projects in India.[10] The issue has, however, not
been taken up for dispute settlement in the WTO context. The JNSSM scheme is
only one of the many solar energy incentive schemes (others exist at the state level),
and the others do not have similar local content provisions.

Consultations with the private sector in India, Pakistan and Bangladesh have
revealed that private-sector power producers in all these countries regard access

BOX 14.1 *LOCAL CONTENT PROVISIONS UNDER THE JAWAHARLAL NEHRU NATIONAL SOLAR MISSION (JNNSM) IN INDIA*

Under the Jawaharlal Nehru National Solar Mission (JNSSM) guidelines for selecting new grid-connected solar projects, India has mandated domestic content for solar PV, as well solar thermal projects.

The guidelines for *solar PV* state that 'developers are expected to procure their project components from domestic manufacturers, as far as possible. However, in the case of solar PV projects to be selected in the first batch during FY 2010–2011, it will be mandatory for projects based on crystalline silicon technology to use the modules manufactured in India. For solar PV projects to be selected in the second batch during FY 2011–2012, it will be mandatory for all the projects to use cells and modules manufactured in India'.

For *solar thermal*, the guidelines state that 'It would be mandatory for project developers to ensure 30 per cent of local content in all plants/installations under solar thermal technology. Land is excluded'.

Source: Ministry of New and Renewable Energy, India (www.mnre.gov.in)

to the best technologies and equipment at competitive prices to be important and have welcomed zero duties on wind equipment, as it would have an impact upon both the price and quality of power that they generate. This may, however, be at odds with the interests of local equipment manufacturers that may demand greater protection, particularly against a perceived surge in Chinese imports while they establish themselves competitively; foreign equipment manufacturers, on the other hand, might criticize LCRs as preventing an optimized supply chain. Thus, depending upon the type of firm and activity, it may be possible to hear conflicting views on local content measures.

The chapter now briefly touches upon two domestic RE policy-related themes: subsidies and technical standards and requirements. These measures have the potential to shape trade and investment flows in renewable energy equipment and services. Given that the RE frameworks and the sector in South Asia are still evolving, no conclusions can be drawn on how these policies have shaped RE trade and investment flows in the region. It should be possible to assess this in greater detail in the coming years.

Domestic policies for RE deployment and their trade and investment implications

Subsidies

Subsidies are an important tool used by governments worldwide in the deployment of renewable energy worldwide, and can take the form of grants, capital subsidies, soft loans and tax credits. In fact, the rapid scale-up of grid-connected solar PV in recent years, especially in countries such as Germany, Spain, the US and Japan, even when equipment costs did not decline, can largely be attributed to government support policies for solar PV (Kirkegaard et al., 2010). Such support schemes appear to be needed for renewable energy, at least until a time when it attains 'grid parity' or becomes cost competitive with non-conventional forms of energy.

Subsidies and incentive schemes are usually granted to firms that produce renewable energy or renewable energy equipment. However, if these subsidies, whether *de jure* or *de facto*, are made contingent upon the use of domestically produced equipment, it may also affect manufacturers based abroad who were previously serving the market through exports. Hence, if subsidy provisions based on their design adversely affect foreign manufacturers, they may be deemed to run afoul of WTO subsidy rules. Consideration for trade rules is something that all countries, particularly WTO members, may need to keep in mind as they design renewable energy incentives. Renewable energy production incentives by themselves need not discriminate against foreign equipment. They may do so, however, if they are provided to domestic equipment manufacturers (as distinct from power producers) or if they are provided to power procurers conditional upon the use of domestic equipment.

In response to country incentives, foreign energy-service providers, as well as foreign equipment manufacturers, may alternatively decide to establish themselves in a country to benefit from these subsidies and incentives provided. In Ontario, the feed-in tariffs and incentives lined to local content sourcing prompted Japan to complain that it discriminated against imports. But the same measures also attracted Korean, US and European firms to establish themselves in Ontario (*ITCSD Bridges Weekly*, 15 September 2010). Thus, depending upon a number of factors (e.g. market attractiveness and attractiveness as an export base), incentives and subsidies such as local content schemes may induce inward investment in equipment facilities as opposed to servicing the market directly through exports, although from a purely commercial and market perspective, subsides linked to local sourcing may not be the most optimal decision.

Figure 14.1 provides an illustration of how renewable energy-related subsidies may or may not have a trade impact.

Consultations in South Asia underscored the need for the availability of more financing options for the private sector, particularly smaller firms involved in RE generation. It was particularly important to enable firms to cover large upfront costs (for capital and equipment and, in some cases, land) involved in RE power generation. According to certain firm representatives in India, rates of interest and loan tenures (both important determinants of the internal rate of return) could be better. While revenue from carbon credits was a welcome additional revenue stream for RE projects, the need to make the Clean Development Mechanism a more easily accessible and less administratively cumbersome mechanism was a grievance that was often aired. An interesting aspect highlighted by the private sector in Bangladesh was the need for financing schemes to accompany regulations, where it was required. The Bangladesh government has issued an order making it mandatory for

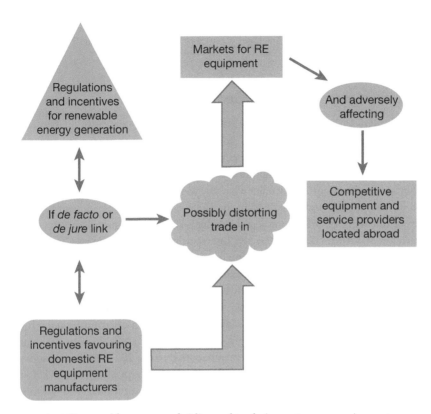

FIGURE 14.1 Renewable energy subsidies and trade impact upon equipment producers

new residential and commercial establishments to install solar electric panels if they want to secure an electric connection. While the aim is laudable, other support-ive aspects required, such as financial incentives for these establishments to cover up-front installation costs, are missing. Incentives may be more effective in enabling such installation than coming out only with mandatory orders.

The type of concessional financing demanded by the private sector during the process of consultations in South Asia should not have trade implications unless linked explicitly to the use of domestically manufactured equipment, or if equip-ment manufacturing itself is heavily subsidized and restricts market opportunities for foreign producers.

From a developing country perspective, it is also important to bear in mind the extent of special and differential treatment on subsidies that they may be entitled to under WTO rules. The WTO Subsidies and Countervailing Measures (SCM) Agreement recognizes three categories of developing country members: least developed country members (the LDCs), members with a gross national product (GNP) per capita of less than US$1000 per year which are listed in Annex VII to the SCM Agreement, and other developing countries. The lower a member's level of development, the more favourable the treatment it receives with respect to subsidy disciplines. In terms of import substitution subsidies, for instance, LDCs have eight years and other developing country members five years to phase out such subsidies.

Technical standards and requirements

Technical standards and requirements are imperative for the successful performance of RE equipment and, consequently, the projects in which they are used. Technical standards are important in conveying confidence and trust between manufacturers, operators, owners, financial institutions and government authorities. They can be either 'design based' or 'performance based' (WTO and UNEP, 2009). In wind, as also in solar, specific industrial certifications are necessary for developers to secure project financing or make a project 'bankable'. The wind industry operates under a series of standards laid down by the International Electrotechnical Commission (IEC) or the International Organization for Standardization (ISO), as well as various private standards. Often, compulsory national standards of varying stringency have also been employed by countries such as Denmark, Japan, Germany, India and the US (Kirkegaard et al., 2009). In India, in order to ensure project quality/performance, the MNRE has been insisting adherence to IEC/international standards for equipment and civil works for solar projects. The subsidy available from the ministry is linked to the use of equipment manufactured to the IEC or other prescribed international standards, and conditions for traceability have also been laid down (JNNSM, Ministry of New and Renewable Energy, India, accessible at www.mnre.gov.in).

The need to comply with different foreign technical regulations and standards involves significant costs for producers and exporters. In general, costs arise from the translation of foreign regulations, the hiring of technical experts to explain foreign regulations, and the adjustment of production facilities to comply with the requirements. Additionally, there is also the need to prove that the exported product meets the foreign regulations. The high costs involved may discourage manufacturers from trying to sell abroad. In the absence of international disciplines that govern the standard-setting process, a risk exists that technical regulations and standards could be adopted and applied solely to protect domestic industries. Hence, reliance on 'international' standards and 'performance-based' standards to the extent possible as provided for under the WTO's Technical Barriers to Trade (TBT) Agreement will avoid discriminatory and trade-restrictive practices. (www.wto.org)

Within South Asia, standards relevant to renewable energy equipment came up as an issue during consultations with the private sector in Pakistan. It was perceived that there was a lack of information and awareness amongst many firms in Pakistan regarding various quality standards and certification for RE equipment. This lack of information and awareness is an area where it was felt that international agencies could step up training activities.

Conclusion

It is clear that the uptake and expansion of renewable energy will be driven mainly through private-sector investment flows. Scaling up renewable energy requires proactive policies on the part of governments in creating the right enabling environment to leverage private-sector investments. While a number of countries in the developed world are at the forefront in terms of putting in place such policies, developing countries have recently started on the same road. Interactions with the private sector in South Asia reveal that in addition to policies on paper, transparency, credibility, stability and effective enforcement lead to a positive perception by the private sector and will encourage investments.

Renewable energy facilitates the transition to a low-carbon growth path and also enhances energy security, in addition to economic benefits such as poverty alleviation, job creation, etc. Countries are interested in deploying reliable renewable power at the lowest cost possible. In addition, they also want the clean energy sector to drive manufacturing and jobs within their economies. Attaining all of these objectives simultaneously may be challenging and may involve balancing various interests, as well as policies. Trade can play an important role in enabling access and rapid deployment of key technologies. Trade-related policies, such as tariffs and local content measures, and domestic renewable energy policies, such as subsidies and technical standards, driven by the pursuit of creating domestic manufacturing jobs and ensuring reliable performance of equipment. If improperly designed, however, they can also have an adverse impact in enabling firms to optimize their supply chains, which have now become global in nature. Policymakers should keep these diverse impacts in mind as they set about designing and balancing their trade and renewable energy policies.

Notes

1 World Wind Energy Association, http://www.wwindea.org/.
2 See http://www.sciencedirect.com/science/article/pii/S0301421510006361.
3 See http://www.sciencedirect.com/science/article/pii/S0301421505002181.
4 See http://www.greenworldinvestor.com/2011/04/06/state-of-solar-energy-in-india-maharashtraupdelhi-formulate-solar-subsidies-though-far-behind-gujarat/.
5 See http://mnre.gov.in/pdf/press-brief-payment-security-mechanism-jnnsm-20062011pdf.pdf.
6 Customs Tariff 2009–2010, Government of India, Central Board of Customs and Excise, www.cbec.gov.in.
7 Tata BP Solar in a letter to the Renewable Energy Ministry, for instance, pointed out that while the government, on the one hand, was encouraging domestic manufacturers, on the other, by imposing a duty on silicon wafers, it was disadvantaging them. According to a firm spokesman, 70 per cent of plants set up in the first phase of the Nehru Solar Mission were based on imported equipment and the number would surge to 80 per cent if the duty was not rolled back ('Excise rollback on silicon wafers may be sought', The Economic Times, 11 March 2011).
8 See http://www.bbc.co.uk/news/world-south-asia-12846672.
9 See http://ictsd.org/i/trade-and-sustainable-development-agenda/84822/.
10 See http://www.business-standard.com/india/news/us-concerned-over-domestic-content-requirements-in-indias-solar-sector/442942/ and also http://india.carbon-outlook.com/news/locke-wants-india-can-ban-foreign-made-solar-panels-indiawestcom.

References

ADB (Asian Development Bank) (2010) *Report and Recommendation of the President to the Board of Directors: Proposed Loan Zorlu Enerji Power Project (Pakistan)*, November 2010

Avato, P. and Coony, J. E. (2009) *Accelerating Clean Energy Technology Research, Development, and Deployment: Lessons from Non-Energy Sectors*, World Bank, Washington, DC

Barradale, M. J. (2010) 'Impact of public policy uncertainty on renewable energy investment: Wind power and the production tax credit', *Energy Policy*, vol 38, pp7968–7709

Barton, J. H. (2007) *Intellectual Property and Access to Clean Energy Technologies in Developing Countries: An Analysis of Solar Photovoltaic, Biofuels and Wind Technologies*, ICTSD Trade and Sustainable Energy Series Issue Paper no 2, International Centre for Trade and Sustainable Development, Geneva, Switzerland

Brown, M. (2001) 'Market failures and barriers as a basis for clean energy policies', *Energy Policy*, vol 29, no 14, pp1197–1207

Cohen, W. and Levinthal, D. (1990) 'Absorptive capacity: A new perspective on learning and innovation', *Administrative Science Quarterly*, vol 35, pp128–152

Hoekman, B., Maskus, K. and Saggi, K. (2005) 'Transfer of technology to developing countries: Unilateral and multilateral policy options', *World Development*, vol 33, no 10, pp1587–1602

Holdren, J. P. (2006) 'The energy innovation imperative: Addressing oil dependence, climate change, and other 21st century energy challenges', *Innovations*, vol 1, no 2 (spring), pp3–23

ICTSD Bridges Weekly (2010) 'Japan challenges Canadian renewable energy incentives at WTO', *ICTSD Bridges Weekly*, 15 September 2010, http://ictsd.org/i/trade-and-sustainable-development-agenda/84822/

Jha, V. (2009) *Trade Flows, Barriers and Market Drivers in Renewable Energy Supply Goods: The Need to Level the Playing Field*, ICTSD Trade and Environment Issue Paper 10, International Centre for Trade and Sustainable Development, Geneva, Switzerland

JNNSM (Jawaharlal Nehru National Solar Mission) (undated) Ministry of New and Renewable Energy, India, www.mnre.gov.in

Kirkegaard, J. F., Hanemann, T. and Weischer, L. (2009) *It Should Be a Breeze: Harnessing the Potential of Open Trade and Investment Flows in the Wind Energy Industry*, Working Paper, World Resources Institute and Peterson Institute for International Economics, Washington, DC

Kirkegaard, J. F., Hanemann, T., Weischer, L. and Miller, M. (2010) *Toward a Sunny Future? Global Integration in the Solar PV Industry*, Working Paper, World Resources Institute and Peterson Institute for International Economics, Washington, DC

Kofoed-Wiuff, A., Sandholt, K. and Marcus-Møller, C. (2006) *A Synthesis of Various Studies on Barriers, Challenges and Opportunities for Renewable Energy Deployment*, Energy Analyses for the IEA RETD Implementing Agreement, May 2006

Mallett, A., Sheridan, N. and Sorrell, S. (2010) *Barriers Busting in Energy Efficiency in Industry*, Report for UNIDO, January 2010

Margolis, R. and Zuboy, J. (2006) *Nontechnical Barriers to Solar Energy Use: Review of Recent Literature*, Technical Report, NREL/TP-520-40116, National Renewable Energy Laboratory, September 2006

Torodova, G. and Durisin, B. (2007) 'Absorptive capacity: Valuing a reconceptualization', *Academy of Management Review*, vol 32, no 3, pp774–786

Vanhaverbeke, W., Myriam, C. and Vareska, V. (2007) *Connecting Absorptive Capacity and Open Innovation*, Hasselt University, Belgium, Eindhoven University of Technology, The Netherlands, and Rotterdam Business School, The Netherlands

Watson, J., Byrne, R., Ockwell, D., Stua, M. and Mallett, A. (2010) *Low Carbon Technology Transfer: Lessons from India and China*, Sussex Energy Group Policy Briefing, November 2010, http://www.sussex.ac.uk/sussexenergygroup/documents/low-carbon-tech-transfer-briefing-nov-101.pdf

Wei, M., Patadia, S. and Kammen, D. (2009) Putting Renewables and Energy Efficiency to Work: How Many Jobs Can the Clean Energy Industry Generate in the US?, Working Paper, University of California, Berkeley, CA, http://rael.berkeley.edu

World Bank (2008a) *International Trade and Climate Change: Economic, Legal and Institutional Perspectives*, World Bank Economic and Sector Work (Environment Department, Sustainable Development Network), Washington, DC

World Bank (2008b) *Potential and Prospects for Regional Energy Trade in the South Asia Region*, ESMAP, Washington, DC

World Bank (2008c) *Renewable Energy (RE) Toolkit: A Resource for Renewable Energy Development*, Energy Sector Management Assistance Program (ESMAP), 30 June

World Bank (2010) *World Development Report*, World Bank, Washington, DC

WTO (World Trade Organization) and UNEP (United Nations Environment Programme) (2009) *Trade and Climate Change: A Report by UNEP and WTO*, World Trade Organization, Geneva, Switzerland

Appendix

TABLE 14.A1

Country	Hydropower potential – economically viable (MW)	Hydropower potential – technical (MW)	Hydropower developed (MW)
Bangladesh		755	230
Bhutan	23,760	30,000	468
India	84,000	150,000	32,300
Nepal	43,000	83,000	600
Pakistan		54,000	6500
Sri Lanka		9100	1250

15

INTERNATIONAL TRANSFERS OF CLIMATE-FRIENDLY TECHNOLOGIES

How the World Trade System Matters

Thomas L. Brewer and Andreas Falke[1]

Purpose and scope

This chapter contributes to the focus of the book on low-carbon technology trans-fers by putting those issues in the context of a wide range of subjects about the interaction of international trade and climate change issues. The chapter defines trade broadly to include not only trade in goods and services, but also direct invest-ment and licensing. Such a broad definition is especially appropriate when discuss-ing international technology transfers, as will be noted in more detail below.

Previous research on policy issues concerning international transfers of low-carbon technologies has been mostly focused on three types of *macro-level* analyses: economic analysis of international trade in goods and associated tariffs; sustainable economic development of countries, with an emphasis on the inter-national economic assistance programmes of developed country governments and capacity-building in developing countries; and legal and policy analysis of intellec-tual property rights issues (see Brewer, 2009b and 2010, for reviews of the literature; see also Brewer, 2008a, 2008b, 2009c; World Bank, 2008; WTO and UNEP, 2009).

Micro-level analyses have been much rarer – and mostly focused on large-scale research and development projects involving government-business partnerships (e.g. de Coninck, 2009, Chapters 5, 8). What have been largely ignored are micro-level analyses of the international business strategies, operations and projects of firms. This chapter includes such micro-level analyses. It uses the wind power industry – which is the largest and fastest growing renewable energy sector – for tangible examples to illustrate the general analytic points. As we shall see, there are diverse types of international technology transfers involving services as well as goods, and direct investments as well as trade and licensing in that industry.

The themes of the chapter are that the world trade system significantly affects international transfers of low-carbon technologies in a wide range of ways and that these effects occur through the strategies, operations and projects of firms.[2] Understanding the issues and options for enhancing international transfers of climate-friendly technologies for mitigation thus requires four different but complementary units of analysis – not only government trade policies, but also industry technologies, firm strategies and project transactions. Before developing those themes, however, it is important to note that issues at the nexus of trade, climate change and low-carbon technology transfer have already emerged on the agendas of governments, international agencies, firms, industry associations and environmental non-governmental organizations (NGOs).

Rhetoric and reality in the nexus of trade, climate change and technology transfer issues

Much of the interest of policy-makers and researchers, to date, at the nexus of international trade and climate change issues has been focused on the possibility that the European Union (EU) and/or the US would impose offsetting border measures to protect against international greenhouse gas (GHG) 'leakage' and the international competitive consequences of establishing emissions trading schemes in some countries, but not in others.[3]

There has been much rhetoric surrounding those issues, especially in the US, where industries such as the steel industry and industry associations such as the National Association of Manufacturers have used international competitiveness issues associated with the proposed establishment of a national cap-and-trade system as a basis for lobbying for the adoption of offsetting border measures (Brewer, forthcoming, Chapters 2, 5; Houser et al., 2008; van Asselt and Brewer, 2009; Hufbauer and Kim, 2010). In the EU, on the other hand, there has been more focus on the environmental issue of potential international leakages of greenhouse gases and the global welfare implications of the adoption or not of offsetting border measures (Gros and Egenhofer, 2010). There has been a special interest in the competitive implications for a small number of industries; the steel and cement industries, for instance, feature both GHG-intensive production processes and vulnerability to international shifts in the location of production as a result of differences in electricity prices and, hence, production costs. While the salience of these issues is likely to wax and wane in the coming years in Europe as the EU Emissions Trading Scheme (ETS) continues to evolve, the issues are dormant in the US because the establishment of a national cap-and-trade system is at least temporarily off the Congressional agenda.

Meanwhile, the current reality about the intersection of international trade, climate change and technology transfer issues is that there have already been a variety of cases in which international transfers of climate-friendly technologies have been issues in trade policies; the cases involve not only the EU and the US, but also Brazil, China, Canada, Japan and other countries.

Illustrative cases

The following eight cases provide tangible illustrations of how international transfers of climate-friendly technologies have emerged during recent years in the context of trade policy issues. The cases include a broad range of industries and technologies in the agricultural and services sectors, as well as the manufacturing sector. In each instance, international transfers of climate-friendly – or, in some instances, climate-unfriendly technologies – have been enmeshed in trade policies. Some of the cases have become subjects of formalized disputes in the World Trade Organization (WTO), while others have been addressed outside the WTO in bilateral negotiations (for information about the substance of the WTO cases and the procedures in the WTO dispute settlement process, see the WTO website at www.wto.org; additional information about all the cases is also available at www.TradeAndClimate.net).

US complaint about Chinese renewable power subsidies in the WTO dispute settlement process

The US filed a request for WTO consultations in December 2010 about Chinese government renewable energy subsidies, including, in particular, its subsidies in the wind power industry and domestic content requirements (WTO Dispute Settlement Case 419). At stake were the quantity and mode of international technology transfer into China, as well as the competitive implications for firms inside and outside China. During the consultation phase of the dispute settlement process in the WTO, before the case reached the phase involving the establishment of a dispute settlement panel, the parties reached an agreement according to which the Chinese government would end the subsidies.

Japanese complaint about Canadian domestic content requirements in the WTO dispute settlement process

After requesting consultations with Canada in September 2010, Japan filed a formal WTO complaint in June 2011 against Canadian treatment of imported equipment for renewable electricity generation in its feed-in-tariff programme due to local content requirements. At issue specifically were the policies of the province of Ontario (the most populous in Canada). In view of the widespread interest in feed-in tariffs in many countries, the outcome of the case was potentially significant for several renewable industries. The case (WTO dispute number DS412) was pending as of early 2012, as was a related case (DS426) filed by the EU against Canada.

EU tariffs on compact fluorescent lights (CFLs) from China and other Asian countries

Although this case did not become a WTO dispute case, it is another example of how trade policy can limit international transfers of low-carbon technologies. In this case, the EU imposed special tariffs on CFLs from China unilaterally on the grounds that Chinese exporters were engaged in 'dumping'.[4] The conflict progressed through several stages, dragged on for a decade, and involved several exporting countries in Asia. At stake in China – and in other Asian countries to which some manufacturing was relocated from China – was the fate of hundreds of manufacturers/exporters of CFLs; at stake in Europe was the energy efficiency of lighting in the commercial and residential sectors of 27 countries. After several years the tariffs were reduced, but they were not eliminated altogether until European retail firms complained to the commission that their desire to sell more low-cost, energy-efficient CFLs was being thwarted by the EU tariffs. The issue of whether there was dumping was not resolved by an independent process because the conflict did not become a WTO dispute.

US limits on rebates for purchasers of hybrid automobiles from certain manufacturers

This case was unusual in that unnamed firms (which were, in fact, Honda and Toyota and which, of course, happened to be Japanese) encountered volume-based caps put on US government rebates to US purchasers of hybrid cars in the US. Thus, the Japanese-based firms, which were leaders in the increasingly popular hybrid technology, reached their caps before their US-based rivals, so the US customers of the Japanese firms could no longer receive the US government tax credits. In response, the Japanese firms shifted production of hybrids from Japanese plants to their facilities in the US in order to placate the US government. Over time, the effects of the rebates on international transfer of climate-friendly technology thus shifted: initially there was less transfer in the form of the *products* (i.e. the hybrid vehicles); but then subsequently there was a transfer of the *production process* technology from the plants in Japan to the Japanese firms' plants in the US. The hybrid vehicles continue to be produced in the US by the Japanese firms. The US government rebate programme eventually expired and was not renewed. The Japanese government did not file a complaint in the WTO.

EU greenhouse gas emissions standards on imported palm oil-based biodiesel fuel from South-East Asia

This case is different from the above cases in as much as the issue was the imposition of barriers to the importation of biofuels produced in greenhouse gas–intensive production processes involving the clearing of rainforests (i.e. carbon sinks) in order to plant palm plantations. The trade barrier thus took the form of environmental standards because the imports were subjected to life-cycle analyses in the EU.

US tariffs on ethanol imports from Brazil

As has been typical with agricultural goods, US tariffs on imports were causally connected to domestic subsidies – in this instance, US domestic subsidies of corn production and corn-based ethanol refineries. The special tariff of 57 cents per gallon on sugar cane-based ethanol from Brazil, in addition to the regular 2.5 per cent was, of course, a particularly troublesome issue in Brazilian–US relations over many years, and it intruded into the WTO negotiating processes on trade in agricultural goods and trade in environmental goods. There was a surprising and unusual turn of events in this case in 2011, however, when the US Congress decided to end the tariffs, as well as the domestic subsidies, as part of a broader movement against government spending, in general, and agricultural and energy subsidies, in particular.[5]

EU objections to US exporters' use of a 'splash and dash' procedure to subsidize US exports of biodiesel fuel

The exploitation of a loophole in US legislation made it possible for US firms to import biodiesel fuel from South-East Asia, add a small portion of US petro diesel to it in order to qualify for a domestic US subsidy of the biodiesel portion of blended fuel, and then export it to Europe (Germany, in particular, where it qualified for additional subsidies). The US firms were thus able to get US government subsidies for a product that was approximately 99 per cent based on imports, with only 1 per cent US content, and then export it without its entering the US market, even though the purpose of the subsidy programme was to expand the US biodiesel industry and increase the use of biodiesel in the US market. The practice ended after coming under pressure from European diplomats and business leaders, and after the US industry association acknowledged that the practice was inconsistent with the intention of the legislation.

Chinese and US reactions to EU inclusion of the aviation industry in the Emissions Trading Scheme (ETS)

When the EU decided to include the aviation sector in its Emissions Trading Scheme in order to create a 'level playing field' for all carriers in the aviation industry, it covered international flights into and out of EU countries – not only the flights of EU-based carriers but also the flights of carriers based outside the EU. US airlines and the US Air Transport Association complained to the European Court of Justice – which rejected the complaint – and also threatened legal action in the International Court of Justice on the grounds that the Chicago Convention concerning international cooperation in civil aviation was being violated. The Chinese government similarly objected to the prospect of its airlines' flights into EU airspace being covered by the EU ETS; the Chinese government put pressure on the EU by preventing a Chinese airline from entering into an agreement with the European manufacturer Airbus to buy nearly US$4 billion worth of new A380 superjumbo planes. Thus, the case involved the use of a trade sanction by China on manufactured goods (airplanes produced in Europe), as well as the imposition of a restriction by the EU on international trade in services (commercial air transport services provided by Chinese airlines).[6] At stake was not only the multi-billion dollar international trade transaction, but the effectiveness of the EU ETS as a stimulator of technological innovation and diffusion. Indeed, a rationale for the establishment of a cap-and-trade system and, thus, the creation of a price for greenhouse gas emissions is, of course, precisely to foster technological innovation and diffusion based on price incentives.

It is clear from these cases, then, that a new era of the intersection of trade and low-carbon technology transfer issues has already arrived. In order to gain a fuller understanding of such issues and their implications for governments' policies and firms' practices affecting international technology transfer, it is necessary to analyse them on the basis of four different, but complementary, units of analysis.

Understanding the issues: Four units of analysis

Government trade policies

Government trade policies obviously comprise an important part of the policy framework within which firms transfer climate-friendly technologies internationally. In particular, it is essential to include political–legal–institutional analysis of the multilateral, plurilateral, regional and bilateral arrangements, as well as unilateral national and sub-national policies concerning international trade. International transfers of technologies are directly affected by policies and agreements concerning trade, investment and licensing in services as well as goods. This chapter builds on and extends previous studies by emphasizing the importance of trade in services as well as goods; this is a natural extension since know-how is often transferred internationally (and domestically) via service transactions. Government trade policies and the role of the WTO are discussed below in the fourth section, and examples in the wind power industry are discussed in section five.

Industry technologies

Industries are often defined in terms of the technologies embodied in products and production processes, and they are often the basis of firms' competitive positions within industries. This is true, for instance, of core technologies in the solar power industry (whether photovoltaic or concentrated) and the wind power industry (whether small turbines for onshore projects or large turbines for offshore projects). International technology transfer issues are therefore specific to particular industries, industry segments or stages in the value-adding process.

Firm strategies

Strategic business decisions about entering and serving foreign markets and, more generally, decisions about the international location of production facilities are at the core of micro-economic analysis of the international direct investments and the strategies of multinational firms.[7] Firms' strategic choices determine the modes by which they transfer technologies internationally. The principal mode of transferring technology internationally is international direct investment, often involving international joint venture partnerships with an international direct investment in the partnership by the foreign firm and a domestic investment in the partnership by a local firm. International direct investment is a particularly common strategic choice in service industries because of the relative importance of face-to-face contacts for services exchanges. The partnership thus represents a partially owned foreign subsidiary for the foreign parent firm and a partially owned domestic subsidiary for the local firm. The proportion of foreign ownership is often limited to less than 50 per cent or some other proportion by host government laws.

Firms, of course, often enter into international licensing agreements – sometimes as a stand alone mode for entering and serving a foreign market, sometimes as one element in a package of arrangements along with international direct investment and trade arrangements. Although the foreign licensee may be an independent firm so that the licensing agreement is an 'arm's-length' agreement, the foreign licensee is often a wholly-owned foreign subsidiary of the licensor; sometimes the licence is between a joint venture as licensee in the recipient host country and the licensor in the source home country. Whatever the arrangements, the enforcement of intellectual property rights (IPRs) is, of course, an issue in both countries and for all firms involved (more about IPRs below).

Project transactions

At the most basic and tangible level, international technology transfers consist of project-specific bundles of services and goods, often through international direct investment as well as trade and licensing. They thus need to be understood as being the results of operational business and engineering management decisions, as well as firms' strategic decisions. The cumulative results of such diverse and numerous individual transactions at the most micro-level add up to the aggregate patterns and trends that are the usual concerns of industry-level and macro-level policy-making by governments.

Government policies and the role of the WTO

While the World Trade Organization (WTO) is the multilateral core of the formally institutionalized trade system, it should be noted that many other institutions and agreements – at all levels, from multilateral to bilateral – are important elements of the international trade system that interacts with climate-friendly technology transfer issues (see www.TradeAndClimate.net for further information and analysis of other elements of the trade system and its interactions with a wide range of climate change issues, including technology transfer issues).

Until now, much of the discussion and the literature about trade and environment issues has focused on the compatibility of WTO rules with potentially trade-restrictive environmental rules. Specifically, the WTO legality of border measures and how to accommodate them under existing WTO rules have been discussed intensely, as noted earlier. In addition, however, a number of WTO agreements and rules are pertinent to the transfer of climate-friendly technologies. Indeed, many of the issues that have been discussed in the stalled – now possibly defunct – Doha Round of trade negotiation have a bearing on facilitating technology transfer; these discussions have also highlighted those rules that impede technology transfers. The relevant WTO agreements include the Agreement on Technical Barriers to Trade (TBT), the Agreement on Trade Related Aspects of Intellectual Property Rights (TRIPS), the Agreement on Trade Related Investment Measures (TRIMS) and the Agreement on Government Procurement.

The disciplines in TRIMS are weak, as they relate primarily to local content and trade balancing requirements and do not cover other types of government policies that affect international direct investments, such as restrictions on the nationalities of members of boards of directors. Furthermore, comprehensive investor protection is not part of the TRIMS, but is instead provided in the hundreds of bilateral investment treaties that are outside the WTO system.

Another area where the WTO is weak – in fact, lacking any agreement or rules – is competition policy (i.e. anti-trust in the US). As a result, the increasing concentration of some industries' structure receives attention from only unofficial observers, if at all. Nor are firms' anti-competitive behaviours subject to WTO scrutiny (except through a focus on predatory dumping practices).

Compulsory licensing under TRIPS for climate-friendly technologies along the lines of pharmaceuticals or similar flexibilities is probably not going to develop because the emergency nature that characterized the pharmaceutical issues has no parallel here (Hufbauer and Kim, 2010, pp29–33).

There is a potential for facilitating the transfer of low-carbon technology through the reduction or elimination of tariff and non-tariff barriers for environmental goods, as discussed in the Doha Round negotiations. Such an agreement could mirror the Information Technology Agreement concluded under the Uruguay Round in going for sector-specific liberalization with great spill-over potential (World Bank, 2008). A sticking point in those negotiations has been the definition of environmental goods and services. The inclusion of biofuels such as Brazilian ethanol in the list, for instance, has been opposed by the US as any inclusion of biofuels would undermine the US domestic subsidy regime of the less efficiently produced US ethanol. Even the conversion of applied tariffs to bound tariffs, as envisaged in a small tariff package for the Doha Round, would ease market access for low-carbon technologies in emerging market countries such as India. Emerging market countries sometimes apply lower tariff levels than they have agreed to be 'bound' under the General Agreement on Tariffs and Trade (GATT) tariff schedules, but reserve the right to increase the applied tariffs up to the bound rates should competitive pressures from imports increase. Another step would be for emerging market countries to join the plurilateral agreement on government procurement or at least agree to transparency measures in this area (van Asselt et al., 2006). Some authors have argued for a 'green round' as a substitute for a failed Doha Round (Hufbauer et al., 2010; Deutsch, 2011).

How WTO agreements, rules and negotiations tangibly affect international transfers of low-carbon technologies at the level of firms' technologies, strategies, operations and projects is illustrated by the wind power industry in the next section.

The wind power industry and international technology transfers

As mentioned in the introduction to the chapter, the wind power industry is the largest and fastest growing part of the renewable energy sector. It experienced an annual growth rate of 21 to 29 per cent each year from 2003 to 2008 (Kirkegaard et al., 2009, p36, Table 1). Although its growth rate declined during the recession years of 2009 to 2010, it fared better than other renewable energy industries (REN21, 2010).

Industry segments

The industry can be segmented by whether installations are onshore or offshore, and as to the latter, whether deep-water offshore or not. Industry segmentation can also be based on turbine size, with larger turbines (up to 5MW or even 6MW) being favoured for deep-water and other offshore installations (Breton and Moe, 2009; Hassan et al., 2010). As of 2011, the offshore turbine segment of the industry was centred in Europe, with the Danish firm Vestas being particularly prominent. Denmark was the world leader in the deployment of shallow offshore facilities. There is a North Sea pilot project being deployed as a joint Siemens–Statoil undertaking in the form of a floating deep-water facility known as Hywind off the coast of Norway (Siemens, 2009; Statoil, 2010).

However, the overwhelming majority of installations, to date, in most countries have been of 3MW turbines or less and for onshore wind farms. Although turbine sizes have been generally increasing over time, the worldwide average size of installed turbines was only slightly more than 1.5MW in 2008 for the first time (Kirkegaard et al., 2009). In the 27 countries of the EU, as of mid-2011, there were only about 3GW of offshore capacity, compared with more than 50GW of onshore capacity (EWEA, 2010; 2011).

Manufactured goods

As for the specifics of wind power technologies, core components of the turbine include generators, gear boxes, bearings and power converters – all located in the

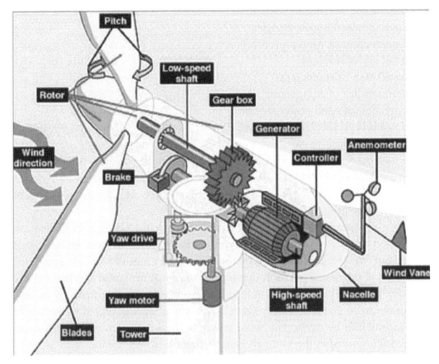

FIGURE 15.1 Components of a wind power facility
Source: USITC, 2005, pp4–3

'nacelle' at the top of the tower. Altogether, there are as many as 7000 component parts in a turbine. Some of them are quite small, but nevertheless crucial – for instance, computer control components; of course, there are also large, conspicuous towers and blades as well. In addition, there are upstream goods such as steel forgings used in the fabrication of steel towers as well as composite materials used in blade fabrication. Figure 15.1 provides a simple depiction of the components of a wind power installation.

Services

The importance and diversity of service transactions in wind power technology transfers are suggested by the following list of services (compiled by the authors from USITC, 2005, p43), which are associated with a wind farm project:

- assessment of wind resources (i.e. potential for producing electricity);
- site analysis;
- project development;
- real estate services;
- project financing;
- project licensing and legal services;

- project engineering and design;
- environmental impact analysis;
- construction of wind power facilities;
- retail sale of turbines;
- installation of equipment;
- maintenance of equipment;
- operation of wind power facilities;
- transmission, distribution and sale of electricity generated by wind power.

These and other services are provided by a large number of firms based in many countries (USITC, 2005, pp4–10).[8] Spanish firms Iberdrola and EHN Acciona Group, and Florida Power & Light in the US are the world's largest operators of wind power facilities.

Other key operators of wind power facilities include the UK firm PPM, Scottish Power and the Japanese firm Eurus. Providers of other services include large energy and engineering companies that supply a wide range of vertically integrated products and services, wind farm developers, and small firms that specialize in the provision of niche wind energy services. Turbine manufacturers frequently participate in the wind power services market by providing services related to the sale of their turbines, or by developing wind power facilities at which their turbines are installed. For example, the German firm Siemens provides services such as training, repair and monitoring services in conjunction with the sale of its turbines, while the Japanese firm Mitsubishi supplies services such as design, construction and installation to its customers. There are many firms that focus on the development of wind power projects: BlueSkyWind, Evergreen Wind Power, Windland and Atlantic Renewable Energy Corp in the US; Sea Breeze Power and Western Wind in Canada; and Airtricity, Energia Hidroelectrice de Navarra (EHN), National Wind Power, Renewable Energy Systems and WindKraft Nord AG in various European countries.

Capacity, production and trade patterns

Slightly more than half of the world's installed wind power capacity is in Europe, with another fifth in the US and about one tenth in each of China and India. Substantial growth is expected in all of these countries for the next several years. The international patterns of production are similar, with European firms having about 60 per cent of the world market share. Firms based in China and India are gaining world market shares and increasingly significant market presences outside their home countries – in particular, the Chinese firms Sinovel, Goldwind and Dongfang, and the Indian firm Suzlon (see especially Lewis, 2007, on China and India). Indian and Chinese firms have also been expanding internationally through international direct investments, including in Europe and the US.

Germany accounts for about two-fifths of world exports of wind turbines and Denmark about one fourth. Elsewhere in Europe, the Spanish firm Gamesa exports turbines to the US and other countries. The Japanese firm Mitsubishi also exports turbines, including to the US. Among firms based in the US, GE has a presence in many foreign markets. Among importing countries, the US is by far the largest, with almost half the world total.

International direct investments

International acquisitions through direct investments by Suzlon of India have enabled it to establish a significant presence in Europe and to use its European affiliates as sources of technology transfer back to India. Suzlon has a wholly owned subsidiary in Germany – namely, REPower, which it acquired in 2007 and which has long been a major firm in the German wind power industry. The Chinese firm Goldwind also has a German subsidiary, Vensys, which it bought in 2008 after having had an international licensing arrangement with it from 2003. *Thus, in these instances, technology transfer to 'developing' countries from 'developed' countries has occurred through international direct investments in developed countries by firms based in developing countries.* There are other patterns and examples of international direct investments in the industry – arrangements that often involve two-way technology transfers. For instance, two German firms – Siemens and Nordex – manufacture turbines in Denmark through their subsidiaries there. Other firms in the industry also rely on international direct investments for access to other countries' technology and markets.

International direct investment is therefore often more important than trade in goods as a firm strategy in shaping the industry's international structure and its international technology transfer practices. It has been observed, for instance, that 'Cross-border investment rather than trade is the dominant mode of the industry's global integration' (Kirkegaard et al., 2009, p1). The importance of international direct investment as a key strategic choice of firms and thus the principal mode of international technology transfers is especially true in services. These observations could be generalized to other low-carbon industries as well (see UNCTAD, 2010, on low-carbon industries).

Firms' international technology transfers – whether through trade, investment or licensing – are, of course, subject to a variety of government policy barriers, including many covered by the WTO agreements enumerated above. The most obvious are tariffs on goods.

Tariffs[9]

An extensive study of the wind power industry by Kirkegaard et al., (2009, p42, Table 8) found that applied tariff levels on wind turbine imports in 25 countries plus the EU-27 varied from 0 to 10 per cent. India was an especially interesting case, as it unilaterally lowered its applied rate from 25 to 12.5 to 7.5 per cent over the period of 2002 to 2008. China had 8 per cent applied tariffs throughout the 2002 to 2008 period. Some countries had 10 per cent tariffs: Colombia, Indonesia,

Mexico (for non-NAFTA countries), Taiwan and Thailand. On the other hand, EU tariffs at 2.7 per cent, US tariffs at 1.3 per cent and Japan's tariffs at 0.0 per cent were all obviously at the low end of the scale. Brazil has countered the general worldwide pattern of low tariffs; it increased its applied tariff rates on small turbines (up to 2.6MW) from 0 to 14 per cent beginning in 2010.

A World Bank study (2008, p63, Table 3.6) of 18 high GHG-emitting developing counties found that half had tariffs on 'wind technology' below 10 per cent. These levels on wind technology were 'much lower' than the average industrial goods levels in the same countries. However, while tariffs on turbines may not be a significant obstacle to technology transfers in the wind power industry in many countries, there are important exceptions, where rates are 10 per cent or above. In any case, non-tariff barriers are often more problematic than tariffs.

Non-tariff barriers

In the same World Bank study noted above (2008, p63, Table 3.6), non-tariff barriers (NTBs) to international transfers of wind power technology to the 18 high GHG-emitting developing countries were much higher than the equivalent tariffs. Computed as *ad valorem* tariff equivalents, the NTBs were 59 per cent for Malaysia, 60 per cent for Zambia, 70 per cent for Egypt, and 87 to 89 per cent for Brazil, Nigeria and the Philippines. In a separate study of the wind industry, in particular, Kirkegaard et al., (2009, p1) concluded that 'the principal barriers to global integration are non-tariff trade barriers and formal and informal barriers that distort firms' investment decisions'. In particular, the following NTBs were found to be significant barriers (Kirkegaard et al., 2009, pp20–24):

- *Local content requirements.* For instance, before it was eliminated in 2009, China had a 70 per cent local content requirement for foreign firms investing in wind turbine manufacturing facilities, a requirement that reduced imports from the foreign firms' home countries and suppliers in third countries and that led to the more rapid development of domestic suppliers in China. Brazil had a 60 per cent local content requirement for wind turbines for several years, but rescinded it after complaints from firms and governments.
- *Industrial standards and certification requirements.* Although there are widely accepted international standards, such as those of the International Organization for Standardization (ISO), there are also significant differences in national standards that pose important barriers to international technology transfers in the wind power industry. These differences pertain not only to goods but also to services.
- *Government procurement requirements and other political pressures.* Many national and sub-national governments impose domestic purchasing requirements for wind farm projects. Some national governments have not signed up to the (plurilateral) WTO Government Purchasing Agreement at all or have not included the relevant agencies in their schedules.

As for IPR issues in the wind power industry – as in other industries – there is evidence to suggest that IPRs are not so much of a barrier to international transfers of climate-friendly technologies as is commonly assumed, at least in developed countries (Kirkegaard et al., 2009, p128, Box 3). On the other hand, there are instances where it has been an obstacle (World Bank, 2008, pp60–61, Box 3.4). In the case of India, Mallett, et al., (2009) found that IPR issues slowed down the rate of diffusion but did not prevent it. There will no doubt be additional research on IPR issues in the wind power and other industries, and it will hopefully further clarify what is rhetoric and what is reality.

Conclusions

Barriers to international transfers of low-carbon technologies are already on the active agenda of the international trade system – that is a reality. However, common assumptions about the barriers to such transactions are not borne out by the information accumulated to date: despite the rhetoric, neither tariffs nor intellectual property rights are generally the most significant impediments to such transfers worldwide, though there are important country-specific and technology-specific exceptions to the general pattern. The most problematic barriers tend to be in the form of non-tariff barriers such as government policies that constrain international direct investment in manufacturing and services, or other barriers to trade in services. In the wind power industry, these barriers include restrictions on government procurement programmes, technical standards and local content requirements for international direct investment projects.

Whatever barriers may be present are potentially of interest in the WTO, which has the legal competence to address most of them. In fact, there have already been dispute cases concerning low-carbon industries – including wind power – brought to the WTO. Furthermore, the wide range of WTO agreements in terms of their coverage of industries and their coverage of types of international transactions of relevance to international technology transfers indicates that WTO agreements are an important part of the institutional framework within which such transfers occur – or do not occur.

Yet, there are reservations to this generalization: the absence of a comprehensive WTO agreement covering international direct investment beyond the limited coverage of the TRIMs agreement prevents the WTO from being relevant to some key issues affecting international transfers of low-carbon technology. Furthermore, the absence of a WTO competition policy agreement means that industry structure and behaviour that can inhibit competition and technological innovation and diffusion are not subject to any multilateral monitoring or disciplines. It seems unlikely that either direct investment or competition will be added to the WTO's legal competence in the near future.

Moreover, of course, the Doha Round WTO negotiations have been at an impasse for many years now, with no prospect for a breakthrough in the near future. It may be, therefore, that in the future progress in reducing trade-related and investment-related government policy barriers to international transfers of low-carbon technologies will take place outside the WTO in other venues. Nevertheless, the WTO will continue to play a central role in the enforcement of its existing rules that directly affect the strategies, operations and project transactions of firms as they decide whether and how to undertake technology transfers.

Notes

1 The authors are indebted to the Schöller Foundation of Germany for a generous grant to Professor Brewer for research on international technology transfer and climate change issues. We are also indebted to Maria Drabble for her excellent research assistance. This chapter is one result of on-going research focused on the intersection of international trade and climate change issues. For additional information, please see www.TradeAndClimate.net.

2 There are also international technology transfers through government agencies, NGOs, academic institutions and foundations. The emphasis of this chapter is private-sector transactions.

3 International 'leakage' refers to the potential phenomenon whereby greenhouse gas emissions can increase in some foreign countries as their production increases, while emissions in the country (countries) with an emissions trading system decrease. It is possible, in fact, that the increases in some countries would exceed the decreases in other countries, so that there would be a net global increase in emissions. The imposition of an emissions trading system in a region or country, without comparable efforts in other countries, can therefore be counter-productive at the global level. There is a project under way organized by the Energy Policy Forum of Stanford University that compares the results of diverse computable general equilibrium modelling studies of the implications of offsetting border measures.

4 Dumping refers to selling goods abroad at lower prices than at home and/or selling them at below the cost of production. See www.wto.org for further information about the practice of dumping and the WTO Agreement on Dumping.

5 The Brazilian ethanol, which is based on sugar cane, is not only cheaper than US ethanol, which is based on corn; the Brazilian ethanol is more effective than the US ethanol in reducing greenhouse gas emissions. In fact, complete life analyses of US corn-based ethanol indicate that there is an increase in greenhouse gas emissions because of the carbon-intensive production methods that are used.

6 The final resolution of the China–EU conflict, like the US–EU conflict, was pending at the time of writing.

7 There is a vast literature on theoretical explanations of firms' strategic choices in international business, including, in particular, the central role of international technology transfer issues. See especially the contributions to Rugman (2009).

8 The following paragraph is based on excerpts from the report (USITC, 2005, pp4–10), with substantial editing.

9 Specific data about trade in wind power goods is often limited to assembled turbines, which have the Harmonized System six-digit code HS 8502.31. Towers and blades for wind power installations and their material inputs do not have unique HS codes, but rather are included as multi-use goods in codes with other types of towers and blades and materials for other uses.

References

Breton, S.-P. and Moe, G. (2009) 'Status, plans and technologies for offshore wind turbines in Europe and North America', *Renewable Energy*, vol 34, issue 3, pp646–654

Brewer, T. L. (2007) 'International energy technology transfers for climate change mitigation: What, who, how, why, when, where, how much ... and the implications for international institutional architecture', Paper prepared for CESifo Venice Summer Institute Workshop: Europe and Global Environmental Issues, Venice, Italy, 14–15 July 2007, CESifo Working Paper no 2408, www.usclimatechange.com

Brewer, T. L. (2008a) 'Climate change technology transfer: A new paradigm and policy agenda', *Climate Policy*, vol 8, pp516–526

Brewer, T. L. (2008b) 'The technology agenda for international climate change policy: A taxonomy for structuring analyses and negotiations', in C. Egenhofer (ed) *Beyond Bali: Strategic Issues for the Post-2012 Climate Change Regime*, Centre for European Policy Studies, Brussels, pp134–145, www.ceps.eu and at www.usclimatechange.com

Brewer, T. L. (2009a) 'Technology transfer and climate change: International flows, barriers and frameworks', in L. Brainard and I. Sorkin (eds) *Climate Change, Trade and Competitiveness*, The Brookings Institution, Washington, DC

Brewer, T. L. (2009b) 'The trade regime and the climate regime: Institutional evolution and adaptation', *Climate Policy*, vol 3, no 2003, pp329–341

Brewer, T. L. (2009c) 'The WTO and the Kyoto Protocol: Interaction issues', *Climate Policy*, vol 4, no 2004, pp3–12

Brewer, T. L. (2010) 'Trade policies and climate change policies: A rapidly expanding joint agenda', Introduction to a symposium on trade issues and climate change issues, *The World Economy*, vol 33, no 6, pp799–809

Brewer, T. L. (forthcoming) *The United States in a Warming World: The Political Economy of Government, Business and Public Responses to Climate Change*

Brewer, T. L. and Lunden, S. (2006) 'Environmental policy and institutional transparency in Europe', in L. Oxelheim (ed) *Corporate and Institutional Transparency for Economic Growth in Europe*, Elsevier, New York, NY

Brewer, T. L. and Young, S. (2000) *The Multilateral Investment Regime and Multinational Enterprises*, Oxford University Press (updated paperback edition), Oxford

Brewer, T. L. and Young, S. (2009) 'The multilateral regime for FDI: Institutions and their implications for business strategy', in A. M. Rugman (ed) *Oxford Handbook of International Business*, Oxford University Press, Oxford, pp282–313

de Coninck, H. (2009) *Technology Rules! Can Technology-Oriented Agreements Help Address Climate Change?*, Vrije Universiteit Amsterdam, Amsterdam

Deutsch, K. G. (2011) *Doha oder Dada*, Deutsche Bank Research, Frankfurt

EWEA (European Wind Energy Association) (2010) *Annual Report 2010: Powering the Energy Debate*, www.ewea.org, accessed 22 August 2011

EWEA (2011) *Offshore Statistics*, www.ewea.org, accessed 22 August 2011

Gros, D. and Egenhofer, C. (2010) *Climate Change and Trade: Taxing Carbon at the Border?*, Centre for European Policy Studies, Brussels

Hassan, G. et al., (2010) 'Floating offshore wind energy: A review of the current status and an assessment of the prospects', *Wind Engineering*, vol 34, no 1, pp1–16

Houser, T., Bradley, R., Childs, B., Werksman, J. and Heilmayr, R. (2008) *Leveling the Carbon Playing Field: International Competition and US Climate Policy Design*, World Resource Institute, Washington, DC

Hufbauer, C. and Kim, J. (2010) *Climate Change and Trade: Searching for Ways to Avoid a Train Wreck*, Centre for Trade and Economic Integration, Geneva

Hufbauer, C., Schott, J. and Wong, W. F. (2010) *Figuring Out the Doha Round*, Peterson Institute of International Economics, Washington, DC

Justice, D. and Philbert, C. (2005) *International Energy Technology Collaboration and Climate Change Mitigation: Synthesis Report*, Organisation for Economic Co-operation and Development, Paris

Kirkegaard, J. F., Hanemann, T. and Weischer, L. (2009) *It Should Be a Breeze: Harnessing the Potential of Open Trade and Investment Flows in the Wind Energy Industry*, World Resource Institute, Washington, DC

Lewis, J. I. (2007) 'Technology acquisition and innovation in the developing world: Wind turbine development in China and India', *Studies in Comparative International Development*, vol 42, pp3–4

Mallett, A. et al., (2009) *UK–India Collaborative Study: Barriers to the Transfer of Low Carbon Energy Technology: Phase II*, Report for UK Department for Energy and Climate Change (DECC), London

REN21 (2010) *Renewables 2010: Global Status Report*, REN21, Paris

Rugman, A. M. (ed) (2009) *Oxford Handbook of International Business*, Oxford University Press, Oxford and New York

Siebert, H. (ed) (2003) *The Economics of International Environmental Problems*, Springer Verlag, Heidelberg

Siemens (2009) 'Hywind: Siemens and StatoilHydro install first floating wind turbine', 10 June, http://www.siemens.com/press/en/pressrelease/?press=/en/pressrelease/2009/renewable_energy/ere200906064.htm, accessed 23 November 2010

Statoil (2010) 'Hywind – the world's first full-scale floating wind turbine', http://www.statoil.com/en/TechnologyInnovation/NewEnergy/RenewablePowerProduction/Offshore/Hywind/Pages/HywindPuttingWindPowerToTheTest.aspx, accessed 23 November 2010

UNCTAD (United Nations Conference on Trade and Development) (2010) *World Investment Report 2010: Investing in a Low-Carbon Economy*, United Nations, New York and Geneva

USITC (United States International Trade Commission) (2005) *Renewable Energy Services: An Examination of US and Foreign Markets*, USITC, Washington, DC, www.usitc.gov

van Assell, H. and Brewer, T. L. (2009) 'Addressing competitiveness and leakage concerns in climate policy: An analysis of border adjustment measures in the US and the EU', *Energy Policy*, vol 38, no 1, pp42–51

van Assell, H., van der Grijp, N. and Oosterhuis, F. (2006) 'Greener public purchasing: Opportunities for climate-friendly government procurement under WTO and EU rules', *Climate Policy*, vol 6, pp217–229

World Bank (2008) *International Trade and Climate Change*, World Bank, Washington, DC

World Bank (2010) *World Development Report 2010: Development and Climate Change*, World Bank, Washington, DC

WTO and UNEP (World Trade Organization and United Nations Environment Programme) (2009) *Trade and Climate Change*, WTO, Geneva

PART VII

Moving Forward: New Directions for Policy and Practice

16

LOW-CARBON ENERGY TECHNOLOGY DIFFUSION

A UK Practitioner's Perspective

David Vincent

This chapter presents a personal view about low-carbon innovation and low-carbon economic development in developing countries. It is drawn from my perspective working largely within a developed country context, but increasingly in recent years having engaged with organizations and initiatives in developing countries to share insights and learning. For over 25 years now I have worked on energy efficiency and low-carbon technology innovation programmes, as well as policy formulation and execution for successive UK governments. This included, for example, the UK government's first renewable energy technologies research, development and deployment programme in 1978 and the Energy Efficiency Best Practice programme, which I helped to design in 1988 and subsequently directed from 1994 to 2001. From 2001 to 2011, I was one of the directors at the Carbon Trust[1] – an independent not-for-dividend company set up by the UK government with support from business to accelerate the move to a low-carbon economy.

At the Carbon Trust I worked on a range of interventions to improve energy efficiency and develop low-carbon technologies. During the last few years, while at the Carbon Trust, I have been working with the UK Foreign and Commonwealth Office and the UK Department for International Development to help them promulgate UK knowledge and experience on energy efficiency and low-carbon development overseas. Of particular relevance to this chapter is my involvement with the development of ideas for low-carbon technology innovation and Climate Innovation Centre models to help developing countries find and implement low-carbon ways to develop their economies. The Carbon Trust's publication on low-carbon technology innovation and diffusion centres in relation to developing countries has contributed towards understanding the challenges of this complex topic.[2] My hope is that the insights and personal opinions presented in this chapter will both stimulate thought and lead to fruitful areas for debate, collaboration and action on climate and energy security in both developed and developing countries.

Framing the low-carbon challenge

There are a number ways of framing this huge and unique global challenge depending upon our perspectives, experiences and vocabularies. For me, the climate change mitigation and energy supply challenges for developing countries need to be put in context with the other huge challenges that developing countries are facing, such as poverty, malnutrition, disease and lack of education. The Millennium Development Goals (MDGs)[3] provide a summary of priority goals which many developing countries share. Goal 7, to 'ensure environmental sustainability', makes a small reference to carbon emissions as a sub-area that must also be discussed and ultimately addressed. Reduction targets don't feature in those goals – and for very good reason. Many developing countries are already very low carbon owing to their low levels of economic activity relative to many developed nations. This highlights important human development challenges, such as improving access to energy for poor and marginalized communities, which understandably often take priority over carbon mitigation concerns. Add to this the fact that developing countries didn't create the problem of greenhouse gas (GHG) emissions, and the question arises: why low-carbon economic growth should be a priority for them. Why not just growth by whatever energy source is the cheapest and most easily available? My response is in three parts.

First, irrespective of whose greenhouse gases they are, these gases are accumulating at an unprecedented rate. And in a world where internationally traded goods carry embedded greenhouse gases across traditional borders which define the boundaries for national greenhouse gas emission inventories, 'whose greenhouse gases' becomes a moot point. The Carbon Trust's research into international carbon flows provides a detailed analysis.[4] The Intergovernmental Panel on Climate Change's (IPCC's) *Fourth Assessment Report*[5] makes clear that the connection between rising atmospheric greenhouse gas concentrations and rising global temperatures which drive the observed changes to our climate and extreme behaviours in global weather systems is 'unequivocal'.

Second, developing countries are often in the climate change frontline – more so than many developed countries – and are often less able to respond adequately.[6]

Third, energy imports are going to get more expensive as prices and price volatility rise. It is better, then, to develop alternative, renewable, sustainable sources of energy and make these technologies available to developing countries on an affordable basis – for example, by analogy with the British pharmaceutical company GlaxoSmithKline's (GSK) decision to supply the expensive rotavirus vaccine, used for prevention of severe diarrhoea in children, at cost price.[7]

The scientific evidence is clear – for those who wish to see it. On land, in our oceans and our atmosphere, climate change is happening now. The Association of British Insurers said in 2004 that 'Managing the impacts of climate change is a major challenge for society – we already live with its effects every day'.[8] It presents a serious threat to the environment, to biodiversity, to habitats and to the economies and well-being of all countries and peoples. For developing countries and their peoples, the impacts are being particularly felt.[9] In the end, however, we will all experience climate change impacts in one form or another – physical changes to our environments, less robust supply chains, costs of raw materials, climate-induced migrations of displaced populations (as I think we are beginning to see in parts of Africa). As has been explained by Lord Stern and his team in the *Economics of Climate Change*,[10] action now will be less onerous and less disruptive than action deferred.

Political will to underpin a low-carbon vision, shared amongst key stakeholders (such as other political parties, business and civil society), is the first essential ingredient. Unless a country or jurisdiction wants to achieve this goal, and commits to it over time (i.e. decades) across the political parties and stakeholders, no amount of low-carbon aid, innovation centres or whatever will be effective. In my view, committing to a goal is not the same as committing to achieving that goal over time with the aid of policies, measures, finance and delivery agencies. The policy and delivery landscape will evolve in the light of experience and how successive governments decide how best to take on the challenge. I have seen a number of national energy or carbon abatement plans with targets, but fewer sets of coherent, consistent policies, few delivery plans and few delivery agencies with a long-term remit to transform economies from high to low carbon over a 30- to 40-year timeframe. By contrast, consider how long environmental protection and associated legislation, and the various environment protection agencies, have been around – decades, most with long-term remits and secure positions in the national governmental infrastructure. The political commitment to a goal or a target is incomplete unless it is accompanied by a delivery plan and a delivery agency – or agencies – with weight and resources, interim targets and a means whereby progress can be reliably and independently tracked and reported.

Given the political will to assign a high priority to low-carbon economic development, where does a country start on its low-carbon economic development journey? I suggest the starting point should be a mapping exercise to cover:

- the policy and market landscape to identify strengths and weaknesses in relation to the low-carbon economic development goals;
- the existing institutions to assess which, if any, would be able to execute the policies and deliver the goals; and
- the availability of local skills and capacities to deliver the goods and services required to achieve the policy goals.

Where a relevant institution exists, explore with them whether they could become the focal point for low-carbon innovation (including technologies, policies and capacity building, training, etc.). Where no such institution exists, explore with stakeholders and others (e.g. from other countries) whether a new institution – e.g. a low-carbon innovation centre – would help to facilitate and accelerate their low-carbon journey.

It is worth bearing in mind that the various concepts of innovation have been influenced by Western cultures and sophisticated markets where streams of incremental change punctuated by occasional disruptive change have driven, and continue to drive, technological development, diversity and economic growth. Some innovation is consumer and market driven (e.g. consumer electronics). This market-led innovation requires little if any public policy intervention. Low-carbon technology innovation, on the other hand, is neither consumer nor market driven. Public policy intervention support is essential to encourage the market to invest in low-carbon innovation. This relationship with public policy puts innovation centres and other delivery agencies into a special category within the policy and market framework. Working out how low-carbon change entities can be established, flourish and deliver desired outcomes in what are quite different cultures is no trivial challenge. We should think about the policy, market and cultural contexts in which these entities will seek to achieve their goals and, importantly, develop as they and their operating environments change. By definition, successful change agencies will not only change the environments in which they seek to have impact, but will also themselves change in order to be effective in their new environments.

'Technology' is important but it is only one aspect of building a low-carbon developing economy. Having a supportive and stable policy and market framework over time; developing the right business and financial models; having education and training schemes in place; and developing the skills to encourage low-carbon consumer behaviour are also important. Thus, for example, we are debating in the UK whether we can achieve prosperity without growth,[11] and how we can learn about the creation of working low-carbon communities, whereas in some of the larger emerging economies, the new middle classes are wondering how best to demonstrate 'arrival' through material goods with their newfound net disposable incomes – and confidence to spend that money. By 'confidence' I mean that money alone doesn't put people in the social group that buys certain goods and services or adopts certain patterns of behaviour. Changing these patterns and behaviours to make purchasing decisions which reflect sustainable consumption as the norm will, throughout the developed and developing world, take time – unless, of course, people can be encouraged, or constrained through choice editing or enforced regulation, to be economical with resources and purchases. This is where the role of the state becomes important and attitudes towards state intervention become crucial.

Lastly, we should bear in mind that for many emerging economies, the urgent challenges of today usually outweigh the important challenges of tomorrow. Some say that future challenges are for tomorrow's children; that 'technology' will find a way to allow future generations to have their cake and eat it; and, best of all, that these challenges can be deeply discounted in the eyes of today's people and some of today's economists. Others argue that current generations have a stewardship responsibility to respect and protect environments and natural assets for successive generations; and that the sheer global scale of the mitigation and adaptation challenges may not be soluble by 'technology' alone. The question for us, I suggest, is that while a single drought or crop failure cannot be ascribed entirely to climate change, we are observing and monitoring climate change impacts around the world. What is already in store for us, and for generations yet to come, is locked up in the inertia of climate change. It's not just developing countries which are seeing climate change impacts. The Arctic, with its thinning sea ice, is also in the climate change frontline. It is, perhaps, ironic that the thinning sea ice could make Arctic shipping trading routes a reality and will enable the exploitation of huge oil deposits which have been found recently.

Each year, we are learning more and more about what we are seeing. Although it is difficult to determine when sets of events around the world become statistically important trends, we know enough to be able to say with confidence that the combination of rising populations and a 'business-as-usual' model of economic growth is unsustainable and will lead to long-term, possibly irreversible, damage to our world with consequences for which tomorrow's children, our children, will not thank us.

The challenge

For me, the challenge of addressing the human drivers of climate change is an environmental, economic and social challenge. The common factor which links these interconnected aspects is energy supply and use in a world which is increasingly industrializing, urbanizing and growing in population. And, as if this challenge were not big enough, we have the additional dimension of urgency. Climate change is happening now and the longer we leave concerted action at scale, the greater the risk that we will not keep atmospheric greenhouse gas concentrations within 'safe' limits – though safe for whom and for what is a moot point.

The challenge for developing countries wanting to start their low-carbon economic development journey has three distinct elements:

1. How to obtain and deliver the energy needed to meet economic and social development aspirations, having proper regard to the huge 'catching-up' exercise in which the emerging economies are engaged, without adding significantly to global greenhouse gas emissions? What clean energy supply and energy-efficient technologies exist ready for adaptation and deployment in, and for, developing countries? At what cost can these technologies be adapted and deployed over time as part of economic and social provision – energy for industry, and energy to improve the quality of life for hundreds of millions of people? What new technologies and techniques need to be developed and who should be shouldering the costs of developing them? Under what terms should they be made available to developing countries? How can developing countries acquire the use of these technologies and thereby avoid

becoming 'locked' into currently economically cheaper but environmentally more expensive fossil-fuel technologies and infrastructures?

2. How to develop the essential economic and societal building blocks, such as trust and governance, financial mechanisms and hard infrastructures (e.g. pipes, wires and roads; or decentralized energy systems), and soft infrastructures (e.g. education and training, professional networks, contract law) and the sustained sense of common purpose over time from governments to governments? These are the essential prerequisites to moving to sustainable, equitable, growing low-carbon economies which will deliver what people need (and, in due course, want).

3. How to balance the urgency of now with the importance of tomorrow? Most developing countries (and some might say, not just developing countries) are struggling with this challenge when they think about political priorities and allocation of scarce resources. Perhaps beyond all other factors this is the one that shapes the debate and the allocation of time, political capital, resources and professional capabilities. It is also the one which carries the greatest risk, in my view, of blowing countries off course and on to high-carbon economic pathways. What is cheaper and 'easier' today is likely to be more expensive and more difficult to unravel tomorrow. Finding the right balance between the urgency of 'now' and the importance of 'tomorrow' is a challenge we all face.

Technology innovation

Technology innovation means different things to different people. For example, consider developed countries. If there is an aspiration to develop an indigenous low-carbon sector (e.g. Germany's drive to build a solar photovoltaic industry or Denmark's determination to become a world player in wind turbine technology), technology innovation is synonymous with strong research, development and demonstration (RD&D) capabilities; and sustained policies and measures to create attractive market conditions to invest in the development, manufacture and deployment of these technologies. For a developing country, I think technology innovation means something different, certainly from the perspective of needs, capabilities and aspirations. Not only are strong RD&D capabilities and policies and measures in short supply, but it is also worth considering whether R&D is really necessary. Many developing countries are more interested in adapting (as appropriate) and deploying existing technologies because this offers the fastest route to increasing energy supply capacity. For them, R&D is not the immediate priority. Over time, perhaps, aspirations may change and some countries may wish to become technology providers. If and when that comes to pass, the need for R&D capability can be reconsidered, and if it is needed, whether it is best provided by the developing country directly or via a partnership with developed countries. An innovation centre would be in a good position to facilitate, inform and specify the R&D.

For many developing countries, innovation is more to do with pilot scale trials, demonstrations and building the skills base. Trials and demonstrations would be best set up in the developing country concerned. My reason for saying this is because technology and system operators (e.g. an energy utility) need to find out in what way technology originally developed in, and for, developed countries needs to be adapted in order to work well in the home country concerned. Only by testing technology out in real conditions will the operators know how to make technology work properly and how much it will cost to operate and maintain. Knowing the operating cost base of technologies in the energy supply fleet is vital to the commercial viability of the utility operator.

Technology innovation is only part of the story

The above consideration makes it clear, I suggest, that 'technology innovation' is necessary but not sufficient to make the transition to a low-carbon developing economy. A set of policies and measures, appropriate to the starting point for each country to encourage low-carbon development and discourage investment in high-carbon assets, a sense of common purpose, and education and training all have important parts to play alongside technology innovation. When we are thinking about how to work with developing countries, we should consider in addition to fitness for purpose of particular technologies: the need for hard and soft infrastructures; the policy and cultural landscapes; the existing institutional and delivery capacities; and the will to deliver change.

I think we should not lose sight of the fact that the way in which we describe 'technology' and 'innovation' draws heavily on their roots in developed societies. They presume certain starting points: a society with adequate numbers of suitably trained and experienced people – managers, supervisors and skilled workers; traditions of good governance, organization and trust; public–private cooperation, proven organizational business models, intellectual property protection, licensing, and finance to develop and deliver technologies to meet identified needs. Even in developed countries, we are finding it difficult to finance the timely innovation and commercialization of key clean energy technologies such as carbon capture and storage, offshore wind and marine technologies. So how can we expect developing countries to fare? Capital and credit are often in short supply, as are professional skills (engineering, project management, systems maintenance, education and training in the management of new technologies). Good governance and political will cannot always be taken for granted and the 'soft infrastructure' referred to above might be fragmented and vulnerable. In my view, how to enable access to low-carbon technologies by developing countries is not just a question of technology development and deployment. It is also a much bigger multi-faceted question about social equity, governance, education, training and consumer education.

Moving to a low-carbon economy is a huge challenge for any economy, developed or developing. It's the nature of the challenges that are different. It requires the necessary institutions and agencies to map out a strategy, design an effective delivery plan, implement it and track progress in a reliable and transparent way. Where these are weak or lacking, what are the options to help? What are the respective roles of governments, business, civil society and the international community and institutions? These are questions that I hope this chapter and this book, more generally, will stimulate us to think about.

Accelerating low-carbon growth in a developing world: Is a new approach necessary?

For the reasons outlined above, I don't think a 'one size fits all' solution is the best way to approach the low-carbon challenges that developing countries face. Each developing country is starting from a different place, with different resources and with different ideas about the best pathway to reach its goals.

Some developing countries have some of the institutional and delivery jigsaw pieces in place – or at least on the table. But many do not. If countries have long-term political commitments to the vision of low-carbon economic development, and to its timely delivery, it may be possible for them to harness what they have got, acquire the missing pieces through international cooperation and start their journey. International knowledge sharing and learning is important here, including that facilitated via international bodies showcasing what countries are doing to put their low-carbon economy aspirations into practice. The United Nations Environment Programme (UNEP)-inspired Sustainable Energy Finance Alliance is one example of an initiative which has the potential to provide impartial information on public finance instruments which have been designed to support and accelerate low-carbon technology innovation.[12]

Some countries may have little in the way of institutional infrastructure. To promote low carbon up the list of priorities will require massive injections of intelligent finance, appropriate know-how, good governance and skilled people. There is a case to consider new models for engagement – models which are designed and resourced to facilitate and build confidence between governments, business, and sources of finance, academia and entrepreneurs to identify and tackle the barriers to achieving low-carbon economic development goals, whether they are technological, institutional, financial or societal. The respective roles of governments, business, finance and civil society are crucial to consider as integral parts of the framework in which a low-carbon technology innovation centre model would operate.

A low-carbon innovation centre network model

The Carbon Trust published a study in 2008 which outlined a low-carbon technology innovation centre concept: *Low Carbon Technology Innovation and Diffusion Centres: Accelerating Low Carbon Growth in a Developing World*.[13] Largely based on the Carbon Trust's own experiences of low-carbon technology innovation in the UK, it has been widely shared with countries, agencies and individuals around the world. What is significant about this technology innovation model is not that it offers a template for developing countries to replicate, but that it stimulates thought on preparing the guiding principles whereby developing countries can design, build and operate their own innovation centres to suit their own starting points, resources, institutions and circumstances. Only in this way can developing countries have ownership and sovereignty over their 'entity of change'. I have drawn on my own experience to propose a set of guiding principles which I hope will help inform the debate:

1. The first step to helping a developing country design its innovation centre is to consider if it needs an innovation centre at all. Map the landscape of existing priorities, policies, institutions and resources; determine where these align well with the low-carbon economy objective, where they do not and where there are gaps. For example, alleviating poverty aligns well with the provision of local low carbon, clean energy supplies and highly energy-efficient products such as light-emitting diode (LED) lamps. Building coal-fired power stations to fuel high-carbon economic growth may help economic development in the short term, but in the long run will serve to put it at risk.
2. If an innovation centre 'gap' is identified, determine the scope of activities, based on the mapping exercise, needed to fill the gap and achieve low-carbon economic development objectives For example, in which part of the low-carbon technology value chain does a particular developing country wish to be active – and therefore where are its priorities for development? Does it want to develop its own clean energy technology sectors or buy technologies or components and then create systems to suit?
3. Give the centre a 'licence to operate' which permits: coordination and action across traditional boundaries, and the right to learn by doing (i.e. work at the edge of the comfort zone of the stakeholders by sharing costs and risks in order to advance learning and understanding). The Carbon Trust's research consortium to explore and develop novel organic and dye-sensitized photovoltaic materials is one example of an initiative designed to share the costs and risks of technology development.
4. Ensure political commitment to make available sufficient funds over a sufficient timeframe to be material to the challenge. The transition to a low-carbon global economy will take several decades and cost hundreds of billions of dollars. A given innovation centre may be required for 10 to 20 years but would change in the light of developing needs and the environment in which it is operating.
5. Develop and encourage good governance practice to earn trust and confidence with stakeholders. By this I mean making sure that the transformation from a high- to a low-carbon economy takes place in a climate where stakeholders can trust each other. For example, if a company says that it has met its carbon emissions reduction target, then stakeholders should be entitled to believe the company because the system of company carbon reporting is independently verified. Or if consumers see a carbon footprint label on a product, they can believe what it is saying about the carbon footprint of that product because the label has been independently certified – as is the case with the Carbon Trust's

product carbon label. How this is achieved will depend upon the custom and practice of the country in question. In the UK, for example, good governance is part of the way we do business. Companies have boards and their own governance, decision-making and accounting and reporting procedures. Only with good governance in place will the entity be trusted to act independently and make its own decisions about the allocation of resources to meet agreed objectives.

6. Recruit a leadership team with transformational experience and vision. Support them with a strong board and good, knowledgeable staff who have wide technology, business and innovation experience, and are committed to making a significant contribution to the low-carbon economy objective.

7. Establish an independent process for evaluating the impact of the innovation centre.

8. Embed in the innovation centre the capacity to codify and disseminate learning and experience with other similar centres by an international network facility. The network aspect of the innovation centre model is, in my view, essential. Not only will it help to avoid reinventing wheels, but it will also generate confidence and courage to try new things and new approaches – in other words, real innovation and sharing of good practice. Independent funding should be made available for networking both virtually and through face-to-face discussion and knowledge sharing.

Innovation centres: Scope of interest beyond 'technology'

Whereas most ideas about innovation focus on bringing in new technologies or business models, climate innovation centres need to go beyond the technologies and models and consider the wider impacts of low-carbon technology innovation on society. The impact upon society of new technology and ways of working or organization can be significant – even life changing. It is better, therefore, to think about the wider societal scope of interest of an innovation centre in parallel with 'technology' in order to be ready for the societal changes which innovation could bring about. In Western society, for example, we have seen a revolution in communications technology which is having profound societal impacts via the explosion of social networks. In rural and farming communities, the introduction of mobile phones is helping farmers to find the best prices for their produce. If innovation centres accelerate the introduction of clean energy supplies, thereby increasing the total capacity of energy supplies for a given country, we are going to see a large proportion of that energy used for manufacturing. That is the way industrial revolutions have happened before, and unless there is a counter-force to the economic driver to find and build the cheapest manufacturing model, based on energy, we could see labour for manufacturing progressively displaced by mechanization over time (although notable exceptions to this trend have been observed – see, for example, the Indian initiative run by TERI to light up 1 billion lives).[14] How is the displaced labour to find a place in the new economy? I think via education, training and the discovered need for new goods and services not previously existing or imagined. For example, new energy technologies will need managing and maintaining and the new opportunities created by more energy will need more, and differently skilled, labour.

My point here is that in the same way that innovation centres seek to bring forward new technologies and business models, so should they also look to help societies make the transition to a world of clean energy supplies and efficient use of energy. If we don't consider the impacts of innovation, we will more likely than not find that there have been societal consequences which then need managing. I recall a case related to me many years ago by a consultant tasked to assess the impact of a selection of welfare projects in developing countries. One was a project to build a village well with the aim of saving women the onerous task of walking several miles each way to a source of drinking water. Before the well was constructed, the men of the village worked the land. What happened, apparently, was that the well in the village released the women and teenage girls from the arduous water collection task (which, in fact, was team driven and social – they chatted and sang to make the journey go by faster). Instead, they were put to work in the fields (back-breaking, less eye contact and more individual), thereby releasing the men from their toils in the fields to spend more time doing much less arduous work and, so the story goes, enjoy increased leisure. The women were, quite rightly, indignant – and not a little irritated. This was not quite the end result for which the project originators were aiming. It is, however, an example of how well-intentioned and valuable changes can have unexpected consequences (please understand, I am not disputing for one moment the value of making nearby water supplies available).

Role of external support: Finance

External financial support to help developing countries design, build and operate their innovation centres will be essential. But it has to be the right kind of financial support which acknowledges that:

- The aim of the support package should be to enable developing countries to take over responsibility for their innovation centres over time.
- No two developing countries will have the same needs – one 'size' of external support will not fit all. The particular needs have to be identified in order to determine what kind of innovation centre model would be appropriate.
- Money alone will not achieve low-carbon goals. Money needs to be part of a package which also includes experience, knowledge and skills appropriate to the needs of the particular developing country.

At present, there are many sources of external funding – for example, the United Nations Development Programme (UNDP)/the United Nations Industrial Development Organization (UNIDO), the World Bank, the Inter-American Bank, the European Bank for Reconstruction and Development, the European Commission, and so on – all with their own rules, procedures, criteria and scopes of interest. These, and the other funding sources, support all kinds of activity – for instance, from small, one-off energy efficiency in industry projects in the region of US$500,000 to US$5 million, to larger capital projects such as hydro-electric power in the region of hundreds of millions of dollars. In December 2009, the United Nations Framework Convention on Climate Change (UNFCCC) Conference of the Parties – COP15 – at Copenhagen produced an accord which, among other things, said: 'Developing Countries, especially those with low emitting economies, should be provided incentives to continue to develop on a low emissions pathway.'[15] At Copenhagen, developed countries entered into a collective commitment to provide US$30 billion by 2012 and US$100 billion by 2020 to address the needs of developing countries. The accord went on to say that 'in order to enhance action on development and transfer of technology, it decided to establish a Technology Mechanism'. This progressed further at COP16 in Cancun (December 2010) to include the idea of a Climate Technology Centre and Network (CTCN) across developing countries.

A number of governments and agencies, such as the UK Department for International Development and infoDev (Information for Development programme), a special programme of the World Bank, are exploring which engagement models and approaches would be most effective. So now we have another source of funding emerging. To add to this complex landscape, a new US$100 million initiative was launched by the World Bank in June 2011 under its Market Readiness Initiative. Initial grants were given to Chile, China, Colombia, Costa Rica, Indonesia, Mexico, Thailand and Turkey to help design, plan and implement market-based instruments to limit greenhouse gas emissions. The initiative eventually aims to support around 15 countries to construct market-based schemes in their countries, such as emission trading schemes or international crediting mechanisms.

To get the most out of these and other sources of funding for developing countries, I think there needs to be a simplification of this complex funding landscape. At present, the nature and scale of the low-carbon challenge, and the need to take a holistic view of that challenge, do not map well on to the patchwork of funding sources. There is insufficient coordination of funding sources and insufficient framing of the low-carbon transition challenge (of which technology innovation is a part). I think the lack of strategic coordination is a significant defect in the compartmentalized funding landscape. For example, some funding agencies are geared up to lend substantial funds for the purchase of capital equipment but don't have funds to help the host country develop a delivery strategy and vehicle to maximize the impact of funds disbursed and, importantly in my view, work out a means whereby the initial initiative can be made self-sustaining without external financial support. What I think is needed is strategic coordination between funding agencies and with respective host governments.

The framing of the low-carbon transition challenge and the mapping analysis would reveal the nature and scale of the challenge, where funds are needed, how much and for how long, and how their use could be made more effective through the development of the policy and market framework. Presenting the low-carbon challenge as a holistic endeavour over time during which different scales of funding would be required for different activities is one half of the equation. On the other side, I see the need for an initiative by the funding agencies to explore how to work together (and with host governments) to create intelligent budgets to be used sequentially to fund progressive parts of the low-carbon transition plan (including the preparation of the plan itself).

However, it is not just a matter of money. Money alone is insufficient. What are also needed are skills, knowledge and experience. These are hard won and in short supply. A review of external international financial mechanisms to see how they can be made 'fit for purpose' along the lines of the above would be worthwhile. This kind of review could play a role in delivering an effective Technology Mechanism as part of the UNFCCC process.[16]

Role of external support: Business

An exciting, relatively untapped (some might say unlikely, but I think that would be unfair) source of help are the enlightened elements of the international business community – companies such as Marks & Spencer, Astra Zeneca, Vodafone and Unilever. The world is beginning to change so far as the roles of business are concerned. We are seeing the beginnings of a new, enlightened phenomenon known as shared economic and societal value creation described by, for example, Michael Porter and Mark Kramer.[17] Companies who reshape their business model to deliver both corporate economic value and community societal value are going to be more successful, they argue. Tackling climate change and moving to a low-carbon global economy must surely be the biggest and most challenging example of the need for corporate economic and community societal value creation that we face today. The opportunities for business to achieve both economic and societal value in developing countries are immense and deserve serious consideration.

Closing remarks

Making the move to a low-carbon economy is not simply a matter of low-carbon technology innovation and/or transfer. Many developing countries are already 'low carbon' but for reasons of low levels of economic growth and slow progress in human and economic development. If the international community is to succeed in reducing global greenhouse gas emissions and to achieve sustainable economic development, there has to be more cooperation between developed and developing nations and between governments and businesses. The international community, I suggest, should look at 'low carbon' as part of a move to a world where sustainability and prosperity walk hand in hand, respectful of the specific starting points and resources which characterize each developing country. Most developing countries simply do not have the resources to start their low carbon, sustainable economic journey alone. Some are committed to taking such action as they can, but others (as with many developed countries) lack political will or are unconvinced of the necessity for such a journey, or both. Some do not think it is, or should be, their problem. Some lack the capacity to develop and/or deploy low-carbon technologies appropriate to their circumstances; and some are already (as are many developed countries) locked into high-carbon technologies – energy supplies, in particular.

As part of an integrated approach, the Climate Innovation Centre concept has great potential; but it will need to be tailored to meet individual country needs. However, it is not only the innovation centre concept that will need tailoring. The external funding mechanisms which are currently available require overhaul so that they can more easily provide effective support in the wider sense – intelligent money strategically delivered, in other words. In addition, those elements of the international forward-thinking business community have the potential to lead a whole new approach to bear on the climate change challenge facing developing countries.

In conclusion, I invite you to read and reflect on the following summary points:

- Developing countries have a number of economic and social development priorities which may or may not include low-carbon economic development.
- Political will to underpin a low-carbon vision, shared amongst key stakeholders (other political parties, business, civil society), is the first essential ingredient of low-carbon development. Unless a country or jurisdiction wants to achieve this goal, and commits to it over time (i.e. decades) across the political parties and stakeholders, no amount of funding and assistance, innovation centres or whatever will be effective.
- Given the political will to assign a high priority to low-carbon economic development, a good starting point to help design an appropriate innovation centre (or develop an existing organization) would be a mapping exercise of the policy, delivery and market landscape to identify strengths and weaknesses in relation to the low-carbon economic development goal.
- 'Innovation' will have different meanings, content and goals for different countries at different stages of development.
- 'Technology' is important, but it is only one aspect of building a low-carbon developing economy. Having a supportive and stable policy and market framework over time; developing the right business and financial models; having education and training schemes in place; and developing the skills to encourage low-carbon, sustainable consumer behaviours are also important.
- Over time, and with the knowledge and understanding of the host country/ jurisdiction, responsibility for low-carbon technology innovation should be transferred from dependency upon externally sourced aid and skills, through co-funding and co-sourcing, and ultimately into fully funded, locally managed and staffed climate innovation centres. The time this could take should not be underestimated.
- Finally, it should be borne in mind that for many developing countries, the urgent challenges of today can outweigh the challenges of tomorrow. Finding the right balance between the urgency of 'now' and the importance of 'tomorrow' is a challenge faced by developed and developing countries alike.

Notes

1 See http://www.carbontrust.co.uk/.
2 Carbon Trust (2008) *Low Carbon Technology Innovation and Diffusion Centres: Accelerating Low Carbon Growth in a Developing World*.
3 The Millennium Development Goals: http://www.un.org/millenniumgoals/.
4 Carbon Trust (2011) *International Carbon Flows*, http://www.carbontrust.co.uk/policy-legislation/international-carbon-flows/pages/default.aspx.
5 IPCC (Intergovernmental Panel on Climate Change) (2007) *Fourth Assessment Report*, http://ipcc.ch/publications_and_data/ar4/syr/en/contents.html.
6 New Economics Foundation (2004) *Up in Smoke? Threats from, and Responses to, the Impact of Global Warming on Human Development*.
7 Interview with Andrew Witty, chief executive officer of GlaxoSmithKline on BBC Radio 4 Today programme, 6 June 2011, http://www.gsk.com/.
8 John Parker, the ABI's head of general insurance; ABI News Release, 8 June 2004.
9 Interview with Andrew Witty, chief executive officer of GlaxoSmithKline on BBC Radio 4 Today programme, 6 June 2011, http://www.gsk.com/.
10 *The Economics of Climate Change* by Lord Stern and his team published by HM Treasury in the UK in October 2006.
11 Tim Jackson (2009) *Prosperity without Growth*, University of Surrey, UK.
12 UNEP Sustainable Energy Finance Alliance. For information on this international initiative, visit www.sefalliance.org.
13 Carbon Trust (2008) *Low Carbon Technology Innovation and Diffusion Centres: Accelerating Low Carbon Growth in a Developing World*.
14 The Lighting up a Billion Lives initiative: http://labl.teriin.org/support/.
15 UNFCCC Copenhagen Accord, 2010.
16 UNFCCC Copenhagen Accord, 2010.
17 Michael Porter and Mark Kramer (2011) 'Creating shared value', *Harvard Business Review*, January–February.

17

TECHNOLOGY TRANSFER AND GLOBAL MARKETS

Jon C. Lovett, Peter S. Hofman, Karlijn Morsink and Joy Clancy

Introduction

A key element of climate change mitigation and adaptation is the transfer of more effective and efficient low-carbon technologies between developed and developing countries. Although several policy mechanisms for technology transfer are in place, most observers agree that these have not been very effective in accelerating the rate of diffusion of energy-efficient and renewable-based technologies. There is a need for market-oriented approaches in order to diffuse efficient technologies more rapidly and to reduce high transaction costs, which are a major factor explaining the low effectiveness of existing mechanisms (Michaelowa and Jotzo, 2005; Jung, 2006; Hofman et al., 2008; Lovett et al., 2009; Byigero et al., 2010; Timilsina et al., 2010).

At the 2007 G8 Summit in Heiligendamm, it was recognized that an 'expanded approach to collaboratively accelerate the widespread adoption of clean energy and climate friendly technology' was needed (G8, 2007). In successive outputs from the United Nations Framework Convention on Climate Change (UNFCCC) negotiations, such as the 2007 *Bali Action Plan*, the 2008 Poznan Strategic Programme on Technology Transfer, the 2009 Copenhagen Accord and the 2010 Cancun Technology Mechanism, the requirement for scaling up technology transfer features prominently. The problem is that project-based funding mechanisms, such as those under the Global Environment Facility (GEF) and Clean Development Mechanism (CDM), can never do more than provide a fraction of the resources needed to transfer sufficient environmentally sound technologies to permit economic advancement of developing countries while minimizing greenhouse gas emissions; for example, at the Poznan meeting, the G77 and China proposed funding in the range of 1 per cent of gross domestic product (GDP) from developed countries (Lovett et al., 2009). The often repeated call is for greater access to technologies through open intellectual property rights (IPRs) and more financial support for technology transfer. In practice, the only viable answer is to meaningfully engage the private sector and associated global markets that transfer technology. The challenge then becomes how to put in place the appropriate institutions and regulatory environment in order to gear markets towards rapid delivery of a range of technologies that have proven to be efficient and affordable for several uses but yet have limited market shares.

A further key element of the transfer process is the build-up of capacity in the recipient countries, such as the knowledge, skills and organization necessary for effective implementation of clean technologies and, ultimately, the emergence of domestic production. Although enhanced environmental regulation, such as that envisaged under the UNFCCC agreements, can promote innovation and competitiveness in line with the 'Porter Hypothesis' (Porter and van der Linde, 1995), and to some extent financing can be provided through UNFCCC agreements; the transfer and uptake of the technology requires something more. The cases outlined here are from a range of sectors, including energy-efficient lighting, solar panels, energy-efficient cement production and high-efficiency electric motors. These cases are chosen because the uses they represent make up a significant share of global energy production and because some of these technologies, such as efficient motors, have as yet limited penetration. Effective technology transfer mechanisms need to take into account the specific nature of the selected technologies, the various forms of distribution and delivery relevant for their diffusion, and the local context for successful adoption and implementation of these technologies by users. This chapter reviews existing mechanisms and develops ideas for more effective technology transfer mechanisms for these selected technologies. Key elements of such mechanisms need to take into account effective global access to environmentally sound and energy-efficient technologies, while also ensuring that appropriate technological capabilities are developed at the local level. In conclusion we demonstrate that neither access to technologies nor financing are the real limiting factors, but rather it is creation of the appropriate enabling environment to allow technology markets to work. We suggest multi-stakeholder partnerships as one possible way forward (Morsink et al., 2011).

Technology transfer mechanisms

The role of technology transfer (TT) has been part of the UNFCCC and its negotiations since its creation in 1992 and it continues to play a central role. For developing countries, in particular, the transfer of environmentally sound technologies (ESTs) from 'North to South' has been an important component for their commitment to climate agreements in recognition of the principle that mitigation efforts do not impair economic growth. In other words, the view is that deployment of new clean technology will enable countries to decouple greenhouse gas emissions (GHG) from development. There are some clear barriers to EST transfer: they are usually more expensive than conventional technologies and many developing countries do not have the installed manufacturing capacity for ESTs. More hidden are the lack of regulatory frameworks that would encourage private-sector engagement with developing country markets to promote ESTs and the incentives in place for retaining inefficient or high greenhouse gas-emitting technologies.

As noted above, a number of financial mechanisms are in place to encourage EST transfer and to help overcome the higher cost of clean technology, most notably the CDM and GEF. However, it is generally agreed that much more technology transfer is needed for climate change mitigation and adaptation. At COP13 in Bali (December 2007), it was decided 'to elaborate a strategic programme to scale up the level of investment for technology transfer to help developing countries address their needs for environmentally sound technologies, specifically considering how such a strategic programme might be implemented along with its relationship to existing and emerging activities and initiatives regarding technology transfer and to report on its findings to the twenty-eighth session of the Subsidiary Body for Implementation for consideration by Parties (Decision 4/CP.13)' (UNFCCC, 2008a). Developing countries take the position that any commitment to specific GHG reduction goals can only happen if accompanied by very significant expansion of technology transfer and support facilities. Consequently, in September 2008 China and the G77 put forward a proposal for a technology mechanism to accelerate the 'development, deployment, adoption, diffusion and transfer of environmentally sound technologies among all Parties, particularly from Annex II parties to non-annex I Parties, in order to avoid the lock-in effects of non-environmentally sound technologies on developing country Parties, and to promote their shift to sustainable development paths' (UNFCCC, 2008b). At the 2008 UNFCCC COP14 in Poznan, one of the few major decisions made was the Poznan Strategic Programme on Technology Transfer, which recognized limitations in the Global Environment Facility approach and the need for major private-sector involvement to cover the shortfall in funding needed (UNFCCC, 2008c; Lovett et al., 2009). Development of a Technology Mechanism was initiated at the 2009 COP15 in Article 11 of the Copenhagen Accord: 'In order to enhance action on development and transfer of technology we decide to establish a Technology Mechanism to accelerate technology development and transfer in support of action on adaptation and mitigation that will be guided by a country-driven approach and be based on national circumstances and priorities' (UNFCCC, 2009). IPRs continued to be a stumbling block, overcome to some extent by the proposed formation of a network of 'Climate Innovation Centres' to 'develop and deploy appropriate technologies to mitigate and adapt to climate change' (Sagar, 2010). The Technology Mechanism was agreed at COP16 in Cancun in 2010, to become operational in 2012 (UNFCCC, 2010). However, developing effective clean technology transfer is not straightforward. Existing mechanisms are often said to have high transaction costs and lack effectiveness as they are unable to mobilize the investment potential of the private sector and widespread adoption beyond the initial projects selected for support (Egenhofer et al., 2007; Forsyth, 2007). For example, the necessary institutional, technical and economic capability may be lacking for CDM projects in Africa (Jung, 2006; Timilsina et al., 2010), and an absence of the necessary regulatory frameworks can further prevent private-sector involvement (Michaelowa and Jotzo, 2005). Indeed, sub-Saharan Africa is perceived as a high risk for foreign direct investment (FDI) (linked, for example, to poor energy infrastructure, political instability and corruption) which is considered to have an influence on CDM investment (Byigero et al., 2010). Moreover, in most of Africa the relative lack of industrialization means that greenhouse gas baselines are low, so there is limited opportunity to mitigate emissions through CDM projects. On the other hand, any acquisition to the most modern technology enables technological leapfrogging.

The elements of more effective technology transfer lie in creating in-country capacity to manufacture and market the EST products. The intergovernmental Panel on Climate Change (IPCC) defines technology transfer as 'a broad set of processes covering the flows of know-how, experience and equipment for mitigating and adapting to climate change amongst different stakeholders such as governments, private sector entities, financial institutions, NGOs and research/education institutions. It comprises the process of learning to understand, utilize and replicate the technology, including the capacity to choose and adapt to local conditions and integrate it with indigenous technologies' (IPCC, 2001, p101). This definition has also been adopted by the GEF (GEF, 2010). Transfer can take many forms, but one of the most dominant forms has been through FDI. The importance of FDI in the successful development strategies of Asian newly industrializing countries is often stressed. Key elements of this success were the strategy and ability to imitate and replicate technology indigenously and the parallel development of local skills and knowledge. Others forms are transfer through the provision of products incorporating the technology (e.g. energy-efficient lighting or photovoltaic panels for off-grid electrical supply, or licensing the capability to produce such products, perhaps to an indigenous firm, through co-development of domestic and foreign firms or through a joint venture) (Barton, 2007). A further form is the support of national capability to research and produce the products independent of a foreign company. A final form is technology transfer through official development assistance (ODA). Especially for low-income developing countries, this is the dominant form of technology transfer, while technology transfer to low-income developing countries through FDI is rather limited.

Experiences with technology transfer have led to understanding the concept of technology transfer as a process that includes a flow of knowledge, as well as goods and which has to be paralleled by processes for learning and capacity-building in developing countries. It is therefore necessary to see technology transfer as part of a broader process of sustained low-carbon technology capacity development in recipient countries (Ockwell et al., 2008). Others have in a similar vein stressed the importance of adequate absorptive capacity and technological capabilities in recipient countries (Mytelka, 2007). Key constraints for effective technology transfer of low-carbon technologies to developing countries are therefore often related to the lack of transfer of capital goods, equipment and knowledge in combination with the lack of an appropriate host environment for technology transfer.

In the next section we present four case studies to illustrate the diversity of enabling environments, incentives and barriers for the transfer of ESTs. In some cases, such as energy-efficient lighting and solar photovoltaics (PV), new regulations, the availability of funding and other incentives such as tax concessions are promoting their spread. In other cases, such as cement manufacture and efficient motors, the diffusion of ESTs is somewhat slower.

Four case studies

We have selected four technologies – lighting, solar panels, cement production and electric motors – for further analysis of technology transfer through global technology markets. Widespread uptake of these technologies would lead to significant reduction in global carbon emissions as the end-uses and industrial uses they represent take up a significant share of global energy and electricity use. Each section briefly reviews the current situation and efforts being made to encourage transfer of the technology.

Lighting

Lighting is an important element in the daily life of the majority of the world population. It is also a major contributor to the climate change problem as it represents around 19 per cent of the world's electricity consumption and emits 1900 million tonnes of CO_2 on an annual basis, which is equivalent to around 8 per cent of world emissions (IEA, 2006, p25). During the past few decades, energy-efficient lighting has become available, with energy savings of between 70 and 80 per cent relative to incandescent lamps, most notably by compact fluorescent lighting (CFL), which is now commercially available, while light-emitting diode (LED) technologies are expected to deliver light even more efficiently than CFL.

The global market in lighting has long been dominated by three leading multinational lamp manufacturers: Philips (based in The Netherlands), Osram (Germany) and General Electric (US). These three companies have a presence in almost all global markets and have a significant share in global trade in lighting products, facilitated by a high degree of standardization between international lighting markets (IEA, 2006, p251). Although the 'Big Three' all have manufacturing facilities in China, they account for only a small proportion of the large Chinese market (IEA, 2006, p251).

In terms of the technology involved, energy-efficient lighting is more than substituting traditional light bulbs for CFLs. 'Lighting energy can be saved in many ways, including (i) improving the efficiency of the light source; (ii) improving the efficiency of the specific component of lighting system, typically the ballast;

(iii) improving the efficiency of the luminaries; (iv) improving the efficiency of the control gear deployed; and (v) making better use of daylight inside built environment' (Figueres and Bosi, 2006, p2). In terms of the process of technology transfer, this implies that effective TT involves also the build-up of knowledge and skills to facilitate the appropriate use and implementation of energy-efficient lighting within the specific local user contexts.

A number of developing countries have implemented relatively successful energy-efficient lighting programmes. Major CFL substitution programmes have been implemented in Brazil, Mexico, Peru, South Africa, Guadeloupe and Martinique. China has implemented an ambitious Green Lights Programme and has become the world's largest CFL market. The Green Lights Programme was originally initiated by the Chinese government in 1996; yet its successor (2001 to 2005), in which UNEP and the GEF have an active role as supervisors and funding partners, was significantly more far reaching and ambitious. The project had as its main objective to 'reduce lighting energy use in China in 2010 by 10 per cent relative to a constant efficiency scenario' and as a secondary goal 'to increase exports of efficient quality lighting products, aiding the Chinese economy and helping to reduce energy use and GHG emissions worldwide' (UNDP, 2000, in Lefevre et al., 2006). However, creation and operation of successful efficient lighting initiatives is more dependent upon organization and political priority than direct economic advantages (IEA, 2006). Key factors for successful programmes have included the following: the price differential of CFLs compared to incandescent lamps has been minimized by direct subsidy or soft-financing; there has been a proactive promotional campaign; the quality of CFLs has been ensured; and there has been pressure on the energy system, such as a power crisis.

Especially with regard to the quality of CFL, there is a need to ensure that good quality is guaranteed. According to a recent study 'analysis shows that one out of two compact fluorescent light-bulbs (CFLs) available in many areas of the world is of shoddy quality. Unless this issue is addressed in the near term, we will fall far short of energy saving goals, turning consumers against CFLs in the process' (USAID, 2007). A key component in China's relatively successful CFL programme has been the set-up of a national standardization organization that is responsible for the quality of the CFL products and for which all producers are obliged to test their products against a number of minimum quality standards developed by the organization. An innovative approach to improving product quality is the Lighting Africa Quality Assurance Product Awards Ceremony, where businesses are recognized for their efforts to improve quality (Lighting Africa, 2011).

The Global Environment Facility has played a major role in the efforts to support a global phase out of incandescent light bulbs and develop a lighting market transformation strategy benefiting all economies, including the developing world. This initiative was triggered by an increasing number of countries announcing their intention to phase out incandescent lighting. One of the first countries was Cuba, which banned the sale of incandescent lighting in 2005 and started a programme of replacing traditional light bulbs with CFL, a process said to be finalized in 2008, making Cuba effectively the first country where incandescent lighting is phased out. Since early 2007, almost all Organisation for Economic Co-operation and Development (OECD) governments began to develop policies aimed at phasing out inefficient incandescent lighting. Australia was among the first to pronounce a time schedule and regulations for phasing out incandescent lighting by 2012; other OECD countries (the European Union, the US and Japan) followed with similar time paths. Industries have also expressed their willingness and readiness to phase out incandescent lighting, Philips being the first in December 2006, followed by other firms. The EU has legislative processes for phase-out, while various EU countries developed national measures for phasing out incandescent lighting ahead of the EU time schedule. For OECD countries, the process of phasing out has been set in motion and is likely to be finalized around 2012. A number of developing countries have already started a process of phasing out, China being the most well known. Another example is Ghana, where a policy was introduced by the government to ban imports of incandescent light bulbs and other high-energy consumption lamps. In Ghana the shift to low-energy lighting is driven by a power crisis. The government is providing 6 million CFL bulbs free of charge to replace incandescent bulbs, saving 430GWh of electricity a year and reducing peak demand by 124MW (Energy Foundation Ghana, 2011). This has the combined effect of stabilizing grid supply and avoiding the installation of additional generating capacity.

In addition to CFL, light-emitting diodes (LEDs) and electron-stimulated luminescence (ESL) lamps are alternative lighting ESTs. Technical advances in LEDs have resulted in a doubling of efficiency and light output every three years since their introduction in the 1960s, giving rise to a wide range of applications; and they are increasingly being considered as a lighting solution in developing countries (Pode, 2010). In their assessment of LED technology transfer to India, Ockwell et al., (2006) found four key barriers to TT:

1. Financial: manufacturing of LED chips is capital intensive and requires large investments beyond the scale of the relatively small Indian manufacturers.
2. IPRs: LED technology is highly protected with patents and Indian companies have been unable to obtain licences, choosing instead to import LED chips.
3. Market barriers: large lighting and LED manufacturers have not invested in India due to the small domestic market, and no joint ventures are established.
4. Human capital: although India has highly skilled engineers, expertise (and academic education) in LED technology is scarce (Ockwell et al., 2006).

Solar panels

A crucial component in mitigating climate change is a transition to a more renewable-based electricity system. Solar energy, captured through photovoltaic panels and/or thin-film solar technologies is expected to be a key component of such a transition. While the installed base of solar PV is predominantly in the industrialized world, particularly Germany and Japan, solar panels form a clean and renewable source of energy that can contribute to improving energy access and health conditions in low-income areas of developing countries. For example, solar PV is widely used in Kenya, particularly by the rural middle class of small business owners, school teachers, civil servants and cash-crop farmers who use it to power televisions, radios, mobile phones and to help with children's education and evening work, such as marking and accounting (Jacobson, 2007). In its June 2011 budget, the Kenyan government granted duty remission on raw materials for the production of solar panels in order to encourage local production and to help meet demand driven by policies to increase household use of solar PV and water heating, thereby reducing grid load. The International Finance Corporation and World Bank are also promoting commercial off-grid markets, including solar PV and other technologies, in sub-Saharan Africa through the Lighting Africa programme, with the aim of providing off-grid lighting to 2.5 million people by 2012 and 250 million people by 2030 (Lighting Africa, 2011).

Production of solar panels is concentrated in a limited number of countries and dominated by a select number of multinational companies. The production of PV panels is expensive and requires large-scale precision manufacturing capability. It is a moderately concentrated industry; the four leading firms produce about 45 per cent of the market (Barton, 2007). From a value chain perspective, the number of companies involved becomes lower when travelling higher up the PV value chain. The upper level of the value chain involves the production of silicon, the main resource for the solar cell, and this requires substantial know-how and investment, as does the production of wafers (EPIA/Greenpeace, 2008). With regard to the intermediate level of cell and module producers, know-how and investment needs are smaller than for silicon and wafer production, and the number of firms in the market is higher. With regard to the installation of solar panels, at the end of the value chain, these installers are often found to be small locally based businesses (EPIA/Greenpeace, 2008).

By 2000 around 1.3 million solar home systems were installed in developing countries; but organizational, financial and technical problems created difficulties for effective implementation, and market transparency is limited, leading to a lack of knowledge and information for potential users about cost-effective systems (Nieuwenhout et al., 2001). More recently, the IFC (2007) evaluated various solar energy projects in developing countries and, while recognizing the large potential market, highlighted problems of identifying the market segment most likely to take up the technology – although as mentioned above, in Kenya this is the rural middle class (Jacobson, 2007). In Bangladesh and India, poor people are constrained from obtaining solar lighting through financial exclusion, weak governance and passive non-governmental organization (NGO) and customer participation (Wong, 2011).

Although it is generally recognized that costs of solar panels still form a barrier, there are also indications that with the help of proper domestic incentives it is possible to move the PV market forward where a sufficiently strong commercial supply chain network has been developed (van der Vleuten et al., 2007). Examples of successful commercial markets for Solar Home Systems (SHS) can be found in Kenya, Morocco, Sri Lanka, on the Tibetan plateau in western China and in Zimbabwe. Estimates suggest that commercial markets in these locations have reached penetration of up to several megawatts of installed peak power per country or up to approximately 5 per cent of the rural population (van de Vleuten et al., 2007, p1439). Similarly, Otieno (2003) reports positive results in Kenya. Decentralized (off-grid) rural electrification based on the installation of standalone systems in rural households or the setting up of mini-grids – where PV can be combined with other renewable energy technologies or with LPG/diesel – enables the provision of key services such as lighting, refrigeration, education, communication and health. During 2007, around 100MW of PV solar energy was installed in rural areas in developing countries, enabling access to electricity for approximately 1 million families (EPIA/Greenpeace, 2008).

Key elements that need to be taken into account for effective technology transfer of solar panels are establishing effective platforms for interaction, facilitating standardization through appropriate organizations, and increasing awareness and access to information by building regional or local knowledge centres (Shum and Watanabe, 2008). Because developing countries play rather different roles in the current solar panel value chains, strategies need to be differentiated. For some of the more advanced Asian countries that have gained access to the solar panel production value chain, the focus can be on facilitating access to technology in the form of co-development programmes (such as for multi-crystalline panels, but also for the emerging thin-film technologies) and sharing IPRs. The focus can also be on expanding global silicon production for PV (silicon, of course, is a major input for several industries, such as semiconductors and metallic alloys) with the participation of developing countries. Another type of approach should focus on supporting the build-up of regional platforms for the interaction of key stakeholders – knowledge centres that apply lessons learned from the many solar home systems that have been installed in relatively poor rural areas in developing countries and that act as catalyst for standardization processes.

Energy efficiency in the cement industry

The cement industry holds a key position in contemporary society as it creates the raw material for bridges, buildings, dams and other infrastructure. But the cement industry is also a major contributor to the climate change problem as cement production is roughly responsible for 5 to 8 per cent of global GHG emissions (Batelle, 2002; Müller and Harnisch, 2008; Worrell et al., 2009). Cement production occurs all over the globe, but production is now predominantly located in developing countries with 74 per cent of world production (Roy, 2008; Worrell et al., 2009). Production has expanded rapidly by 60 per cent in past decade, particularly in developing countries, with China responsible for 44 per cent of world production in 2004 (Price et al., 2006; Roy, 2008).

Cement is among the industries with the largest mitigation potential, together with steel, and pulp and paper industries (Roy, 2008). With 1930 million tonnes of CO_2, the cement industry emits 4.6 per cent of global anthropogenic GHG emissions (Watson et al., 2005), though estimates range up to 8 per cent (Müller and Harnisch, 2008). Around 50 per cent of cement emissions arise from the chemical process of converting limestone to lime in order to produce clinker, which accounts for around 90 per cent of cement emissions if powered by fossil fuels. Offsite electricity and transport emissions account for the remaining 10 per cent of emissions. Developing countries account for 70 per cent of global cement emissions, a figure which is set to rise as developing countries continue to have higher demand for their construction and infrastructure sectors.

China is by far the largest producer of cement, with its production more than the next 20 largest countries combined. Western Europe is the second largest producer at 11 per cent, followed by South and East Asia at 8 per cent. The industry has undergone significant consolidation over the past decade to the point where the five largest companies represent 42 per cent of global capacity and the ten largest 55 per cent. However, the cement sector also comprises a vast number of small firms. For example, estimates for the total number of firms in China are from 5000 to 8300, and the top five cement producers in Russia account for only 10 per cent of production capacity (Watson et al., 2005). Cement is primarily consumed close to where it is produced for two key reasons. The first is that raw materials for cement production are widely available. The second is that cement is a costly product to transport relative to its value, particularly over land. Only 5.8 per cent of production is traded, with 40 per cent of this traded between regions. The largest exporters are Western Europe, Japan and India, while the US is the largest net importer of cement, importing 8 per cent of its consumption (Watson et al., 2005), primarily from China, Canada, Columbia, Mexico and the Republic of Korea.

The set-up of an agreement in the cement industry will be a challenge because production is spread among many plants and companies across the globe, while the level of international trade is rather low. Furthermore, a process of increased consolidation of the traditionally fragmented cement industry is under way through mergers and acquisitions, and through growth of large national players in emerging economies such as China and India. This increasing consolidation process may be accompanied by the establishment of a global cement industry institution, and thus better enable the cement industry to become a strong partner in sectoral agreements (Watson et al., 2005). According to Watson et al., (2005), an emissions intensity agreement could be a way for the cement industry to move forward, with efficiency gains potentially leading to 16 per cent lower emissions.

The cement sector has reasonably good conditions for international cooperation as portions of the cement industry have also organized themselves under the World Business Council for Sustainable Development's Cement Sustainability Initiative (CSI). The key challenge will be how to involve China as the CSI includes 16 companies representing about 50 per cent of global cement production outside of China. Key components of the initiative are 'climate protection and CO_2 management', where monitoring and reporting of CO_2 emission has been mainstreamed under members by setting up a common approach and monitoring and reporting protocol. The initiative also aims to develop public policy and market mechanisms for reducing CO_2 emissions, but this has not been specified yet and fixed emissions targets are unlikely to be popular given the central economic role of the industry (Bradley et al., 2007). A more likely approach is a focus on technology and financial assistance towards developing countries such as China and other countries where significant growth is expected.

Unlike lighting or solar PV, there is no simple technological entry point for reducing GHG emissions from the cement industry. Rather, it is a matter of simultaneously tackling a range of issues within an overall approach to reaching a consensus among producers. Key elements for cement agreements need to focus particularly on the options available to increase the amount of blended cement, replacing old plant with energy-efficient technology, and on diffusing best practices with regard to energy management, which also includes developing low-carbon or renewable biomass-based energy provision for this rather energy-intensive industry (de Coninck et al., 2007; Müller and Harnisch, 2008). Pilot projects for carbon capture and storage (CCS) could also be connected to the energy requirements of the cement industry, although CCS has yet to move beyond the demonstration stage (Russell et al., 2011) and has not gained much credibility in the UNFCCC negotiations as an effective tool for mitigation as there are few additional benefits compared to alternative carbon capture approaches. In particular, forestry offers a wide range of livelihood opportunities.

Energy-efficient electric motors

Electric motor systems are considered to be responsible for up to 40 per cent of industrial electricity demand worldwide (Brunner, 2007) – thus a major source of greenhouse gas emissions. It is estimated that uptake of high-efficiency motors (HEMs) could improve the efficiency of motor systems by 25 to 30 per cent on average. Motor system components are widely traded commodity goods that are currently subject to different testing standards and performance and labelling requirements (SEEEM, 2006). As a result, there are substantial variations in the market penetration of high-efficiency motors and motor systems around the globe. Countries that have implemented minimum energy performance standards at relatively high efficiency levels have market shares for high-efficiency motors of over 70 per cent, whereas the market share in countries without them hovers below 10 per cent, despite voluntary programmes (SEEEM, 2006). The International Energy Agency estimates that up to 7 per cent of global electricity demand could be saved by more energy-efficient motors and motor systems. At present, both markets and policy-makers tend to focus exclusively on individual system components, such as motors or pumps, with an improvement potential of 2 to 5 per cent instead of optimizing systems (McKane et al., 2008).

According to McKane et al., (2008), the barriers to uptake of more efficient motors are foremost institutional and behavioural, rather than technical. The fundamental problems are lack of awareness of the energy-efficiency opportunities by firms, suppliers and consultants; there is a lack of understanding on how to implement energy-efficiency improvements, and a consistent organization structure for energy management within most industrial facilities is often absent. At the institutional level, there is a general lack of standards. Without performance indicators that relate energy consumption to production output, it is difficult to document improvements in system efficiency (McKane et al., 2008).

Energy efficiency in motors and drives is generally considered to be cost effective (i.e. the more efficient motors and systems pay the additional cost (more material in steel and copper, additional power electronic components, additional labour for design and engineering, testing, etc.) within less than two years). Especially in new equipment, there is no reason whatsoever not to buy and install optimally designed and highly energy-efficient motors and electronic adjustable speed drives where feasible (SEEEM, 2006). There is a wide variety of electric motor producers with some of the largest players from the OECD, while China also has a significant number of large producers. Motors are sold from the manufacturer to three different channels: large industrial end-users, distributors who sell them to small end-users, and original equipment manufacturers (OEMs) who put them into complete systems. Only in the first channel is there a direct link between initial cost and quality and energy savings that will result from the installation of efficient motor systems.

A number of national and international activities are in place to promote energy-efficient motors and provide standards (SEEEM, 2006; Saidur, 2010): the International Electrotechnical Commission (IEC), the International Conference on Electrical and Electronics Engineering (ICEEE) and other organizations have provided standards for testing the energy efficiency of electric motors. CEMEP in Europe, NEMA in the US and many other organizations have launched labelling schemes and voluntary standards for high-efficiency motors. Mandatory minimal energy performance standards have been enacted in the US, Canada, Australia, New Zealand, Brazil, China and Mexico. The Motor Challenge campaign has increased awareness, competence and acceptance of efficient electrical motors in the US and Europe. Similar campaigns have been started in Australia, China and other countries. An elaborate database for energy-efficient motors is provided by EuroDEEM. The Collaborative Labelling and Appliance Standards Programme (CLASP) has worked internationally to harmonize standards and labels. APEC–ESIS has worked on harmonized and effective energy efficiency standards in Asian countries.

Discussion

The four cases suggest that significant reductions of GHG emissions can be realized by a more rapid diffusion of energy-efficient and renewable-based technologies. However, the path to more rapid acceleration of these technologies is not straightforward. First of all, the diversity of the technologies calls for approaches tailored to the specific characteristics of the technologies, users and the global value chains that they represent. Those technologies fulfilling the needs of households, such as solar panels and lighting, require a different set-up of mechanisms relative to those oriented to large industries, such as cement and high-efficiency motors. Moreover, energy-use patterns within developing countries differ widely between the poor and middle classes, and mechanisms will need to exploit various delivery models to reach those different groups effectively. The technologies differ with respect to the type and nature of investments they require and their complexity; and this has to be reflected in the proposed mechanisms. The nature of global value chains and the integration of developing countries within those value chains also differ by region. Whereas several Asian countries have gained access to some of the value chains for the identified technologies, the role of other developing countries in global value chains is much more limited.

Second, adequate technological capabilities and absorptive capacity are key requirements for the successful spread of the technologies outlined in the case studies here. 'Technology flows can be embodied in foreign direct investment (FDI), intermediate goods, capital equipment, or licensing, but may have little or no effect on development or growth without absorptive capacity' (Narula, 2004, p3). Host environments differ widely across developing countries. The East Asian newly industrializing countries (South Korea, Taiwan, Singapore and Hong Kong) have developed strongly supportive enabling environments and effective absorptive capacity for the acquisition and exploitation of a range of technologies, including high-tech manufactures. They have built relatively effective 'national systems of innovation' and play an increasing role in strategy technology alliances, partnerships and agreements between countries and firms from the North and South (Archibugi and Pietrobelli, 2003). China made significant progress in this respect during the last two decades. For example the level of R&D as a percentage of China's GDP rose from 0.7 to 1.1 per cent during 1997 to 2002; China aims to increase this to 2.2 per cent by 2015, while in most African countries (with the exception of South Africa) declining trends were observed, with starting levels of R&D spending already well below 0.5 per cent of GDP (UNIDO, 2005). China has become a major recipient of FDI, while for the cases analysed here, Chinese firms take up an increasing share in domestic and international production of energy-efficient lighting, solar panels, high-efficiency motors and cement production (and by acquiring an increasing number of technology licences).

Nevertheless, barriers of access to environmentally sound technologies and the availability of financing are still cited as the major constraints for further upgrading and delivery of technologies for mitigation and adaptation. For example, the G77 and China have proposed to set up a technology mechanism under the UNFCCC, cited at the beginning of this chapter, which particularly aims at increasing access to, and financing of, environmentally sound technologies, including through co-development of technologies and intellectual property rights sharing (UNFCCC, 2008b).

A number of conclusions can also be drawn more specifically for the selected technologies. Energy-efficient lighting is the most straightforward case because it represents a not too complex form of substitution of existing products by a new generation of better products. There is also broad political and industrial support for the phase-out of incandescent lighting which offers a route to a global agreement on energy-efficient lighting and the phase-out of incandescent lighting. Apart from global partners (OECD countries, developing countries and the lighting industry), such an agreement needs to be accompanied by stakeholder collaboration at national and regional levels. At national levels it is important to develop strategies and regulations to prevent inefficient lighting from resurfacing, and it is crucial to safeguard the quality of energy-efficient lighting by setting up standardization and testing organizations. Programmes such as Lighting Africa can also act as vehicles for diffusion, while taking into account the specific needs of various user groups, such as poor households and micro-businesses. Large companies such as Philips are setting up multi-stakeholder partnerships with developing countries to jointly manufacture energy-efficient lighting (Morsink et al., 2011) and are developing ways of selling their products to the 'bottom billion' to open up new markets. Governments support widespread uptake of lighting EST because it lowers peak grid power demands and substantially reduces the need for increasing generating capacity.

For solar panels, it is important to distinguish between the group of countries that are actively involved in solar panel production and the group that are mainly involved in application of solar panels in local electricity systems. For some of the more advanced Asian countries that have gained access to the solar panel production value chain, the focus can be on facilitating access to technology in the form of co-development programmes (such as for multi-crystalline panels, but also for the emerging thin-film technologies) and sharing IPRs. The focus can also be on expanding global silicon production with the participation of developing countries. Another type of approach should focus on supporting the build-up of regional platforms for the interaction of key stakeholders and knowledge centres that apply lessons learned from the many solar home systems that have been installed in relatively poor rural areas in developing countries and that act as catalyst for standardization processes.

With regard to cement production, technology agreements need to focus particularly on the options available to increase the amount of blended cement, and on diffusing best practices with regard to energy management, which can also involve developing low-carbon energy provision for this rather energy-intensive industry. This requires the build-up of energy management expertise and of effective channels to bring support to the many smaller cement companies in developing countries. For the larger cement industry operations, pilot projects for carbon capture and storage could also be part of a mitigation package.

High-efficiency motors include a range of technologies and applications across all industries. Mechanisms need to focus on two key aspects. First, there is still much to be gained by replacing less efficient motors by high-efficiency electric motors. Key components are awareness-building and information campaigns and incentives to lower the higher costs of these high-efficiency motors relative to the less efficient ones. Second, there is a need for energy management expertise and for improving energy management across the board of industries.

As mentioned in the introduction, an important element for effective technology transfer is the creation of an enabling environment in the recipient country, which is defined as 'government actions, such as fair trade policies, removal of technical, legal and administrative barriers to technology transfer, sound economic policy, regulatory frameworks and transparency, all of which create an environment conducive to private and public sector technology transfer' (IPCC, 2001, p26). This implies that country-specific social and institutional contexts need to be taken into account. For many cases, the lack of fine-tuning to the specific socio-cultural and institutional context has been a contributing factor to the failure of technology transfer. Incorporation of a needs assessment and active national and local stakeholder involvement and contribution is one way to prevent this. Such a multi-stakeholder approach also holds promise for the build-up of local technological capabilities that are deemed crucial for any effective technology transfer. Enabling factors such as infrastructure and a supportive political, legal and regulatory framework are highly relevant; but most critical seem to be partnerships with multiple stakeholders from the private and public sectors and from civil society because these will bring in local knowledge, leverage and human and infrastructural resources (Morsink et al., 2011). With these capabilities a multi stakeholder partnership can influence the creation of an enabling environment for technology transfer.

A further general perception is that effective technology transfer to developing countries is hampered by limited access to knowledge and the proprietary nature of technologies (i.e. IPRs). However, the role of IPRs as a barrier is far from clear. A general review suggests that middle-income countries prefer relaxed IPRs, whereas low- and high-income countries benefit from good IPR protection (Falvey et al., 2006a, 2006b); and it is clear that there are conflicting discourses about the role of IPRs in development versus technology diffusion (Ockwell et al., 2010). Nonetheless, development of a more open innovation structure could facilitate participation from developing countries. One part of such a strategy could be the creation of a global Knowledge Fund, an idea developed by Lynn Mytelka in which patents of technologies critical to fundamental human needs (e.g. for food, drugs and ESTs) can be deposited (Mytelka, 2007). Financial resources should be put into the fund to ensure that appropriate local capabilities are developed in enterprises, knowledge institutes and the public sector in developing countries ensure that when patents are being utilized, the tacit knowledge required to work these patents in a local context is transferred (Mytelka, 2007). While the Knowledge Fund is basically designed as a global organization, national and regional knowledge centres, such as climate innovation centres (Sagar, 2010), could play a key role in facilitating the further application and diffusion of the knowledge, skills, and engineering practices and technologies to local entrepreneurs, firms and other relevant stakeholders.

Acknowledgements

This chapter was originally published in 2008 as a working paper commissioned by The Netherlands Directorate-General for International Cooperation (DGIS) to gain further insight into potential mechanisms for technology transfer of environmental sound technologies in the light of the climate change challenge (Hofman et al., 2008). We gratefully acknowledge support from the DGIS during the preparation of this chapter.

References

Archibugi, D. and Pietrobelli, C. (2003) 'The globalization of technology and its implications for developing countries – Windows of opportunity or further burden?', *Technological Forecasting and Social Change*, vol 70, pp861–883

Barton, J. (2007) *Intellectual Property and Access to Clean Energy Technologies in Developing Countries: An Analysis of Solar Photovoltaic, Biofuel and Wind Technologies*, Issue Paper no 2, ICTSD Programme on Trade and Environment, Geneva, Switzerland

Batelle (2002) *Toward a Sustainable Cement Industry*, Report commissioned by the WBSCD, Batelle, Colombus, OH

Bradley, R., Baumert, K. A., Childs, B., Herzog, T. and Pershing, J. (2007) *Slicing the Pie: Sector-Based Approaches to International Climate Agreements*, WRI Report, Washington, DC

Brunner, C. U. (2007) *International Harmonization of Motor Standards Saves Energy*, Presentation at APEC workshop, Beijing, China

Byigero, A. D., Clancy, J. S. and Skutsch, M. S. (2010) 'Clean Development Mechanism in sub-Saharan Africa and the prospects of the Nairobi Framework Initiative', *Climate Policy* vol 10, no 2, pp181–189

de Coninck, H., Bakker, S., Junginger, M., Kuik, O., Massey, E. and van der Zwaan, B. (2007) *Agreement on Technology? Exploring the Political Feasibility of Technology-Oriented Agreements and Their Compatibility with Cap-and-Trade Approaches to Address Climate Change*, ECN report ECN-E-07-092, Amsterdam, The Netherlands

Egenhofer, C., Milford, L., Fujiwara, N., Brewer, T. L. and Alessi, M. (2007) *Low-Carbon Technologies in the Post-Bali Period: Accelerating Their Development and Deployment*, European Climate Platform Report no 4, Centre for European Policy Studies, Brussels, Belgium

Energy Foundation Ghana (2011) http://www.ghanaef.org/, accessed 11 August 2011

EPIA/Greenpeace (2008) *Solar Generation V*, Joint report by EPIA and Greenpeace, Brussels/Amsterdam, Belgium/The Netherlands

Falvey, R., Foster, N. and Memedovic, O. (2006a) *The Role of Intellectual Property Rights in Technology Transfer and Economic Growth: Theory and Evidence*, UNIDO, Vienna, Austria

Falvey, R., Foster, N. and Greenaway, D. (2006b) 'Intellectual property rights and economic growth', *Review of Development Economics*, vol 10, pp700–719

Figueres, C. and Bosi, M. (2006) *Achieving Greenhouse Gas Emission Reductions in Developing Countries through Energy Efficient Lighting Projects in the Clean Development Mechanism (CDM)*, World Bank, Washington, DC

Forsyth, T. (2007) 'Promoting the "development dividend" of climate technology transfer: Can cross-sector partnerships help?', *World Development*, vol 35, no 10, pp1684–1698

G8 (2007) *Summit Declaration, Heiligendamm 2007*, para 54, http://www.g-8.de/Webs/G8/EN/Homepage/home.html, accessed 11 August 2011

GEF (Global Environment Facility) (2010) *What Is Technology Transfer?*, http://www.thegef.org/gef/KM/tech_transfer, accessed 11 August 2011

Hofman, P. S., Lovett, J. C., Morsink, K. and Clancy, J. (2008) *Making Global Technology Markets Work for Combating Climate Change and Contributing to MDGs*, Report to the Climate and Energy Section of the Dutch Foreign Ministry, Twente Centre for Studies in Technology and Sustainable Development, University of Twente, Enschede, The Netherlands

IEA (International Energy Agency) (2006) *Light's Labour Lost – Policies for Energy-Efficient Lighting*, IEA, Paris

IFC (International Finance Corporation) (2007) *Selling Solar: Lessons from More than a Decade of IFC's Experience*, International Finance Corporation, Washington, DC

IPCC (Intergovernmental Panel on Climate Change) (2001) *Annex II, Summary for Policymakers: Methodological and Technological Issues in Technology Transfer*, Special IPCC Report by B. Metz, O. Davidson, J. W. Martens, S. N. M. Van Rooijen and L. V. W. McGrory, UNEP, Nairobi, Kenya

Jacobson, A. (2007) 'Connective power: Solar electrification and social change in Kenya', *World Development*, vol 35, no 1, pp144–162

Jung, M. (2006) 'Host country attractiveness for CDM non-sink projects', *Energy Policy*, vol 34, pp2173–2184

Lefevre, N., de T'Serclaes, P. and Waide, P. (2006) *Barriers to Technology Diffusion: The Case of Compact Fluorescent Lamps*, COM/ENV/EPOC/IEA/SLT(2006)10, IEA, Paris

Lighting Africa (2011) http://www.lightingafrica.org/, accesssed 11 August 2011

Lovett, J. C., Hofman, P. S., Morsink, K., Balderas Torres, A., Clancy, J. S. and Krabbendam, K. (2009) 'Review of the 2008 UNFCCC Meeting in Poznan', *Energy Policy*, vol 37, pp3701–3705

McKane, A., Williams, R., Perry, W. and Li, T. (2008) *Setting the Standard for Industrial Energy Efficiency*, IEA Workshop, Paris

Michaelowa, A. and Jotzo, F. (2005) 'Transaction costs, institutional rigidities and the size of the clean development mechanism', *Energy Policy*, vol 33, pp511–523

Morsink, K., Hofman, P. S. and Lovett, J. C. (2011) 'Multi stakeholder partnerships for transfer of environmentally sound technologies', *Energy Policy*, vol 39, pp1–5

Müller, M. and Harnisch, J. (2008) *How to Turn around the Trend of Cement Related Emissions in the Developing World*, Report prepared for the WWF–Lafarge Conservation Partnership, WWF International, Gland, Switzerland

Mytelka, L. (2007) *Technology Transfer Issues in Environmental Goods and Services*, Issue Paper no 6, ICTSD, Geneva, Switzerland

Narula, R. (2004) *Understanding Absorptive Capacities in an 'Innovation Systems' Context: Consequences for Economic and Employment Growth*, Prepared for the ILO, Background paper for the World Employment Report, ILO, Geneva, Switzerland

Nieuwenhout, F. D. J., van Dijk, A., Lasschuit, P. E., van Roekel, G., van Dijk, V. A. P., Hirsch, D., Arriaza, H., Hankins, M., Sharma, B. D. and Wade, H. (2001) 'Experience with solar home systems in developing countries: a review', *Progress in Photovoltaics: Research and Applications*, vol 9, no 6, pp455–474

Ockwell, D. G., Watson, J., MacKerron, G., Pal, P. and Yamin, F. (2006) *UK–India Collaboration to Identify the Barriers to the Transfer of Low Carbon Energy Technology*, Final Report, University of Sussex in collaboration with TERI and IDS, Sussex, UK

Ockwell, D. G., Watson, J., MacKerron, G., Pal, P. and Yamin, F. (2008) 'Key policy considerations for facilitating low carbon technology transfer to developing countries', *Energy Policy*, vol 36, pp4104–4115

Ockwell, D. G., Hauma, R., Mallett, A. and Watson, J. (2010) 'Intellectual property rights and low carbon technology transfer: Conflicting discourses of diffusion and development', *Global Environmental Change*, vol 20, pp729–738

Otieno, D. (2003) 'Solar PV in Kenya', *Refocus*, September–October 2003, pp40–41

Porter, M. E. and van der Linde, C. (1995) 'Toward a new conception of the environment–competitiveness relationship', *Journal of Economic Perspectives*, vol 9, pp97–118

Pode, R. (2010) 'Solution to enhance the acceptability of solar-powered LED lighting technology', *Renewable and Sustainable Energy Reviews*, vol 14, no 3, pp1096–1103

Price, L., de la Rue du Can, S., Sinton, J., Worrell, E., Zhou, N. and Sathaye, J. et al., (2006) *Sectoral Trends in Global Energy Use and Greenhouse Gas Emissions*, LBNL- 56144, Lawrence Berkeley National Laboratory, Berkeley, CA

Roy, J. (2008) *IPCC Results on Mitigation in the Industry Sector: Some Results from India*, International Workshop on Sectoral Approaches for International Climate Policy, Hosted by the International Energy Agency, Paris, 14–15 May, 2008, France

Russell, S., Markusson, N. and Scott, V. (2011) 'What will CCS demonstrations demonstrate?', *Mitigation and Adaptation Strategies for Global Change*, DOI: 10.1007/s11027-011-9313-y

Sagar, A. (2010) *Climate Innovation Centres: A New Way to Foster Climate Technologies in the Developing World?*, International Bank for Reconstruction and Development/World Bank, Washington, DC

Saidur, R. (2010) 'A review on electrical motors energy use and energy savings', *Renewable and Sustainable Energy Reviews*, vol 14, no 3, pp877–898

SEEEM (Standards for Energy Efficiency of Electric Motor Systems) (2006) *Market Transformation to Promote Efficient Motor Systems*, Project Launch Paper, SEEEM, Zurich, Switzerland

Shum, K. L. and Watanabe, C. (2008) 'Towards a local learning (innovation) model of solar photovoltaic deployment', *Energy Policy*, vol 36, pp508–521

Timilsina, G., De Gouvello, C., Thioye, M. and Dayo, F. (2010) 'Clean Development Mechanism Potential and challenges in sub-Saharan Africa', *Mitigation and Adaptation Strategies for Global Change*, vol 15, pp93–111

UNFCCC (United Nations Framework Convention on Climate Change) (2001) *Annex: Framework for Meaningful and Effective Actions to Enhance the Implementation of Article 4, Paragraph 5, of the Convention*, FCCC/CP/2001/13/Add.1, Report of the Conference of the Parties on its seventh session, held at Marrakesh from 29 October to 10 November 2001, Morocco

UNFCCC (2008a) *Report of the Global Environment Facility on a Strategic Programme to Scale Up the Level of Investment for Technology Transfer*, FCCC/SBI/2008/5, 29 May 2008, UNFCCC Subsidiary Body for Implementation, Twenty-eighth session, Bonn, Germany

UNFCCC (2008b) 'Proposal by the G77 and China for a Technology Mechanism under the UNFCCC', http://unfccc.int/files/meetings/ad_hoc_working_groups/lca/application/pdf/technology_proposal_g77_8.pdf, accessed 11 August 2011

UNFCCC (2008c) *Development and Transfer of Technologies, Decision 2/CP.14*, FCCC/CP/2008/7/Add.1., p3, Report of the Conference of the Parties on its 14th session, held in Poznan from 1 to 12 December 2008, Poland

UNFCCC (2009) *Copenhagen Accord*, FCCC/CP/2009/11/Add.1., pp5–7, Report of the Conference of the Parties on its 15th session, held in Copenhagen from 7 to 19 December 2009, Denmark

UNFCCC (2010) *IV. Finance, Technology and Capacity-Building. B. Technology Development and Transfer*, FCCC/CP/2010/7/Add.1, p18, Report of the Conference of the Parties on its 16th session, held in Cancun from 29 November to 10 December 2010, Mexico

UNIDO (United Nations Industrial Development Organization) (2005) *Industrial Development Report 2005: Capability Building for Catching Up*, UNIDO, Vienna, Austria

USAID (*US Agency for International Development*) (2007) *Confidence in Quality: Harmonization of CFLs to Help Asia Address Climate Change*, USAID Regional Development Mission for Asia, Bangkok, October 2007

van der Vleuten, F., Stam, N. and van der Plas, R. (2007) 'Putting solar home system programmes into perspective: What lessons are relevant?', *Energy Policy*, vol 35, pp1439–1451

Watson, C., Newman, J., Upton, S. and Hackmann, P. (2005) *Can Transnational Sectoral Agreements Help Reduce Greenhouse Gas Emissions?*, Background paper for the meeting of the Round Table on Sustainable Development, 1–2 June 2005, World Bank, Paris

Wong, S. (2011) 'Overcoming obstacles against effective solar lighting interventions in South Asia', *Energy Policy*, doi:10.1016/j.enpol.2010.09.030

Worrell, E., Bernstein, L., Roy, J., Price, L. and Harnisch, J. (2009) 'Industrial energy efficiency and climate change mitigation', *Energy Efficiency*, vol 2, pp109–123

18

REDUCING THE COST OF TECHNOLOGY TRANSFER THROUGH COMMUNITY PARTNERSHIPS

Tim Forsyth

It is widely agreed that low-carbon technology transfer to developing countries can both reduce greenhouse gas emissions and alleviate poverty. But there are also strong concerns that this kind of technology transfer is costly, time consuming and unattractive to private investors. These concerns need to be rethought.

Some early international agreements on environment have contributed to these beliefs. The United Nations Framework Convention on Climate Change (UNFCCC) (Article 4.5) stated that parties 'shall take all practicable steps to promote, facilitate and finance' technology transfer. And Chapter 34 of Agenda 21 – while acknowledging the need to protect intellectual property rights – proposed that technology transfer should be promoted 'on favourable terms, including on concessional and preferential terms'.

These statements seem to imply that technology transfer can be done and, indeed, *should* be coordinated by governments – even if it means subsidizing costs – and that this is the responsibility of more industrialized countries. There is clearly a case for Annex I countries to assist poorer countries to adopt new technologies. But these statements – both written in 1992 – today seem inappropriate because experience has shown that technology transfer cannot occur if it is considered a cost by investors. Subsidizing technologies also seems at odds with new economic thinking and the role of trade laws.

Instead, there are ways of reducing the costs of technology transfer to developing countries that offer more optimistic possibilities than subsidization. In particular, this chapter focuses on the role of partnerships between international investors and local communities as a way of reducing costs. Costs can be reduced by spreading tasks for technology transfer between different stakeholders. Moreover, careful participation and local benefits can increase the long-term success of technology transfer by ensuring that technologies are appropriate to local needs.

But for this to happen more fully, three changes are required. First, the role and diversity of partnerships needs to be discussed more fully in international negotiations as a means of implementing climate change policy. At present there is a tendency to discuss partnerships in rhetorical terms. Instead, partnerships can be firm institutions based on contracts and the delivery of services.

Second, there is a need for international organizations and the UNFCCC to identify the role of community partnerships in both accelerating low-carbon technology transfer and in enhancing local development. More assistance and guidance for achieving partnerships can help to mitigate climate change rather than simply rewarding companies for mitigating emissions alone.

And third, there is a need to consider more carefully the 'development dividend' possible from international technology investments. The term 'development dividend' was used to describe the requirement under the UNFCCC's Clean Development Mechanism (CDM) to provide local development benefits as well as climate change mitigation. Development dividends might include local benefits such as employment, energy and other public services, as well as reducing emissions. Allowing local parties to achieve a development dividend from technology transfer is one way of encouraging partnerships with local stakeholders.

But first we start with a discussion of why partnerships can help to reduce the costs of technology transfer.

Low-carbon technology transfer and local development

Much previous experience with technology transfer has shown that technology transfer in developing countries both requires, and benefits from, public participation. For example:

- *Technology appropriateness:* some new technologies do not suit local conditions or present advantages over existing technologies. For example, experience in the Philippines has shown that biogas-fuelled irrigation pumps introduced during the 1960s to 1980s were generally not used by farmers as they still considered rainfall appropriate for their agricultural needs. A Danish biogas generator in Delhi during the 1980s did not anticipate the high organic matter in Indian waste (Forsyth, 2007).
- *Impacts upon local markets:* technology transfer will depend upon how new technologies change markets for commodities and labour. For example, in 1979, the United Nations Children's Fund (UNICEF) introduced a biogas generator in Fateh-Singh-Ka-Purwa, India. This project did not anticipate that introducing new technology would suddenly commodify cow dung, which had previously been considered a free product. The resulting rise in the price for cow

dung made the technology uneconomic (Forsyth, 1999, p135). Similarly, some expected changes might not occur. In Thailand, the introduction of solar panels failed to lead to income generation because local machinery (such as sewing machines) could not operate on this energy (Green, 2004, p755).

- *Costs need to be recovered:* technologies based on subsidies are uneconomic and not trusted by users who fear the subsidies will be withdrawn. Requiring local users to contribute to some of the technology costs will increase the economic feasibility of technologies, and demonstrate that technologies are useful (Gregory et al., 1997; IPCC, 2000; Douthwaite, 2002).
- *Local capacity:* technology transfer is not simply moving equipment to new locations. It also requires training local people to understand and maintain the technology, and to seek ways of using technologies for local purposes. It also means constructing bodies to collect fees to help pay for technologies, which in turn can assist in demonstrating the value to local stakeholders. In terms of energy investment, this is equivalent to creating local utilities (UNFCCC, 2003; Mallett, 2007).
- *Local politics and inequality:* long experience from the Green Revolution has demonstrated that distributing new technologies such as tractors can enhance, rather than reduce, social inequality. These factors need to be considered if the objective of technology transfer is to enhance local development (Heaton et al., 1994).

The problem with these lessons is that they imply that long-term successful technology transfer in developing countries often requires investors to undertake tasks usually outside their commercial interest or professional ability. These tasks might include building capacity for running and maintaining technologies, rather than simply selling technologies to new users. As a result, much low-carbon technology transfer has faltered because of a perceived lack of additional funding for activities such as capacity-building.

Moreover, these lessons seem far away from either international negotiations on climate change or the mechanisms used to implement climate change policies. As discussed above, international agreements such as the UNFCCC and Agenda 21 have focused more on rhetorical statements about North–South responsibilities. Market-based mechanisms such as the CDM, too, have not directly addressed these practical problems of technology transfer, and instead have offered incentives only for mitigation alone.

As discussed in earlier chapters, many developing countries hoped that the CDM would enhance industrial technology transfer because its original text stated that it was for attracting investment for 'sustainable development', in general, as well as climate change mitigation. But the CDM has been criticized for failing to achieve either sufficient technology transfer or the development dividend. Critics have suggested too much investment has gone into projects that deliver fast destruction of gases (notably HFC23), which offer fast returns on carbon credits, rather than into projects that might build local development capacity by providing livelihoods or public services, such as heating and electricity. Moreover, these investments have also been attracted to countries such as China and India, which have received much existing foreign direct investment, rather than countries that do not receive investment, such as in sub-Saharan Africa (Bäckstrand and Lövbrand, 2006; IISD, 2006, p4). In sub-Saharan Africa, for example, one climate adviser noted: 'unfortunately, the CDM in Africa is largely about selling stories for corporate social responsibility rather than profits on climate change credits'.[1]

Moreover, the CDM rules and governance have been considered costly and complex (Boyd et al., 2009). The decision-making process of the CDM Executive Board and national authorities is considered by some investors and policy-makers to be time consuming and bureaucratic, with too many restrictions on what kinds of investment are considered appropriate (Cosbey et al., 2005). The costs of the CDM have also been increased by the decision under the Marrakesh Accords of 2001 to create the UNFCCC Adaptation Fund by extracting 2 per cent of CDM profits from the sale of Certified Emissions Reduction units (CERs) (Taiyab, 2005; Lohmann, 2009). The objective of the Adaptation Fund is to assist the poorest countries in adapting to climate change. But the fund makes investing in the CDM more costly for investors. It also means that there is less pressure for CDM projects to contribute directly to 'development' (as opposed to greenhouse gas mitigation alone) because some of each project's proceeds will lead indirectly to the Adaptation Fund.

Some evaluation schemes for CDM investment provide ways of assessing the development dividend. For example, the Gold Standard[2] is a non-profit organization that certifies carbon credits with particular attention to social as well as environmental benefits. But there are still barriers to encouraging large-scale investments that enhance technology transfer for climate change mitigation to developing countries. The UNFCCC Subsidiary Body for Scientific and Technological Advice (SBSTA) has helped to develop a technology information system (TT:CLEAR[3]), including an inventory of environmental technology and projects. It has also created the Expert Group on Technology Transfer. The Copenhagen and Cancun Conferences of the Parties to the UNFCCC in 2009 and 2010 also started the creation of a Clean Technology Mechanism based on international grant finance and discussion about the Climate Technology Centre and Network.

But there has been little attention to how far technology transfer and the development dividend can be integrated through involving local people. In 2003, the UNFCCC (2003, p16) wrote: 'governments can create enabling environments for technology diffusion and transfer if they endorse the importance of socially and environmentally oriented organizations and mandate social impact assessments for technology transfer projects'. Such statements, of course, indicate the valuable role played by intermediary non-governmental organizations (NGOs), but fall short of acknowledging the commercial needs and interactions that drive non-state actors to engage in technology transfer. Similarly, the UNFCCC further adopts a state-led perspective by writing '*transferring* experience, knowledge, skills and practices is "capacity building"' (UNFCCC, 2003, p4; emphasis added), which suggests that end-users may not have pre-existing capacities that may be strengthened.

Can community partnerships reduce costs?

Community partnerships are a more localized form of collaboration between investors and end-users that can assist in low-carbon technology transfer. Partnerships can help to overcome the problems of low-carbon technology transfer by increasing local guidance about which technologies are appropriate. Moreover, partnerships can help to reduce the costs of investment by sharing tasks between companies and local people (Forsyth, 2010; Morsink et al., 2011).

Most discussion of 'partnerships' in environmental and public policy started during the 1990s. Some early applications of community partnerships were based on classic contractual public–private partnerships, which involved an agreement between the state and an investor to provide an environmental or public service (Plummer, 2002). Since the late 1990s, however, the discussion of community or cross-sector partnerships (CSPs) has increasingly involved local deliberation, which allows local people to shape public policy (World Bank, 2000; Linder, 2000; Ählström and Sjöström, 2005). In turn, deliberative CSPs are similar to other public policy approaches, such as 'public policy partnerships' (Rosenau, 2000), the 'mutual state' (Mayo and Moore, 2001) or 'network' governance (Hajer and Wagenaar, 2003); or what British Prime Minister David Cameron in 2010 called 'The Big Society'.

Proponents of CSPs have argued that they can address three 'policy deficits' (Biermann et al., 2007). The 'regulatory deficit' refers to the lack of established norms in arenas where states lack capacity. The 'implementation deficit' is the inability of states to carry out policy locally at the sub-state level. The 'participation deficit' is the challenge of including less powerful stakeholders in policy processes, such as poorer citizens or those whose needs have not been addressed by policy to date (Otiso, 2003).

But against this, critics have also suggested that partnerships will never be powerful enough to replace the regulation that should come from states. In particular, some research has shown that 'partnerships' might serve more rhetorical purposes of demonstrating consultation with local stakeholders rather than building new and permanent deliberative capacity. Studies from the UK and Canada, for example, have suggested that investors and governments have consulted local communities as token participants in decisions rather than choosing to build long-term deliberative capacity (Jupp, 2000; Sherlock, et al., 2004; Mitchell, 2005).

Moreover, it is simplistic to refer to 'communities' rhetorically to indicate local cohesion. There are divisions within local stakeholders – such as between landowners and tenants, men and women, or those with different livelihoods. Indeed, 'communities' might not be passive recipients of technology and investment, but might include small businesses who are seeking their own commercial activities.

Can community partnerships be used more practically in low-carbon technology transfer? One answer might be to see 'partnerships' less as a form of corporate social responsibility, but instead as a business strategy based on reducing costs and securing services. Institutions theory is one way to theorize these.

Institutions theory and partnerships

Institutions theory is based on the perception of an institution as shared behaviour between different actors (Williamson, 1985). Much debate within institutional economics (or new institutional economics) seeks to predict institutions by making assumptions about how actors will respond rationally to maximize their interests under different circumstances (Ostrom, 2005). Alternative approaches have, instead, emphasized the cultural or less universal norms that make certain behaviours legitimate, or which lead to certain resources being valued above others (Hajer and Wagenaar, 2003). The common theme of this work is to assess which circumstances lead to occasions of shared behaviour by different actors.

'Partnerships' are examples of occasions when different actors work collaboratively and persistently. Institutional theorists have proposed that there are two key factors that underlie successful partnerships: transaction costs and assurance mechanisms (Weber, 1998). As mentioned above, typical transaction costs for long-term successful technology transfer can include local training, setting up institutions to collect fees for technologies, and ensuring that technologies are appropriate for local needs and markets. Failing to address these factors can also be a transaction cost if they result in technologies being rejected, or if lengthy disputes arise about the new technology and its perceived impacts.

Assurance mechanisms may be defined as the contracts, laws or expectations that ensure collaboration or partnerships will provide each party with successful outcomes. These mechanisms can be formal – in the sense of laws or contracts – or informal, such as expectations and personal agreements.

A possible third controlling factor on successful partnerships is the perceived 'development dividend' of projects (Forsyth, 2007). If technologies are not appropriate to local needs or are not seen to be useful, then it is unlikely such technologies will be adopted in the long term – even if they have the potential to mitigate climate change.

Consequently, an ideal partnership between international investors and local communities should have minimum transaction costs, maximum assurance mechanisms and should address how local stakeholders identify their own needs. But the actual success of partnerships, however, can also depend upon partners' willingness to cooperate and trust, the ability to use negotiations and legal knowledge, and the ability for parties to achieve mutually compatible (or, indeed, shared) objectives. The CDM might offer benefits for low-carbon technology transfer. But successful community partnerships might depend upon addressing needs not directly related to climate change.

Different types of partnerships

'Partnerships' therefore exist when different parties find it in their interests to collaborate. But the nature of partnerships might vary according to different forms of interests, and different forms of collaboration.

TABLE 18.1 Simplified classification of cross-sector partnerships

Type of partnership	Partnerships defined more in contractual terms ←			Partnerships defined more in deliberative terms →	
	Substitutive	*Complementary*	*Shared*	*Consultative*	*Conflictual*
Typical roles	Classic 'public–private partnership'. Usually state and private investor.	Parties undertake complementary economic roles, sometimes under contract to each other.	Parties undertake overlapping roles.	One party consults another for advice or permission without contracts.	Parties engage in conflict or activism to influence each other or the state.
Example	State contracts private investor to build plants. Ownership transferred to state after some years.	Investor supplies electricity-generating technology; citizens may collect or segregate waste supply.	Investor and citizens may both seek to benefit from waste recycling.	Investor has meetings with citizens to build trust and gain information.	Not a classic partnership but can create agreements (e.g. NGOs influencing business through newspapers).
Typical assurance mechanisms	Formal contract, such as build–operate–transfer.	Contracts between parties; assumption that parties gain from different roles.	Contracts between parties; shared interests.	Desire to avoid conflict or damage to company reputation.	Fear of criticism or reputational damage; loss of trust.
Typical deliberative forums	Negotiations with the state; public tendering process.	Negotiations between companies and communities, often helped by state.	Negotiations between companies and others, often helped by state.	Meetings with community leaders, etc.	Occasional public and private meetings, advertising campaigns, etc.
Typical costs, or threats, to partnership	Failure of either party to satisfy contract.	Collaboration may be seen as less important than individual roles of parties.	Different objectives of collaborators may undermine shared activities.	Consultation seen as 'greenwash' or fail to build sufficient trust.	Long-term loss of trust; withdrawal of either or both parties.

Source: Adapted from Forsyth, 2007

Table 18.1 summarizes some different forms of cross-sector partnerships that reflect different reasons for collaborating. The table shows a sliding scale of partnerships from those that are more contractual towards more deliberative forms. The contractual forms of partnership refer to those where the objectives of collaboration are well known in advance, and there is little disagreement or uncertainty about the nature of activities. The more deliberative partnerships allow greater role for communities to participate in shaping policies or to achieve the development dividend in terms that they define.

According to Table 18.1, there are five types of CSPs:

1. 'Substitutive' partnerships refer to contracts between the state and non-state actors to provide environmental services that previously would have been provided by the state.
2. 'Complementary' partnerships are based on different parties undertaking different roles. For example, a biogas investor might supply the new electricity-generating technology; local stakeholders might provide waste products to use as fuel for the generators. Both waste collection and electricity generation are different activities, but are complementary for this investment.
3. 'Shared' partnerships involve both parties sharing different tasks.
4. 'Consultative' partnerships are based upon local questioning and discussion, without the issuing of contracts.
5. 'Conflictual' partnerships are not 'partnerships' in the usual sense of the word because they involve public criticism between, for example, NGOs and local businesses. But they can be called a type of partnership if they lead to some kind of agreement.

The objective of Table 18.1 is to demonstrate that partnerships do not have to take one form, and that involving communities does not always have to mean rhetorical acts of consultation. Under this classification, the most likely forms of partnership to include both cost reduction and local participation are the complementary and shared forms of partnership because these include both contracts for service delivery as well as local deliberation.

But how successful are these forms of collaboration in practice?

Examples of cross-sector partnerships in practice

The discussion above is based on a rather idealized, theoretical form of institutional theory. The following case studies are selected from studies conducted during the 2000s in order to analyse the success factors of CSPs in practice (Forsyth, 2005, 2007). The studies, of course, are not exhaustive. But they present some illustrations of how community partnerships emerge, and the challenges for lowering the costs of technology transfer. The examples selected are cases of 'complementary' and 'shared' partnerships from India, Philippines and Thailand.

The cases focus on two questions. What critical success factors underlie the contractual aspects of partnerships (and, in particular, transaction costs and assurance mechanisms)? And what are the lessons for local deliberation about the developmental benefits of low-carbon technology transfer?

Contractual arrangements of partnerships

In theory, contractual arrangements are strongest when transaction costs are low and assurance mechanisms are resilient. Some interesting lessons were revealed by these case studies.

In the rice-producing areas of central Thailand and Bulacan in Luzon in the Philippines, two investors used different approaches for contracting with local people. Both of these locations are rice agricultural zones with high levels of rice production from irrigated fields. Rice farming has a long and prosperous history in both zones, but the use of rice husks for renewable energy has not been well developed.

In Luzon, during the early 2000s, the US investor Enron planned to build a very large 40MW electricity generator based on burning rice husks. This plan, however, failed because of problems of transaction costs and assurance mechanisms. Enron hoped that the large supply of rice in this fertile region would allow the generation of a substantial quantity of electricity. But in order to fuel the power plant, Enron needed to negotiate contracts with some 150 local rice millers, which comprised a large proportion of the farms surrounding the proposed power plant. Unfortunately, the farmers decided to negotiate strongly with Enron once negotiations had started, and then began to raise the price of the rice husks. Enron were held to ransom because there was no alternative source of fuel. Under these conditions, Enron's financiers withdrew their support.

In central Thailand, however, AT Biopower built six smaller 16MW rice husk plants between 2000 and 2004. Contracts were made with just 20 to 30 rice millers per plant (a relatively small percentage of all available millers), and using just 10 to 15 per cent of their total husk production. The contracts included fines if millers did not supply contracted amounts and bonuses if they fulfilled targets. As a result, assurance mechanisms were stronger, and transaction costs lower than in Bulacan, Luzon.

A different example of transaction costs and assurance mechanisms involves investment in methane capture from municipal waste in the Philippines. Here, a smaller US investor (PhilBio) tried to establish a power plant using the technology of biomethanation (or methane capture from organic waste) in the wealthy suburb of Manila of Ayala Alabang during the early 2000s. This investor undertook a lengthy negotiating process with the local government about gaining permission to locate the site in this neighbourhood, and to agree upon a power-purchasing agreement for the electricity to be generated. In addition, the company also sought a contract with a local NGO who took responsibility for organizing local waste-pickers to collect waste material to feed the plant. The advantage for the local government was to gain new sources of electricity and to minimize the need to manage urban waste. The NGO was also happy to support the treatment of waste and to find employment for local waste-pickers. Like many rapidly growing countries, the Philippines has experienced a boom in the production of municipal waste, and especially organic waste from markets and fast-food shops that emit large quantities of methane when it decays.

Unfortunately, this initial plan for the project failed for several reasons. As with the example of Enron in Bulacan, local landowners (including the municipality) increased the rents they asked from the US investor because they believed the project was more profitable than it was. Moreover, the company learned that it could not control the waste stream. The contracts signed gave the US company ownership over municipal waste from the point of collection in waste dumps or garbage collection. In reality, however, the company learned that the truck drivers (and, frequently, the friends and families of truck drivers) were carefully removing recyclable materials such as glass, metal, paper and plastic from the waste stream before it was delivered to the biomethanation plant. Accordingly, the company could not make any profit on recycling waste, which affected their ability to achieve profits in accordance with their business plan. In later projects, this company simplified its contractual arrangements by allowing waste-pickers to control the recycling of waste, but insisting on regular delivery of organic waste for methane capture.

A similar – but grander – approach was adopted by an Asian consortium of investors in Lucknow, India. On this occasion, the biomethanation project planned to produce 5MW of power from 400 to 500 tonnes of municipal organic waste per day – and with the deliberate strategy of providing livelihoods for local waste-pickers as waste collectors and segregators (an activity encouraged by Uttar Pradesh's local government). The investor also consulted an NGO (Exnora) that specializes in community waste management. Unfortunately, this ambitious CSP eventually closed in 2004 because it could not secure reliable sources of organic waste (Krishna, 2005). Negotiations to secure a more reliable flow are continuing.

A more positive example might be the Philippines city of General Santos in Mindanao. Here, the local government has strong control over investors and citizens and a national flow of funding to support economic development. The combination of waste supply and a local government friendly towards biomethanation and the employment of urban poor (including waste-pickers) at the plant has meant that a move towards biomethanation has been relatively unproblematic.

These cases show that different investors used various forms of assurance mechanisms to reduce transaction costs. Seeing local communities as commercial operators, with the ability to negotiate and to supply useful commodities, was crucial to success. In some locations, however, local practices as encouraged by the state were important for creating an investment environment that assisted successful CSPs.

Deliberative capacity of partnerships

The deliberative capacity of partnerships refers to the ability of different stakeholders to negotiate and shape the objectives of different investments and technology transfer. Research has shown that there were several barriers in achieving a shared understanding of technologies and investments.

In one site in Suphan Buri, central Thailand, the same Thai investor encountered an unusual level of resistance among local people to a rice husk power plant. Various organizations and citizens were concerned (inaccurately) that the power plant was run by an unpopular local politician. In turn, these opponents then spread misinformation that the generator would reduce rainfall or even sterilize people walking underneath power cables. After these tactics, the investor cancelled the project and moved to new locations. In these sites, the company invested in careful public communication and consultation (consultative partnerships).

Similarly, in the Philippines, the technology of biomethanation encountered a surprising level of resistance from NGOs, including from Greenpeace because activists did not understand the technology. Many people interviewed during the 2000s believed biomethanation to be the same as unsophisticated incineration of municipal waste – a belief probably caused by the fact that biomethanation is another form of waste to energy (albeit without incineration). Moreover, Greenpeace had campaigned successfully to win a national ban on incinerating municipal waste in the Philippines in 2000, and feared that biomethanation might undermine this ban.[4]

Biomethanation was also opposed for other reasons. Local NGOs said that they mistrusted the technology because it removed the organic material from waste flows and gave the opportunity for compost production to the biogas company. One woman showed her mistrust of new technologies by saying: 'you do not have to use complicated methods in converting organic waste back to compost … Filipinos have been converting waste to compost for many thousands of years now.'

In another case in Baguio, northern Luzon, the US investor PhilBio was criticized for allegedly removing livelihoods from local waste-pickers – who earned money by digging through municipal waste dumps for recyclable material. In an interview, a representative of PhilBio was asked: 'What can you do to reassure waste-pickers about your plans?' The answer was: 'Offer them jobs!' But it later emerged that some waste-pickers preferred to keep their claims to land next to waste dumps (where they had been living for some years) than exchange these for waged jobs at waste separation and methane capture plants. Clearly, here, the investor was engaging with long-term patterns of land tenure and development that could not be altered. Negotiating specifically with the waste-pickers themselves and their representative NGO also proved to have high transaction costs because it involved lengthy meetings with large numbers of people, and discussions about various factors of local development (such as water, education and transport) that the company could not influence.

In all of these cases, investors appreciated that the scope for being misunderstood – or of being misrepresented – was high. All investors responded by anticipating local concerns, communicating with local officials and NGOs, and by seeking ways to empower local actors.

These experiences show that building deliberative capacity, or the ability to communicate on shared or complementary visions of technology transfer, is still difficult. But these examples do not prove that collaborating with communities is impossible. They point to the need to anticipate problems and to work to establish mutual understandings, even if the local 'development dividend' that emerges is somewhat incidental to either producing energy or climate change mitigation.

Conclusion

Can community partnerships reduce the costs of low-carbon technology transfer? This chapter argues that they can. But evidence suggests that the potential is still largely unfulfilled. Three key findings seem apparent.

First, 'partnerships' are not just rhetorical acts of corporate social responsibility. Rather, there are different types of partnerships, each with potential for reducing costs of technology transfer. Table 18.1 shows a typology of different community partnerships (or cross-sector partnerships) that can range from strict contractual agreements to more deliberative and negotiated forms of consultation or public debate. So-called 'shared' or 'complementary' forms of partnerships – where communities perform both contracted services and deliberation – offer the most potential to achieve technology transfer and local development benefits simultaneously. But consultation is also useful as a step towards contracts, or as a way of overcoming some potential conflicts in contractual partnerships.

Second, successful community partnerships also requires considering the 'development dividend' of climate technology transfer carefully. There is a need to create dialogue to identify local concerns and to communicate the benefits of technologies. There is also a need to realize that 'development dividends' might actually lie in providing employment, livelihoods and services that are not directly related to the specific technology or, indeed, to climate change mitigation. For example, in the Philippines, local citizens were happy to have waste collected more efficiently – even if this waste then fed low-carbon biogas generators.

And third, these acts are still beginning. The studies presented in this chapter indicate that there is still much work needed on building and explaining the benefits of community partnerships for low-carbon technology transfer. Much international assistance for climate change policies tends to focus on incentives for mitigation alone – rather than on building the social and economic conditions necessary for investment to be successful. The 'development dividend' tends to be identified through national or international discussions under the UNFCCC or the designated national authorities of each country for the CDM. Yet, allowing new technology investments to address local needs – by offering employment and useful local services – might be the best way to achieve climate change mitigation and local development at the same time.

Notes

1 Natasha Calderwood, Carbon Neutral Company, pers comm, 2008.
2 See http://www.cdmgoldstandard.org/.
3 See http://ttclear.unfccc.int/ttclear/jsp/.
4 Greenpeace later confirmed that they do not oppose methane capture and did not endorse the strong rejection of biomethanation expressed in Thailand and the Philippines at the time of research.

References

Ählström, J. and Sjöström, E. (2005) 'CSOs and business partnerships: Strategies for interaction', *Business Strategy and the Environment*, vol 14, pp230–240

Bäckstrand, K. and Lövbrand, E. (2006) 'Planting trees to mitigate climate change: Contested discourses of ecological modernization, green governmentality and civic environmentalism', *Global Environmental Politics*, vol 6, no 1, pp50–75

Biermann, F., Chan, M., Mert, A. and Pattberg, P. (2007) 'Multi-stakeholder partnerships for sustainable development: Does the promise hold?', in P. Glasbergen, F., Biermann and A. Mol (eds) *Partnerships, Governance and Sustainable Development: Reflections on Theory and Practice*, Elgar, Cheltenham, UK, pp239–260

Boyd, E., Hultman, N., Roberts, T., Corbera, E., Cole, J., Bozmoski, A., Ebeling, J., Tippman, R., Mann, P., Brown, K. and Liverman, D. (2009) 'Reforming the CDM for sustainable development: Lessons learned and policy futures', *Environmental Science and Policy*, vol 12, pp820–831

Cosbey, A., Parry, J., Browne, J., Babu, Y., Bhandari, P., Drexhage, J. and Murphy, D. (2005) *Realizing the Development Dividend: Making the CDM Work for Developing Countries, Part 1 Report*, Pre-publication version, May 2005, International Institute for Sustainable Development

Douthwaite, B. (2002) *Enabling Innovation: A Practical Guide to Understanding and Fostering Technological Change*, Earthscan, London

Forsyth, T. (1999) *International Investment and Climate Change: Energy Technologies for Developing Countries*, Earthscan and the Royal Institute of International Affairs, London

Forsyth, T. (2005) 'Enhancing climate technology transfer through greater public–private cooperation: Lessons from Thailand and the Philippines', *Natural Resource Forum*, vol 29, pp165–176

Forsyth, T. (2007) 'Promoting the "development dividend" of climate technology transfer: Can cross-sector partnerships help?', *World Development*, vol 35, no 10, pp1684–1698

Forsyth, T. (2010) 'Panacea or paradox? Cross-sector partnerships, climate change and development', *Wiley Interdisciplinary Reviews: Climate Change*, vol 1, no 5, pp683–696

Green, D. (2004) 'Thailand's solar white elephants: an analysis of 15 years of solar battery charging programmes in northern Thailand', *Energy Policy*, vol 32, pp747–760

Gregory, J., Silveira, S., Derrick, A., Cowley, P., Allinson, C. and Paish, O. (1997) *Financing Renewable Energy Projects: A Guide for Development Workers*, Intermediate Technology Publications in association with the Stockholm Environment Institute, London

Hajer, M. and Wagenaar, H. (eds) (2003) *Deliberative Policy Analysis: Understanding Governance in the Network Society*, Cambridge University Press, Cambridge

Heaton, G., Banks, R. and Ditz, D. (1994) *Missing Links: Technology and Environment Implications in the Industrializing World*, World Resource Institute, Washington, DC

IISD (International Institute for Sustainable Development) (2006) *Third Meeting of the Development Dividend Task Force*, 27–28 March 2006, http://www.iisd.org/climate/global/dividend.asp

IPCC (Intergovernmental Panel on Climate Change) (2000) *Methodological and Technological Issues in Technology Transfer: Special Report of Working Group III*, Cambridge University Press, Cambridge

Jupp, B. (2000) *Working Together: Creating a Better Environment for Cross-Sector Partnership*, Demos, London

Krishna, G. (2005) 'Lucknow waste to energy project dubious', *The Indus Telegraph*, 13 January 2005, http://www.industelegraph.com/story/2005/1/13/21550/7898

Linder, S. (2000) 'Coming to terms with the public–private partnership: A grammar of multiple meanings', in T. Rosenau (ed) *Public–Private Policy Partnerships*, MIT Press, Cambridge, MA

Lohmann, L. (2009) 'Toward a different debate in environmental accounting: The cases of carbon and cost-benefit accounting', *Organisations and Society*, vol 34, pp499–534

Mallett, A. (2007) 'Social acceptance of renewable energy innovations: The role of technology cooperation in urban Mexico', *Energy Policy*, vol 35, no 5, pp2790–2798

Mayo, E. and Moore, H. (2001) *The Mutual State: How Local Communities Can Run Public Services*, New Economics Foundation, London, www.themutualstate.org

Mitchell, B. (2005) 'Participatory partnerships: Engaging and empowering to enhance environmental management and quality of life?', *Social Indicators Research*, vol 71, pp123–144

Morsink, K., Hofman, P. and Lovett, J. (2011) 'Multi-stakeholder partnerships for transfer of environmentally sound technologies', *Energy Policy*, vol 39, pp1–5

Ostrom, E. (2005) *Understanding Institutional Diversity*, Princeton University Press, Princeton, NJ

Otiso, K. (2003) 'State, voluntary and private sector partnerships for slum upgrading in Nairobi City, Kenya', *Cities*, vol 20, no 4, pp221–229

Plummer, J. (2002) *Focusing Partnerships: A Sourcebook for Municipal Capacity Building in Public–Private Partnerships*, Earthscan, London and Sterling, VA

Rosenau, P. (ed) (2000) *Public–Private Policy Partnerships*, MIT Press, Cambridge MA

Sherlock, K., Kirck, E. and Reeves, A. (2004) 'Just the usual suspects? Partnerships and environmental regulation', *Environment and Planning C: Government and Policy*, vol 22, pp651–666

Taiyab, N. (2005) *The Market for Voluntary Carbon Offsets: A New Tool for Sustainable Development?*, International Institute for Environment and Development Gatekeeper Series 121, London

UNFCCC (United Nations Framework Convention on Climate Change) (2003) *Capacity Building in the Development and Transfer of Technologies: Technical Paper*, UNFCCC, FCCC/TP/2003/1, UNFCCC, Bonn

Weber, E. (1998) *Pluralism by the Rules: Conflict and Co-operation in Environmental Regulation*, Georgetown University Press, Washington, DC

Williamson, O. (1985) *The Economic Institutions of Capitalism: Firms, Markets, Relational Contracting*, Free Press, New York, NY

World Bank (2000) *Greening Industry: New Roles for Communities, Markets, and Governments*, World Bank Policy Research Report, Oxford University Press, Oxford

19
CARBON TRADING AND SUSTAINABLE DEVELOPMENT

Exploring options for reforming the Clean Development Mechanism to deliver greater sustainable development benefits

Ben Castle

Introduction

The Clean Development Mechanism (CDM) has become a significant instrument for financing global technology transfer aimed at reducing greenhouse gas (GHG) emissions (e.g. Seres, 2008). The scheme emerged out of the Kyoto Protocol negotiations as a way of satisfying two policy goals: providing developed countries with access to the most cost-effective emission reductions (a key focus of the US negotiation position); and increasing developed country financial support for development efforts (a key objective for many developing countries, led by Brazil) (Disch, 2010). Article 12 of the Kyoto Protocol describes the twin objectives of the CDM as to assist developing countries 'in achieving sustainable development and in contributing to the ultimate objective of the Convention' (UNFCCC, 1998, 12.2, p11), where the 'ultimate objective' is preventing 'dangerous anthropogenic inter-ference with the climate system' (UNFCCC, 1992, 2, p5). Thus, the achievement of sustainable development (SD) benefits for host countries is enshrined as a key objective in the founding legal document of the CDM, alongside enabling devel-oped countries to achieve part of their emission reduction targets at lower costs by purchasing Certified Emissions Reductions (CERs) from CDM projects. However, the CDM has been criticized for failing to deliver both sufficient technology trans-fer and SD benefits (e.g. Forsyth, 2007).

To date, the market nature of the scheme has prioritized the delivery of emissions reductions, with stringent measurement, auditing and reporting requirements being developed by the CDM Executive Board (the Board), in consultation with parties and other experts. Since its launch in 2001, the CDM has grown rapidly, with over 3000 projects now registered and 3 billion tonnes of CO_2 equivalent (CO_2e) expected to be saved by the end of 2012 (UNFCCC, 2011). In contrast, the SD benefits have remained unspecified and vague and the responsibility of national governments to define and police through their Designated National Authority (DNA). DNAs have interpreted their responsibilities differently, with wide variation in the definition and prioritization of SD benefits, as well as approaches to enforcement (Disch, 2010). The market nature of CDM incentivizes investment in the most cost-effective projects which may not necessarily be those with the greatest SD benefits. A significant proportion of overall CDM investment has gone towards large industrial projects which offer high volumes of CERs through 'end-of-pipe' solutions, such as HFC-23 mitigation technologies.[1] While such projects deliver GHG emission reductions, the local SD benefits can be minimal. Some CDM projects have even been accused of resulting in negative local impacts. Docena (2010), for example, suggests that CDM landfill gas recovery projects in the Philippines have damaged the viability of local recycling initiatives and efforts to reduce the overall volumes of waste going to landfill. Another key concern is over the distribution of CDM projects, which is highly concentrated in India and China, with only 0.2 per cent of CERs expected to come from projects in least developed countries (LDCs), where development needs are greatest (de Lopez et al., 2009). These issues have led to significant criticism suggesting that the CDM has offered little, if any, genuine SD benefits.

There have been differing views on the importance of achieving SD benefits. Some have suggested the delivery of emissions reductions and some inward investment to host-countries is sufficient; if anything, additional efforts should be limited to minimizing possible negative local impacts (Streck and Lin, 2008). Others point to the wide range of additional social and environmental benefits that CDM projects can potentially offer, including local air quality and health improvements and employment and poverty reduction, which they claim are too often overlooked, representing a significant missed opportunity (e.g. Cosbey et al., 2006; Schneider, 2007). This broad range of additional SD benefits, often referred to in the literature as 'co-benefits' or the 'development dividend' of projects, is the focus of this chapter.

There is a common sentiment shared by many CDM analysts that the 'CDM does not sufficiently fulfil its objective of assisting host countries in achieving sustainable development' (Schneider, 2007, p10). This chapter accepts this analysis as a starting point and assumes that the SD benefits delivered to date have been suboptimal and that it is desirable to improve the performance of the scheme. With the CDM set to expire at the end of 2012, on-going international negotiations will seek to find agreement on how best to extend, reform or replace the CDM.[2]

The following section summarizes different options for improving the SD benefits of the CDM and explores their likely effectiveness. It also assesses the likely political acceptability of reform options for different nations and the prospects for them being adopted in forthcoming negotiations. It is intended to be of use to readers interested in the future relationship between international carbon trading and development, including policy-makers, business representatives, civil society groups and non-governmental organizations (NGOs).

It is assumed that international carbon trading will continue post-2012 and the options below will therefore be of relevance should the CDM continue largely in its present form or should it be more substantially changed or replaced with a similar mechanism. Additionally, while it is recognized that trading in emissions from forestry-based projects is likely to expand in the future, with significant SD implications (Gutierrez et al., 2007), this is considered to be a separate topic and outside the scope of this chapter.

Options for CDM reform

Based on a wide ranging review of relevant literature, seven broad reform options for improving the SD benefits of the CDM have been identified: no reform and voluntary initiatives; removing barriers to smaller projects; limiting the use of technologies; minimum standards; inflating the value of high SD CERs; improving monitoring and compliance; and ensuring civil society participation. These are outlined below.

No reform and voluntary initiatives (option one)

It could be argued that market forces will naturally support the delivery of development benefits. After all, project developers have an interest in retaining local support and avoiding criticism which could harm their reputation. It is also arguably in the host country's interest to demand high development returns through their DNA. Purchasers of CERs are also weary of criticism and may demand higher project standards (Streck and Lin, 2008), leading Liverman and Boyd (2008, p11) to declare that the 'heady days when buyers and investors bought anything they could get their hands on are over and investors are looking for quality projects that provide social development'.

The success of voluntary initiatives such as the CDM Gold Standard is evidence of a market for CERs with high SD benefits. However, such initiatives seem likely to remain relatively niche compared to overall CER volumes and of most interest to public-sector and ethical investors and those companies in the public eye willing to pay extra for brand benefits. The natural desire for purchasers to obtain CERs at minimal cost is still likely to limit the emphasis placed on SD benefits under conventional CDM projects (Peskett et al., 2007). Furthermore, competition between prospective host countries to attract investment could lead to a 'race to the bottom' in terms of national SD standards (Boyd et al., 2009). It is therefore highly questionable as to whether continued reliance on voluntary actions will be sufficient to increase delivery of SD benefits.

Removing barriers to smaller projects (option two)

Analysis of CDM projects has shown that, in general, smaller-scale projects tend to produce greater SD benefits. As Cosbey et al., (2006, p19) put it: 'small-scale projects tend to yield greater development dividends, and very large-scale projects yield comparatively few. The basic relationship holds across all three elements of sustainable development: social, economic and environmental.' Smaller projects are also more likely to fit with the circumstances typically found in many LDCs where there tends to be fewer large industrial installations and more diffuse opportunities, often in rural areas (de Lopez et al., 2009). Furthermore, smaller-scale projects are more likely to be owned by community organizations, helping to ensure that the benefits are captured locally (Cosbey et al., 2006).

There are already a number of simplified project design, monitoring and review procedures and requirements to try and support smaller projects. Despite this, transaction costs for small projects remain high – estimated to be, on average, 82,000 Euros for development and a further 7000 to 18,000 Euros for each monitoring and verification visit (de Lopez et al., 2009). These costs, combined with limited CER volumes, makes small projects less attractive from an investor perspective, and they act as a barrier for community-owned projects which have little access to upfront capital.

It may be possible to further streamline and simplify additionality, validation and verification processes (Leguet and Elbed, 2008). Revising the current thresholds[3] for small-scale projects upwards would enable more projects to qualify for reduced reporting requirements and procedures (de Lopez et al., 2009). A new category of 'micro-projects' could also be introduced with further reduced requirements. Standardized methodologies for small projects may also further reduce costs (Leguet and Elbed, 2008). While such changes could improve the economic viability of smaller projects, market forces are always likely to favour larger volume projects which offer higher returns for investors (de Lopez et al., 2009). Equally, such reforms will not prevent many LDCs from being viewed as too risky by investors, due to poor governance or instability (Kant, 2010). Significant national capacity building is likely to be necessary to overcome these issues, although that could prove resource intensive and a long-term process.

Small projects of the same kind are able to benefit from 'bundling' together to reduce overall transaction costs. However, the current rules prevent bundling where the total quantity of estimated CERs exceeds the limits for small-scale CDM projects.[4] Reviewing such limits could therefore offer advantages (Cosbey et al., 2006). Similarly, it has also been suggested that recent movement towards more programme-based CDM approaches could support smaller, more diffuse projects in the renewable energy, energy efficiency and transport sectors, which typically have high development benefits (Cosbey et al., 2006). However, there remain significant unresolved questions over the methodologies for ensuring additionality and which types of programmes and sectors would be favoured by the market in practice (Boyd et al., 2009). The complexities involved in designing CDM programmes and producing sufficient evidence of additionality are also likely to be resource intensive for host countries and could act to further exclude LDCs with limited specialist capacity (Cosbey et al., 2006).

There are signs that the Board may be open to some of these reforms (Figueres and Streck, 2009). However, some countries are nervous about reforms that could make the processes of validation too easy, which they fear could 'open up a flood-gate to supply which would overwhelm demand and depress prices' and undermine domestic climate change targets and objectives (Figueres and Streck, 2009, p234). Additionally, more industrially advanced developing countries such as India, China and Brazil are likely to resist reforms which may divert investment from them towards smaller projects in less developed countries.

Limiting the use of technologies (option three)

Different technologies tend to offer differing potential development benefits. While domestic energy efficiency improvements may offer significant local economic, air quality and health benefits, the application of technology to industrial systems to reduce HFC-23 emissions is likely to offer little, if any, SD benefits. Indeed, the dominance of HFC-23 projects in the CDM has been singled out as a major reason for the CDM failing to deliver significant SD benefits (Schneider, 2007).

The decision by the Board to suspend issuance of HFC-23 credits in the summer of 2010 is widely seen as likely to increase the social benefits of CDM. However, the decision was taken due to concerns over the additionality of emissions reductions and not on the basis of SD considerations. Any decision to further limit the list of qualifying technologies would represent a major departure from the current CDM operating principles where such decisions are made primarily on the basis of emissions reduction integrity.

The emergence of carbon capture and storage (CCS) as a potential CDM-qualifying technology is likely to add to the pressures to introduce technology-specific limits. CCS is seen by many as strategically vital in delivering future emission reductions from countries with high levels of gas and coal reserves, and the 2010 Cancun negotiations agreed in principle to the inclusion of CCS in to the CDM. Yet, the added local SD benefits are likely to be weak and there are risks of negative local environmental and health impacts from CO_2 leakage due to poor site selection and storage practices (IPCC, 2005). With individual CCS projects likely to deliver many millions of CERs, CCS could begin to dominate the CDM market and marginalize other projects with intrinsically greater SD potential.

The Board is reluctant to intervene with the choice of qualifying technologies as this is likely to increase CER prices. A compromise in order to support CCS through the CDM, while reducing the risk of CCS dominance, could be to impose limits on the proportion of CCS-originated CERs that states or companies can use to achieve their targets (de Coninck, 2008). Such a change in approach could pave the way for further market intervention to limit the use of technologies with limited likely development benefits. However, it will be difficult to reach agreement on what basis such decisions should be made as definitions of SD benefits are highly subjective (see the fourth option, minimum standards, below). The benefits from a project will also depend to a large extent upon where and how the project is designed and implemented. For example, while some geothermal projects have tended to deliver high social benefits, on average, this is because they have tended to be located in areas with low access to energy. Much of the benefit is therefore reliant on the local circumstances (Cosbey et al., 2006).

Minimum standards (option four)

A number of commentators have suggested the need to introduce globally consistent minimum standards for development benefits (e.g. see Huq, 2002). Under such an approach, projects would not be approved until a certain level of SD benefit is demonstrated. The delivery of SD benefits would therefore be treated in a similar way to how emissions reductions are currently, with globally consistent standards and enforcement. This could take the form of an agreed checklist of SD benefits, similar to the way in which the voluntary Gold Standard scheme operates (Huq, 2002; Boyd et al., 2009).

A number of methods for measuring development benefits have been suggested, comprising multiple environmental, social and economic indicators (see Cosbey et al., 2006, Olsen and Fennhann, 2008; Disch, 2010). However, such methods have been developed for the purpose of comparative research of CDM projects. Reaching international agreement on what indicators should be included with what weightings is likely to prove far more difficult (Cosbey et al., 2006). Different local circumstances, values and preferences, and stages of economic development will inevitably produce diverse national priorities, which will, in turn, influence the preferred definition of SD benefits. While some countries may wish to prioritize economic growth, others (and many NGOs), are likely to emphasize environmental and social benefits. Countries are reluctant to surrender such sovereign freedom (Disch, 2010). Forsyth (2007) also cautions against developers or national authorities using a prescription definition of SD benefits, as this would pre-empt the preferences of local stakeholders (see the seventh option, ensuring civil society participation, below). Greater flexibility could be introduced through a scoring system where projects would be awarded points based on a wide set of criteria and having to achieve an overall threshold score in order to be registered (Boyd et al., 2009). However, in order to gain the support of all countries, any such criteria are likely to have to be so broad and flexible that their value would be questionable. Even if an agreed approach could be found, there is likely to be resistance from developed countries due to the likely increase in CER prices that would result.

Inflating the value of high SD CERs (option five)

An alternative to introducing standards could be to differentiate between the value (or the quantity) of CERs awarded for projects with high SD benefits. This could steer investment towards more beneficial projects. This could be done according to priority sectors, technologies or based on the type of host country (Boyd et al., 2009). This last option offers one of the few ways of successfully steering investment towards LDCs.

Adopting any of these options would mark a significant departure from the principles upon which the CDM currently operates. Such changes would restrict the supply of CERs and would therefore be likely to increase the costs for developed countries purchasing CERs. This could also require agreement on the method for determining CER values, which would inevitably mean reaching agreement on a workable approach for measuring SD benefits, which as discussed in options three (limiting the use of technologies) and four (minimum standards), will be fraught with difficulty. Nevertheless, the European Union has recently decided that post-2012 the use of CERs within the EU Emissions Trading Scheme (ETS) will be restricted to those originating from LDCs. This is likely to be fiercely opposed by the likes of China and India, as well as non-LDC developing countries and representatives of carbon traders (see, for example, CMIA, 2011).

Improving monitoring and compliance (option six)

At present there is also no agreement over how DNAs should ensure delivery of stated SD benefits. As a result, there is a wide range of different approaches practised by DNAs, with many simply reviewing project design documents (PDDs) prior to implementation and assessing them against checklists or national regulations (Disch, 2010). With such a lack of post-implementation checks and enforcement, it is unsurprising that there is evidence that claimed SD benefits can often fail to materialize (CSE, 2005).

Agreeing to stricter monitoring and enforced procedures for DNA s would not necessarily require agreement over common SD standards (as for options two to five: removing barriers to smaller projects; limiting the use of technologies; minimum standards; and inflating the value of high SD CERs) and may therefore prove more politically viable. In order to better ensure delivery of stated development benefits, *post hoc* evaluation activities will need to be undertaken by DNAs to compare project outcomes to initial PDDs. This could involve DNAs undertaking spot checks, site visits and interviews with stakeholders. Failure by project developers to deliver development benefits could result in penalties or fines for non-compliance in order to act as an effective incentive. One area of focus for DNAs could be to ensure high standards of consultation, which is discussed in more detail under the final option below.

Ensuring consistent implementation of such approaches is likely to prove challenging considering the diverse national circumstances and the perverse incentive for DNAs to be 'soft' on developers so as not to deter investment. In order to ensure that DNAs deliver their responsibilities to consistently high standards, a system of accreditation, akin to that for designated operating entities (DOEs), could be developed.[5] This would require DNAs to demonstrate that they have the necessary procedures and practices in place to monitor and enforce national SD standards. Should poor DNA practice or a lack of enforcement be identified, penalties could be applied, with the potential risk of DNA de-accreditation. A system for handling complaints against DNAs could also be introduced to increase accountability. With no accredited DNA, countries would effectively lose the ability to host CDM projects, so such a system could act as powerful incentive to improve monitoring and enforcement practices.

Some countries are likely to be reluctant to agree to such a move due to the loss of national autonomy and the cost implications of running more active DNAs and operationalizing such an accreditation system. Furthermore, without some international agreement over the definition of SD benefits, some countries could simply choose to weaken or abandon national requirements, so that, in practice, DNAs would have very little enforcement to do.

Ensuring civil society participation (option seven)

Processes which ensure that the views of a variety of stakeholders are incorporated with the design of projects are likely to improve decision-making and increase the likelihood of SD benefits being delivered (Eddy and Wiser, 2002; Lovbrand et al., 2009). Case studies of waste-to-energy projects in India, the Philippines and Thailand support this view, and even suggest that in some circumstances deliberative processes can increase SD benefits, while also reducing overall investor costs by making local actors more willing to support project implementation (Forsyth, 2007).

Good participatory practices in Peru have been credited with delivering CDM projects with higher-than-average SD benefits (Disch, 2010). In this case, consultation and engagement activities were coordinated directly by the Peruvian DNA, with the outcomes informing final project approval decisions. This has the advantage of ensuring impartiality and uniformity of consultation approaches across different projects. However, such a centralized approach is highly resource intensive for the DNA and will therefore be resisted by many governments.

An alternative to a DNA-led approach is for consultation activities to be undertaken by project developers and for this to be monitored by the DNA and DOEs. In fact, the CDM rules technically already require stakeholder consultation by developers, but 'there are no clear rules or guidelines and the implementation varies from country to country' (Disch, 2010, p54). Lovbrand et al., (2009, p94) suggest that as a result of this lack of clarity, the 'concerns and views of those directly affected by CDM projects will automatically be less influential in the project design and implementation than those of project developers, host country governments and investors (domestic as well as foreign)'.

Development of clearer guidance on standards for consultation practices for project developers, DNAs, DOEs and national governments could therefore help to ensure that higher standards are achieved more consistently (Eddy and Wiser, 2002). It will inevitably be difficult to define uniform standards for how participation should be delivered across different projects and locations. This is especially true as experience has illustrated the importance of project institutional and contractual arrangements suiting specific local circumstances and stakeholder needs (Forsyth, 2007). Ensuring consistency in the interpretation and enforcement of guidance will also be challenging (as discussed in option six: improving monitoring and compliance). There are also questions over the legitimacy of different participation methods; in many cases it may only be international NGOs who have the financial capacity and technical skills to participate in a meaningful way (Cosbey et al., 2006; Lovbrand et al., 2009), yet they are likely to focus on their specialist areas of concern and are unlikely to be representative of local stakeholders (Forsyth, 2007). Additionally, in practice, it may also prove hard to distinguish between genuine participation activities and box-ticking gestures which offer little real benefit (Lovbrand et al., 2009).

Despite these challenges, clearer guidance on how best to achieve local participation, especially if combined with improved monitoring and enforcement (as outlined in option six), would seem to offer the potential of improving delivery of SD benefits.

Summary discussion

The options outlined above are not exclusive of each other. The wide range of challenges in delivering improved SD benefits suggests that a number of different reforms could be necessary. Indeed, it is possible for the voluntary initiatives outlined in option one to continue to play a useful role alongside the more significant mandatory reforms outlined in options two to seven. There would also be little point in introducing the minimum SD standards outlined in option four, without also improving monitoring and compliance practices, as suggested in option six. However, there are also tensions between some options. For instance, while the focus of option two is on reducing transaction costs for small projects, the introduction of increased evidence requirements (implied by options four and five) could further add to project development costs and therefore act to weaken the commercial attractiveness of smaller projects. Similarly, differentiation in the value of CERs from different origins (outlined in option five) has the potential to increase investment in LDCs, yet a movement towards more programmatic approaches (one possibility considered in option two) could reduce LDC investment viability due to their limited specialist skills and government capabilities.

While each of the options offer different strengths and weaknesses, the final decision on CDM reform and replacement is likely to be a compromise and determined more by what is politically acceptable to different countries and stakeholders than what is practically most effective. The political barriers to implementation of reforms are significant. Many of the ideas for streamlining the project approval process (as outlined in option two) appear workable. However, concerns over the possibility of this generating too many CERs and suppressing carbon prices, and a reluctance by China and India to tolerate reforms which could reduce investment flows to them, is likely to limit their adoption. The significant challenges involved in reaching international agreement on a definition of SD benefits may prove insurmountable. This would effectively rule out options three and four. Although a variant of option five is being pursued by the EU, there remain significant political barriers to its implementation, as well as uncertainty over its implications for delivery of SD benefits. This suggests that improvements to monitoring and compliance (option six) and participatory approaches (option seven) are likely to be the most politically viable reforms which could deliver greater SD benefits. However, even for these options it will be difficult to ensure flexible yet consistent implementation at costs acceptable to all countries.

Any improvement in the delivery of SD benefits under the CDM will ultimately require a redoubling of political commitment to the delivery of SD. This will require explicit agreement that the SD benefits currently delivered under the CDM are too few and honest recognition of the limits to unfettered market-based approaches. Furthermore, it is also likely to require an acceptance of some increase in the costs of CDM for investors (which will be reflected in the prices of CERs) and/or the administrative cost of the scheme (which will be borne by DNAs and host governments).

Notes

1 HFC-23 is a powerful GHG.
2 This chapter was written in 2011 ahead of the Durban negotiations. However, it is anticipated that the discussion outlined here will continue to be relevant to the debate over the future of carbon trading and its relationship to sustainable development.
3 Thresholds for small-scale projects are currently 15MW per year for renewable energy projects, savings of up to 60GWh per year for energy efficiency projects, or savings of up to 60 kilo tonnes CO_2 equivalent for other projects: see decision 1/CMP.2, para 28.
4 This was the case at the time of writing. See 2007 guidance document F-CDM-SSC-BUNDLE, para 9.
5 DOEs are accredited by the UNFCCC as being able to offer legal, accounting and verification services to CDM projects.

References

Boyd, E., Hultman, N., Timmons, R., Esteve, C., Cole, J., Bozmoski, A., Ebeling, J., Tippman, R., Mann, P., Brown, K. and Liverman, D. (2009) 'Reforming the CDM for sustainable development: Lessons learned and policy futures', *Environmental Science & Policy*, vol 12, pp820–831

Cosbey, A., Murphy, D., Drexhage, J. and Balint, J. (2006) *Making Development Work in the CDM: Phase II of the Development Dividend Project*, International Institute for Sustainable Development, Winnipeg, Manitoba, Canada

CMIA (Carbon Markets and Investors Association) (2011) Open letter to African DNAs from the Carbon Markets and Investors Association

CSE (Centre for Science and Environment) (2005) *Biomass for Nothing*, Centre for Science and Environment India, Down to Earth, 15 November 2005, http://www.indiaenvironmentportal.org.in/content/biomass-nothing

de Coninck, H. C. (2008) 'Trojan horse or horn of plenty? Reflections on allowing CCS in the CDM', *Energy Policy*, vol 36, no 3, pp929–936

de Lopez, T., Ponlok, T., Iyadomi, K., Santos, S. and McIntosh, B. (2009) 'Clean Development Mechanism and least developed countries: Changing the rules for greater participation', *Journal of Environment and Development*, vol 18, no 4, pp436–452

Disch, D. (2010) 'A comparative analysis of the 'development dividend' of Clean Development Mechanism projects in six host countries', *Climate and Development*, vol 2, pp50–64

Docena, H. (2010) *The Clean Development Mechanism Projects in the Philippines*, Focus on the Global South Special Report, Philippines

Eddy, N. and Wiser, G. (2002) 'Public participation in the clean development mechanism of the Kyoto Protocol', in C. Bruch (ed) *The New 'Public':The Globalization of Public Participation*, Environmental Law Institute,Washington, DC, pp203–214

Figueres, C. and Streck, C. (2009) 'The evolution of the CDM in a post-2012 climate agreement', *Journal of Environment and Development*, September, vol 18, pp227–247

Forsyth,T. (2007) 'Promoting the 'development dividend' of climate technology transfer: Can cross-sector partnerships help?', *World Development*, vol 5, no 10, pp1684–1698

Gutierrez, M., Boyd, E. and Chang, M. (2007) 'Small-scale forest carbon projects: Adapting CDM to low-income communities', *Global Environmental Change*, vol 17, no 2, pp250–259

Huq, S. (2002) *Applying Sustainable Development Criteria to CDM Projects: PCF Experience*, Prototype Carbon Fund,World Bank,Washington, DC

IPCC (Intergovernmental Panel on Climate Change) (2005) *Carbon Capture and Storage: A Special Report of Working Group III of the Intergovernmental Panel on Climate Change*, United Nations, Geneva

Kant, P. (2010) *Taking CDM beyond China and India*, IGREC Web Publication no 16/2010

Leguet, B. and Elbed, G. (2008) 'A reformed CDM to increase supply: Room for action', in K. Olsen and J. Fenhann (eds) *A Reformed CDM, including new Mechanisms for Sustainable Development*, Perspectives Series, UNEP Riso Centre, Denmark

Liverman, D. and Boyd, E. (2008) 'The CDM, ethics and development', in K. Olsen and J. Fenhann (eds) *A Reformed CDM, Including new Mechanisms for Sustainable Development*, Perspectives Series, UNEP Risø Centre, Denmark

Lovbrand, E., Ridefjall,T. and Nordqvist,J. (2009) 'Closing the legitimacy gap in global environmental governance? Lessons from the emerging CDM market', *Global Environmental Politics*, vol 9, no 2, pp74–100

Olsen, K. and Fennhann,J. (2008) 'Sustainable development benefits of Clean Development Mechanism projects:A new methodology for sustainability assessment based on text analysis of the project design documents submitted for validation', *Energy Policy*, vol 36, no 8, pp2773–2784

Peskett, L., Luttrell, C. and Iwata, M. (2007) *Can Standards for Voluntary Carbon Offsets Ensure Development Benefits?*, Forestry Briefing 13, Overseas Development Institute

Schneider, L. (2007) *Is the CDM Fulfilling Its Environment and Sustainable Development Objectives? An Evaluation of the CDM and Options for Improvement*, Oko Institut, Germany

Seres, S. (2008) *Analysis of Technology Transfer in CDM Projects*, Prepared for the UNFCCC Registration and Issuance Unit, United Nations, Geneva

Streck, C. and Lin,J. (2008) 'Making markets work:A review of CDM performance and the need for reform', *European Journal of International Relations*, vol 19, no 2, pp409–442

Sun, Q., Xu, B.,Wennerstein, R. and Brant, N. (2010) 'Co-benefits of CDM projects and policy implications', *Environmental Economics*, vol 1, issue 2, pp78–88

UNFCCC (United Nations Framework Convention on Climate Change) (1992) *The United Nations Framework Convention on Climate Change*, UN, http://unfccc.int/resource/docs/convkp/conveng.pdf

UNFCCC (1998) *The Kyoto Protocol to the United Nations Framework Convention on Climate Change*, UN, http://unfccc.int/resource/docs/convkp/kpeng.pdf

UNFCCC (2011) *CDM in Numbers*, http://cdm.unfccc.int/Statistics/index.html, accessed 28 April 2011

INDEX